IUTAM SYMPOSIUM ON SEGREGATION IN GRANULAR FLOWS

SOLID MECHANICS AND ITS APPLICATIONS
Volume 81

Series Editor: G.M.L. GLADWELL
Department of Civil Engineering
University of Waterloo
Waterloo, Ontario, Canada N2L 3GI

Aims and Scope of the Series

The fundamental questions arising in mechanics are: *Why?, How?,* and *How much?*
The aim of this series is to provide lucid accounts written bij authoritative researchers
giving vision and insight in answering these questions on the subject of mechanics as it
relates to solids.

The scope of the series covers the entire spectrum of solid mechanics. Thus it includes
the foundation of mechanics; variational formulations; computational mechanics;
statics, kinematics and dynamics of rigid and elastic bodies: vibrations of solids and
structures; dynamical systems and chaos; the theories of elasticity, plasticity and
viscoelasticity; composite materials; rods, beams, shells and membranes; structural
control and stability; soils, rocks and geomechanics; fracture; tribology; experimental
mechanics; biomechanics and machine design.

The median level of presentation is the first year graduate student. Some texts are mono-
graphs defining the current state of the field; others are accessible to final year under-
graduates; but essentially the emphasis is on readability and clarity.

For a list of related mechanics titles, see final pages.

IUTAM Symposium on Segregation in Granular Flows

Proceedings of the IUTAM Symposium
held in Cape May, NJ, U.S.A.
June 5–10, 1999

Edited by

Anthony D. Rosato
New Jersey Institute of Technology,
Department of Mechanical Engineering,
University Heights, Newark, U.S.A.

Co-edited by

Denis L. Blackmore
New Jersey Institute of Technology,
Department of Mathematics,
University Heights, Newark, U.S.A.

KLUWER ACADEMIC PUBLISHERS
DORDRECHT / BOSTON / LONDON

A C.I.P. Catalogue record for this book is available from the Library of Congress.

IUTAM Symposium on Segregation in Granular Flows (1999 : Cape May, N.J.)
   IUTAM Symposium on Segregation in Granular Flows / edited by Anthony D. Rosato ;
co-edited by Denis L. Blackmore.
      p. cm. -- (Solid mechanics and its applications ; v. 81)
   ISBN 0-7923-6547-X (hc. : alk. paper)
   1. Granular materials--Fluid dynamics--Congresses. 2. Inhomogeneous
materials--Fluid dynamics--Congresses. 3. Bulk solids flow--Mathematical
models--Congresses. 4. Particles--Congresses. I. Rosato, Anthony D. II. Blackmore,
Denis L. III. Title. IV. Series.

   TA418.78 .I89 2000
   620.1'05--dc21

00-060505

ISBN 0-7923-6547-X

Published by Kluwer Academic Publishers,
P.O. Box 17, 3300 AA Dordrecht, The Netherlands.

Sold and distributed in North, Central and South America
by Kluwer Academic Publishers,
101 Philip Drive, Norwell, MA 02061, U.S.A.

In all other countries, sold and distributed
by Kluwer Academic Publishers,
P.O. Box 322, 3300 AH Dordrecht, The Netherlands.

Printed on acid-free paper

Printed in the Netherlands.

TABLE OF CONTENTS

"Segregation" is the term used to describe the ubiquitous phenomenon in which a bulk solid, composed of particulates with differing constituent properties, evolves to a spatially non-uniform state. Because segregation is a universal problem of great importance that cuts across almost all fields, it has attracted worldwide interest from the mathematics, physics, engineering and industrial communities. Indeed, the participants of this symposium hail from 14 different countries and are experts in the fields of mechanical and chemical engineering, applied mechanics, physics, geology, chemistry and mathematics. Most interesting is that the theme of the symposium appears to serve as a common thread connecting these often separated communities with a collective goal of unraveling the mysteries of the phenomenon by integrating the perspectives and approaches of each.

The importance of the segregation problem cannot be understated. This becomes evident when one simply recognizes the fact that particulates are universally found as constituents of most commonly used items, which are produced within an extensive complex of industries, i.e, agriculture, ceramics, chemicals, energy, geological systems, mining, pharmaceuticals, plastics, pollution control systems, and powder metallurgy. A brief look at some production figures will serve to provide an estimate of magnitudes involved. In the United Kingdom, the mining industry produces 1.66×10^8 tons of materials each year, while the annual world production of coal is approximately 4.6×10^9 tons. The 1999 world production of various agricultural commodities (almonds, cereals, lentils, grains, buckwheat, barley, millet, oats, maize, rice, rye, sugar cane, and sorghum) totaled 5.34×10^9 tons. In the US, 2.01×10^9 tons of crushed sand and gravel were mined for consumption during the first half of 1999, while in 1994, polymer utilization was approximately 3.8×10^7 tons. Indeed, particulate-based products and related services contribute one trillion dollars annually to the US economy alone. The fundamental problem when segregation occurs is that it poses a formidable impediment to the general requisite of creating and maintaining homogeneous mixtures. An inability to achieve and maintain a well-mixed condition of a bulk solid throughout its processing history can lead to serious flaws in the properties of an end product, leading to unfavorable economic consequences.

Perhaps the most confounding aspect of the phenomenon is that segregation is often governed by a combination of mechanisms, which themselves depend on the nature of the flow, particle properties and their distributions, and environmental conditions. Consider, for example, those particle features which can influence segregation behavior: particle sizes and distributions, shape, morphology, contact friction, elasticity, brittleness, density, chemical affinities, ability to absorb moisture, and magnetic properties. Additional complexities arise through the interplay between different time and length scales. These scales depend upon the mode by which energy is imparted to the bulk solid causing it to flow, such as through vibrations, shearing, and gravity. As a consequence, individual researchers have often focused their efforts on narrow ranges of conditions in an attempt to identify, isolate or enhance dominant mechanisms. Despite an expanding body of literature in the last ten to fifteen years, relationships between identified mechanisms are ambiguous, experimental data is scarce, and there is no accepted model capable of predicting the occurrence of segregation over the wide spectrum of possible flow regimes. These issues bring to bear a host of scientific as well as technical questions, such as the existence of "universal" mechanisms of segregation, their measurement and quantification, the effects of mean flow and particle fluctuations, the evolution of microstructure, and the feasibility of developing a unified model. It was these questions that this symposium attempted to address.

A. D. Rosato

D. L. Blackmore

ACKNOWLEDGEMENTS

The organizers of this symposium are grateful for the financial support of the International Union of Theoretical and Applied Mechanics, as well as the U.S. National Science Foundation, the Federal Energy Technology Center of the U.S. Department of Energy, Union Carbide Corporation, Merck & Company, and the Particle Technology Center at New Jersey Institute of Technology.

The editors of this volume are especially appreciative for the assistance provided by the IUTAM Bureau, and the Scientific and Organizing Committees. Special thanks are due to many other people whose efforts contributed to making the meeting successful. We would especially like to recognize Pat Brannigan who launched the meeting with a welcoming address on behalf of Dr. Saul Fenster (President, NJIT) and the NJIT community; the Mechanical Engineering Department at NJIT and Barbara Valenti for her time, effort and diligence in attending to the multitude of details needed to organize the symposium; to Vonney Nelson (NJIT) for expediting the disbursement of the participant travel grants; to Ninghua Zhang (NJIT) and Jian Liu (NJIT) for their assistance in maintaining the presentation schedule; to Marie Rosato who attended to all of the registration details during the meeting; and last, but not least, to Joan Haber and the staff at the Marquis de Lafayette Hotel in Cape May, NJ for their hospitality and efforts to make certain that the meeting ran smoothly.

Scientific Committee

A. D. Rosato, Chair (NJIT)	USA
S. B. Savage (McGill University)	Canada
J. B. Bridgwater (Cambridge University)	England
J. T. Jenkins (Cornell University)	USA
Z. Mroz (Inst. of Fund. Tech. Research)	Poland
Y. Tsuji (Osaka University)	Japan
L. van Wijngaarden (University of Twente)	Netherlands
J.-L. Auriault (Universite J. Fourier)	France

Organizing Committee

Denis L. Blackmore (Co-Chair)	NJIT
Eiichi Fukushima	New Mexico Resonance
James T. Jenkins	Cornell University
Masami Nakagawa	Colorado School of Mines
Anthony D. Rosato (Chair)	NJIT
Roman Samulyak	Brookhaven National Laboratory
Stuart B. Savage	McGill University
Pushpendra Singh	NJIT

SEGREGATION MECHANISMS IN CONDENSED GRANULAR FLOW

A Summary of Some Fundamental Experiments

JOHN BRIDGWATER
Department of Chemical Engineering
University of Cambridge
Pembroke Street
Cambridge CB2 3RA
England

1. Introduction

The mechanics of granular materials are much influenced by the void fraction created. The action of the forces arising in the process of a deforming bed can cause there to be consolidation of the bed or, alternatively, there to be an open or even fluidised state. Such behaviour can arise due to either the passage of a fluid or mechanical action. The present purpose is to bring together some of the results of a sequence of experimental studies aimed at elucidating the segregating mechanisms which are at work in granular materials, the particles of which are in general maintaining many points of contact during motion. These studies, conducted mainly in the 1980's, provide evidence of the character of the physical processes that are at work. These studies also give a framework for interpreting behaviour in fundamental terms, both for research and for industrial application. These furthermore give a goal for those developing theoretical and computer-based models.

What was the motivation for these studies? About thirty years ago a government survey carried out in the United Kingdom led to the conclusion that the mixing of powders was an important and under-researched area of knowledge. A centrally-funded programme was launched to tackle the area but this fell significantly short of its goals. The bureaucratic enthusiasm to understand a difficult subject had not been tempered by a realisation that the subject lacked not only effective experimental techniques but also an understanding of mechanisms. It was not appreciated that a secure theoretical base was needed but that this was absent. The initiative stemmed form a perceptive article written by Bourne [1] which makes it amply clear that we will not gain control of processes and products until there is a fundamental understanding. It took some time for the work reported here to develop, but what is said here has many of its roots in the issues raised by Bourne. The difficulties expressed by him at that time included problems in the characterisation of the structure of mixtures, problems of knowing what structure was being created, and uncertainty as to which structure might be needed for particular functions. Thus in many processes the mixing action is incidental to the real

1

A.D. Rosato and D.L. Blackmore (eds.), IUTAM Symposium on Segregation in Granular Flows, 1–10.
© 2000 *Kluwer Academic Publishers. Printed in the Netherlands.*

process requirement which could be one of heat transfer either through an external surface or to a gas flowing through or above the mixture. It could be one of agglomeration or, conversely, one of breakage. It could be one of cohesive powder mixing, a dynamic process in which agglomerates are alternately formed and broken up. The subject of granular mixing is thus one of great intellectual significance in its own right; it gives rise to a family of derivative problems, the better understanding of which is of the greatest importance for the chemical, food, pharmaceuticals, and materials industries, to name a few. The changing shape of the industries and the advances in science and engineering make this area even more important today.

2. Physical Processes

When a bed of particles under normal conditions of packing is subjected to a shear strain under a given normal stress, the shear stress rises and to a maximum and then decays to a steady state value. This maximum shear stress is dependent on the initial packing of the particles, with the maximum being enhanced if the particles are more tightly packed initially.   Such ideas underlie the method of assessing the frictional properties of powders by the procedures developed by Jenike and are central to the methods of critical state soil mechanics. When such bodies of powder undergo shear (Figure 1), the main relative inter-particle motion occurs in failure zones, which are variously reported to be between five and twenty particle diameters in width. If a material has a substantial displacement, it is presumed that these zones exist for a portion of the displacement and then shift positions. There is thus, on this hypothesis, a flow structure in which there are blocks of material having no slip with failure zones between the blocks in which there is a very high rate of strain.

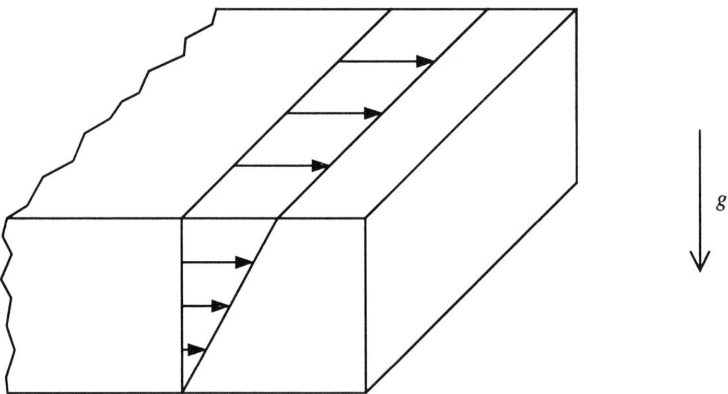

Figure 1. Idealised horizontal failure zone.

The high rates of strain that exist in failure zones give rise to phenomena of particle segregation which have been given the terms *Inter-particle Percolation* and *Particle*

Migration. The first of these is the drainage of a species down through the opening and closing voids created by the particle motion within a failure zone while the second is the motion of a particle larger that the bulk within a deforming region under a strain rate gradient. There is then also the somewhat separate case of behaviour at a free surface, a subject that is now attracting much attention where the separating process is known as *Free Surface Segregation*

3. Inter-particle Percolation

A so-called simple shear cell was built which imposed a linear velocity profile on a bed of spherical particles. The cell had a reciprocating action with rigid rotating side walls to allow high linear strains to be imparted to the material. The particles in the bed thus take up the linear velocity distribution that is shown in Figure 1. A specially designed self-clearing base and top were used to allow an experiment to proceed without the occurrence of consolidation. A particle of different, smaller size traversed the bed repeatedly and the mean shear strain for the particle to cross a bed of height h was found. By varying h the mean percolation velocity could be calculated. One powerful piece of evidence that the bed was behaving uniformly comes from the observation that the mean residence strain $\bar{\gamma}$ is directly proportional to the height of the bed, as Figure 2 reveals. Other checks such as varying the shear strain per cycle show a similar consistency. Cells were used which employed bulk particles of various sizes from 6.03mm to 18.6 mm. The full details of the procedure are well documented [2].

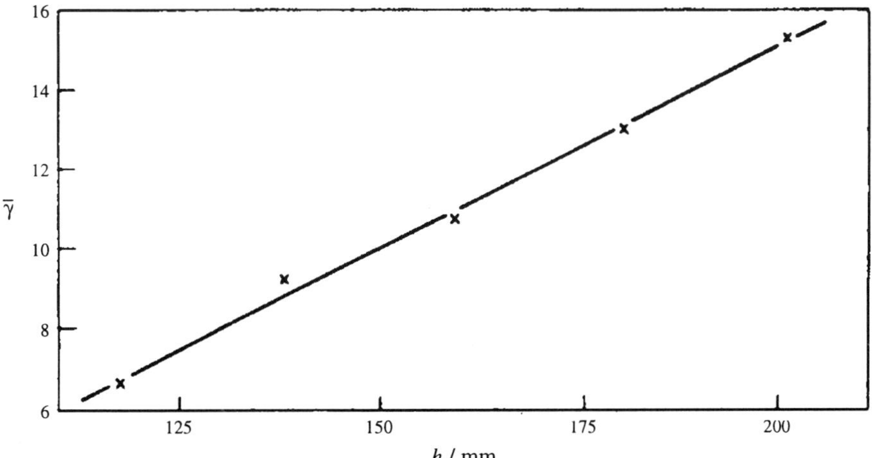

Figure 2. Regression between mean residence strain and bed height h. The diameter of the bulk particles is 18.6 mm. $d_p/d_b = 0.431$, $\sigma = 1.40$ kN m^{-2}, $\omega = 0.4$ s^{-1}.

The results could be presented in dimensionless form thus:

$$\frac{u}{\omega d_b} = f\left(\frac{d_p}{d_b}, \frac{\rho_p}{\rho_b}, \frac{d_b\omega^2}{g}, \frac{\sigma}{E_b}, \frac{E_p}{E_b}, \frac{d_b\rho_b g}{E_b}\right)$$

This equation sets aside the necessary further parameters of friction and particle shape. The significance of the term on the left hand side is that of a dimensionless percolation velocity in which u is the mean vertical velocity, ω is the mean shear strain rate and d_b is the diameter of the bulk particles. The small particles are present in small amounts and are of a diameter d_p. The ratio d_p/d_b is the most significant group; Figure 3 shows the relationship presented in a log-linear plot form. This form is suggested by statistical mechanical arguments [5]. These data can be interpreted to reveal a fluctuation in free space associated with the deforming body of particles having a length scale of about one eighth of a bulk particle diameter. This is the length scale necessary for a slip zone that is eight particle diameters wide to exist.

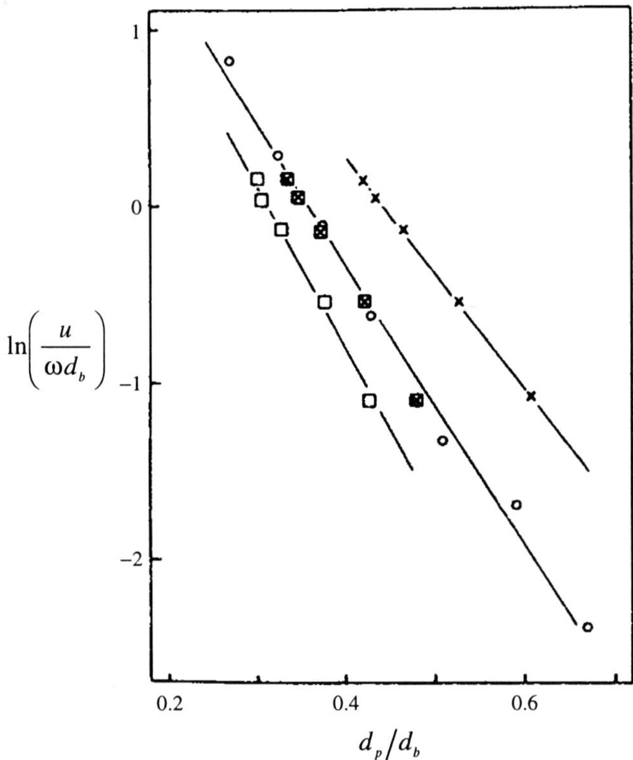

Figure 3.   Plot of $\ln\left(u/\omega d_b\right)$ vs d_p/d_b for the percolation of acrylic resin spheres and cubes through phenolic resin bulk. Spherical percolating particles:

o Cubic percolating particles (side a): □ $d_p = a$; ✕ $d_p = a\sqrt{2}$; ⊠ $d_p = 1.12a$.

Figure 3 also shows results when using percolating particle that are of a cubic shape, the characteristic length being adjusted as indicated. For bulks of different diameter, the dimensionless results of Figure 3 remain the same.

The other groups in the equation did not show a significant effect in the experiments. Thus ρ_p/ρ_b the particle density ratio did not influence behaviour. A weak effect of stress could be assigned to the ratio σ/E_b with E_b, the Young's modulus of the bulk particles, providing a measure of the deformation of the bulk particles due to the applied stress σ. The ratio $d_b\rho_b g/E_b$ gives the effect due to the deformation of the body force due to gravity, not an issue with the large particles studied but one that could arise in the processing of deformable agglomerates. At the rates of strain used, there was no observable influence of strain rate. This is given by the ratio $d_b\dot\gamma^2/g$ which is a measure of the time of existence of a void to the time for a particle to fall into it under the acceleration due to gravity g. The term E_p/E_b measures the effect of percolating particle deformation on the percolation process. It is hard to find out how many of the groups will influence behaviour other than by gaining insight through computer-based modelling.

In further work, the role of interstitial liquid has been studied. This can slow the rate of percolation as it restricts the ability of a particle to move into spaces created in the deforming structure due to the influence of liquid buoyancy and drag. This is consistent with the important practical observation that percolation can be inhibited by the addition of liquid. The experimental work used a fully saturated material; surface tension forces would be a further influence if a material were to be partially saturated.

It is also possible to use the equipment to make evaluations of lateral and axial dispersion coefficients and to determine also the self-diffusion coefficients for bulk material; the issues are fully reported in the literature; see Bridgwater [4] for a summary. Studies using wheat as a bulk material showed there would be very marked effects due to the change of the bed structure itself with strain. Thus at the wall the particles rotated in the direction of the strain but in the bulk, the particles moved to behave as cylinders rolling over one another.

If a mixture is no longer dilute, then the use of a simple shear cell which was effective for single particle percolation was no longer possible. Work has been carried out in an annular shear cell reported by Stephens and Bridgwater [9, 10] and later by Bridgwater et al. [3]. They based their technique upon the use of an annular cell operated so that it developed a shear zone in a horizontal plane. In general they used granular materials of around 2-4mm diameter and, by the insertion of marked tracers initially aligned vertically and carefully sectioning the bed after strain, they were able to find the velocity distribution that arose in a failure zone.

A velocity distribution is shown in Figure 4. In the particular experiment cited, use is made of a mixture of particles. The larger component is 4.00mm glass and the smaller is 1.91mm glass. These have been carefully mixed by hand and placed in the annular cell which was of 307mm outer diameter and 45mm width. The cell has then been rotated and the bed has then been carefully taken apart layer by layer and the fraction of the smaller component is plotted as a function of height. We see that the smaller component is indeed to be found in the lower portion of the bed. However, there is not a clear boundary between the two components. There is thus both a downwards motion of small particles due to percolation and an upwards diffusive flux of this component due to the random forces generated by the particles during the rotation of the cell. In later work, Bridgwater et al. [3] showed how a kinematic flow model, (using percolation rates from Figure 3 as input together with the experimentally found velocity distribution such as that depicted in Figure 4) could describe with some degree of satisfaction the segregation of a dilute mixture in an annular shear cell. They followed the downward motion of layers of small particles. These moved slowly into the failure zone from above, passed through the high strain rate region in the middle of the zone, and then accumulated in the region of decreasing rate of strain at the bottom of the failure zone.

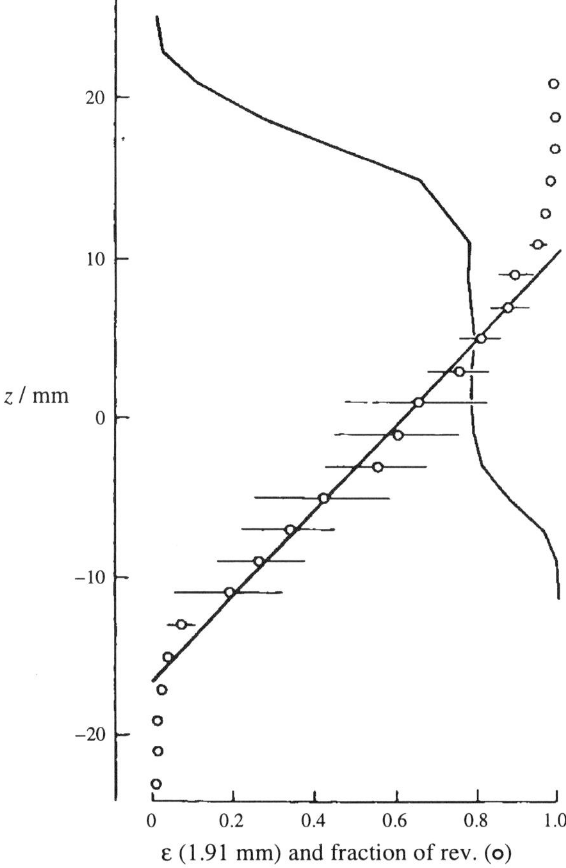

Figure 4. Concentration profile and failure zone identification ia an annular shear cell. 4.00 and 1.91 mm glass beads. Initial conditions: small particles $12 > z > -23$, large particles $z > 12$. Concentration profile after 26 cell revolutions. Velocity profile obtained from last revolution. The cell splits at $z = 0$.

4. Particle Migration

This was examined in a cell with side walls that imparted a strain rate that varied with vertical position. The full details of the technique have been reported in the literature [8]. The cell could be operated so that the strain rate either increased or decreased with the vertical coordinate. A normal stress was applied to the free surface and the material subjected to shear again by use of a reciprocating action. The motion of a particle larger than the bulk was followed by the tedious method of operating the cell for a period and then taking apart the structure to find the position of the large particle. It was found that the particle could move either up or down, its behaviour being unified by the observation that it moved in the direction in which the rate of strain increased. The literature contains many observations of the motion of large through small particles and an array of explanations exist. However, the large particles experience the occurrence of more voids on that side of the large particle where the rate of strain is high. Accordingly the large particle is more likely to move in the direction of the higher rate of strain. The amount of experimental evidence here is not so great, unlike the considerable amount available for inter-particle percolation. It is, however, another good target for computer-based simulation studies since it is much more practicable to explore the significant physics in such a way rather than resorting to difficult experiments on systems where it is hard to characterise the particles and the environment fully.

5. Free Surface Segregation

The work summarised here is reported in detail in two papers by Drahun and Bridgwater [6,7]. They carried out experiments on the occurrence of segregation during the formation of a two-dimensional heap. The overall purpose was to determine the physical mechanisms and to establish, where possible, the scaling laws.

The apparatus deployed is illustrated in Figure 5. The particles fed to the system are placed on a conveyer belt and are fed into a rectangular box that was of 10mm width and gave a slope length that could be varied up to 50mm. The particles can be fed independently or as a layer of material. This layer can be well mixed or may consist of layers one on top of another. The speed of the conveyer determined the rate at which particles were added to the apparatus. The free fall height y of the particles onto the conveyor is held at a constant value for a given experiment but may be varied between experiments. This is achieved by maintaining the conveyer in a constant vertical position but allowing the bed itself to fall in a controlled manner. To minimise the mass of particles to be used, the bed was fitted with an adjustable base, the slope of which could be set by experience to correspond to the angle of repose arising in the experiment. As an experiment proceeded, there were thus two drives in action, one feeding the material to the top of the heap and the other maintaining the free fall height at a constant value. At the end of the experiment, the variation of bed composition with position was determined. To do this, the bed was rotated about hinges in the corner of the bed near the bottom of the heap so that the free surface was now horizontal. Stiff plates were then inserted into the bed at 20mm spacing, the positions of these being determined by slots that had been cut in the perspex walls. The contents of each of these compartments were then withdrawn using a vacuum gun and the mixture analysed by sieving.

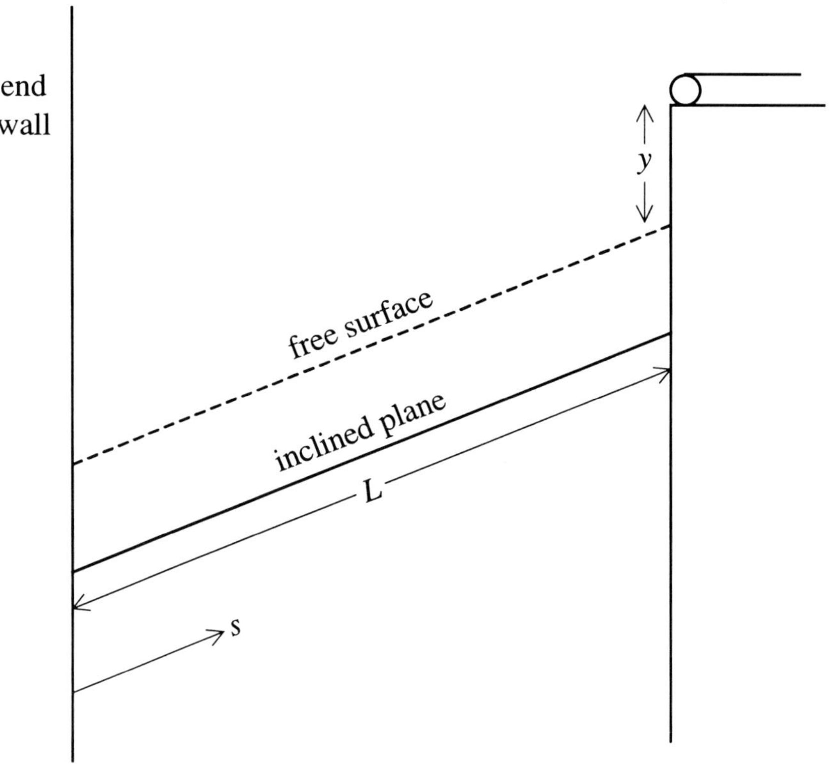

Figure 5. Schematic diagram of the inclined plane showing notation; s denotes tracer displacement.

Consider first the case in which there is one component being fed to the bed in vast excess over the other. The minor component is found in the heap either close to the conveyer or towards the bottom of the slope. The particles of the former type are called sinkers whereas those of the latter type are called floaters. The symbols in Figure 6 show data at three different slope lengths; the data scale with the slope length L. Avalanches were usually observed and thus the effect of these also scaled with slope length. It is likely that the separation step occurs quite quickly near the top of the heap but the resting place of the tracer is determined by the convective effect of the avalanche. If the free fall height y were small, then the tracers of larger size and smaller density were found close to the end wall, Figure 6. If the free fall height is increased, then the larger particles are found near the conveyer end. This is attributed to the greater momentum of the individual impacting particles, which can deform the surface structure as they land and then become sufficiently buried to avoid the avalanching action.

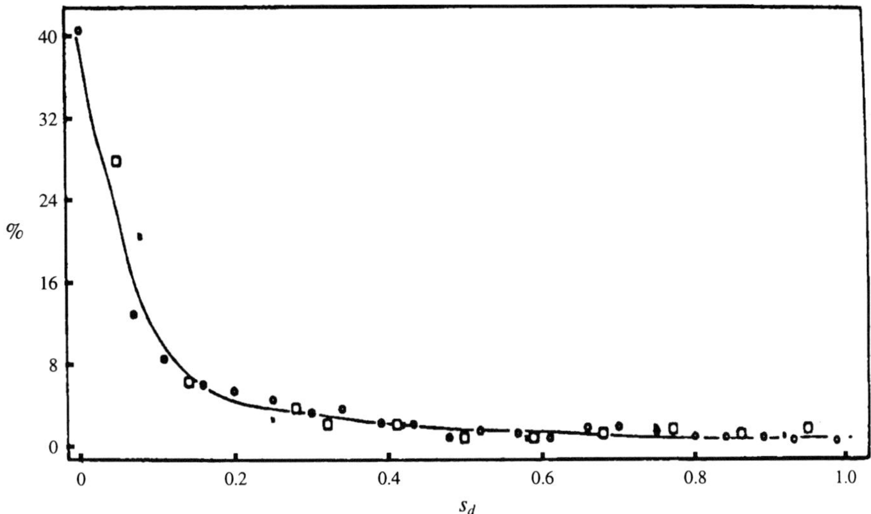

Figure 6. Concentration distribution of tracer vs dimensionless slope position, $s_d = s/L$. d_b=4.00mm, $\rho_b = 2950\text{kg/m}^3 (d_p/d_b = 1.00,\ \rho_p/\rho_b = 0.40)$.

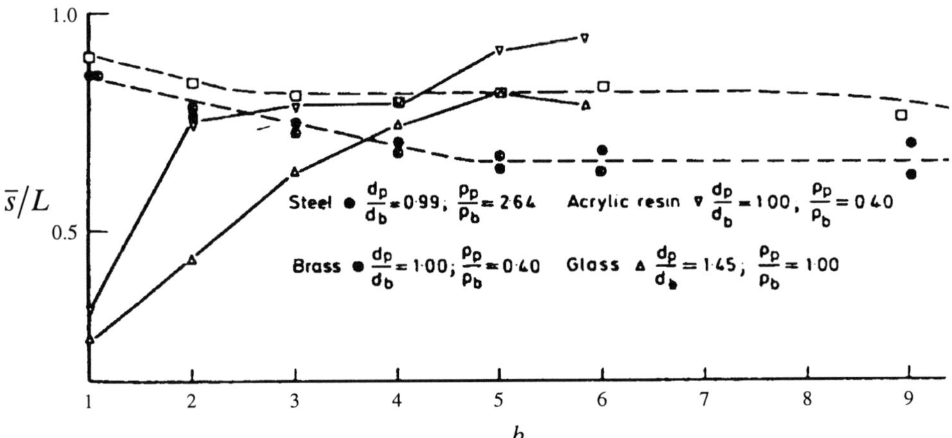

Figure 7. The relationship between dimensionless mean tracer position $\overline{s}/L$ and the thickness of the feed on the conveyer b. Bulk 4.00mm glass beads. □, tracer 2.05mm glass beads.

Smaller particles can be reflected into the space above the free surface and bound over the surface rather with avalanching being irrelevant. If material is fed as a thick layer, then those particles that arrive on the surface itself undergo the behaviour of Figure 6.

However, this is inhibited for ones arriving more deeply buried and we then have the behaviour depicted in Figure 7.

6. Conclusions

Descriptions have been provided of studies designed to elucidate the mechanisms of segregation. Those reported were as carried out by one group with a particular interest in the physics of powder mixing as it related to the processing industries, broadly interpreted. The effects of scale have been accounted successfully in many instances. A fuller overall account of the work has been given in the book edited by Mehta [4]. Significant advances in instrumentation, theory and computation now allow us to make powerful inroads into this research area. The studies reported are give useful targets for such work and remain of value in seeking to understand the behaviour of practical systems.

7. References

1.   Bourne, J.R.: The mixing of powders, pastes and non-Newtonian liquids, *The Chemical Engineer* **181**, (1964), CE202-216.
2.   Bridgwater, J., Cooke, M.H., and Scott, A.M.: Inter-particle percolation: equipment development and mean percolation velocities, *Transactions of the Institution of Chemical Engineers* **56** (1978), 157-167.
3.   Bridgwater, J. Foo, W.S., and Stephens, D.J.: Particle mixing and segregation in failure zones: Theory and experiment, *Powder Technology* **41** (1985), 147-158.
4.   Bridgwater, J. Mixing and segregation mechanisms in particle flow: in A Mehta (ed.), *Granular Matter, An Interdisciplinary Approach*, Springer-Verlag, 1994, New York, pp 161-193.
5.   Cooke, M.H., and Bridgwater, J.: Interparticle percolation: a statistical mechanical interpretation, *Industrial Engineering Chemistry Fundamentals* **18** (1979), 25-27.
6.   Drahun, J.A., and Bridgwater, J.: Free surface segregation, *Institution of Chemical Engineers Symposium Series* **65** (1981), S4/Q/1-S4/Q/14.
7.   Drahun, J.A., and Bridgwater, J.: The mechanisms of free surface segregation, *Powder Technology* **36** (1983), 39-53.
8.   Foo, W.S., and Bridgwater, J.: Particle migration, *Powder Technology* **36** (1983), 271-273.
9.   Stephens, D.J., and Bridgwater, J.: The mixing and segregation of cohesionless particulate materials. Part I. Failure zone formation, *Powder Technology* **21** (1978), 17-28.
10. Stephens, D.J., and Bridgwater, J.: The mixing and segregation of cohesionless particulate materials. Part II. Microscopic mechanisms for particles differing in size, *Powder Technology* **21** (1978), 29-44.

SEGREGATION MECHANISMS AND THEIR QUANTIFICATION USING SEGREGATION TESTERS

PROF. SUNIL R. DE SILVA
Telemark College
Kjoelnes Ring
3914 Porsgrunn, Norway

ARE DYRØY and PROF. GISLE G. ENSTAD
Telemark Technological R & D Centre
Kjoelnes Ring
3914 Porsgrunn, Norway

ABSTRACT

In handling particulate materials with broad size distributions, or mixtures of such materials with components exhibiting differences in physical properties, one often encounters a phenomenon called segregation. Segregation is a tendency for certain sizes, or components with similar properties, to preferentially collect in one or another physical zone of a collective. This tendency to separate into different zones is largely caused by differences in mobility. The extent of the separation is dependent on a number for factors, including the physical characteristics of the particles in the mixture. The extent of differences between these characteristics, the process conditions, such as the rate of fill or the height of fill, and the handling regime that is encountered by the particles in the particular process being considered, will all have an influence on the segregation that will take place.

One of the most common process situations in which segregation occurs is in the filling of heaps. Heaps are formed during, for example, the creation of stockpiles, and in the filling of silos. Separation of particles of similar properties to different points, or zones, in the heap occurs as a result of several mechanisms. The final pattern of segregation depends on which of these mechanisms were dominant or contributory with respect to the nature of the mixture, the process parameters, and the handling regime involved in the situation being analysed.

This paper describes the results of extensive investigations using both two- and three-dimensional testers, which clearly show that the mechanisms which contribute to segregation are very much dependent on all the parameters mentioned above. Some evidence is presented showing that it may be possible to predict a three-dimensional pattern of heap segregation from two-dimensional testing. These indications must, however, be considered preliminary, and further work is shown to be necessary in order to establish a more rigorous link between these results.

A.D. Rosato and D.L. Blackmore (eds.), IUTAM Symposium on Segregation in Granular Flows, 11–29.

1. Introduction

Segregation is a phenomenon that has long been recognised as the cause of several problems in materials handling situations. However, many more problems are due to this phenomenon than industry appears to be aware of! Some of the earliest investigations were related to the handling of coal [1-7]. In 1925, Garve [8] found that small particles were found in the centre of a silo being filled, while larger particles were found near the walls – the classic pattern of heap segregation. Systematic studies of the phenomenon, however, were not carried out until much more recently, with investigators such as Williams [9-11] and Bridgwater [12-17] and their co-workers providing information on a number of mechanisms that may act during heap formation. Others who have contributed significantly are Johanson [18-21], Bagster [22-24] and Standish [25-27]. There also have been many attempts at modelling the segregation process, notably by Shinohara [28-38], Savage [39], Popplewell et al [40], Alonso et al [41], Arteaga and Tuzün [42], Dolgunin and Okolov [43] and Meakin and Jullien [44]. Neither the experiments nor the models have yet proved universal enough to describe all the mechanisms that act during the formation of a heap, and it needs to be questioned whether straightforward deterministic modelling is at all a feasible method of dealing with such complex, and often chaotic, situations.

At Telemark College/Telemark Technological R&D Centre, we have been concerned for a very long time with the problems that are caused by segregation. Our approach has been largely pragmatic, and has been directed towards first identifying, and then minimising, the mechanisms of segregation that are most likely to dominate in any given situation. Initially this was not an easy task because we simply did not know which mechanisms we were dealing with – or even whether the problems that appeared downstream were the result of segregation, or of some other cause [45]! Accordingly, our efforts in the last 5-6 years have been primarily directed at 1) developing testers to identify the various mechanisms that play a role in heap segregation, and 2) solving industrial problems based on the understanding and quantification of the underlying phenomena.

2. Mechanisms of Segregation

Mosby [46] identified ten different mechanisms of segregation. These are: rolling, sieving, push-away, angle of repose, percolation, displacement, trajectory, air current, fluidization and impact. De Silva [47] identified two more: concentration driven displacement and agglomeration. In a more recent work on segregation testers, Salter [48] also identified a mechanism, which he called "embedding."

Various terms are used to describe these mechanisms, and there is also a considerable degree of confusion regarding the differences between mechanisms, processes and handling regimes, as evidenced in the otherwise very useful "User Guide to Segregation" by Bates [49]. In order to make it clear what is meant, the following definitions/examples are to be noted:

- A segregation <u>mechanism</u> is a localized event, which leads to the separation of one type of component or size class from another. Examples were given above.
- A segregating <u>process</u> is a situation in which the mechanisms can become active, for example in the formation of a heap, the filling or discharging of a silo, or the transport of material in a container, or along a chute, belt or pneumatic transport line.
- Handling regimes are circumstances, which promote or reduce the effect of various segregation mechanisms. Examples are air flows set up during the filling of silos, local shear caused by movements or vibrations, rapid or slow filling of heaps, mass- or funnel flow in the discharge of silos, or the introduction of material into processes or systems which preferentially accelerate or retard the different components in a mixture. Very short descriptions of the mechanisms and alternative terms are given below.

Trajectory	-	segregation caused by the fact that air drag reduces the speed of smaller particles, in free flight, to a greater extent than it does that of larger (or heavier) ones. Could possibly be better called "inertia segregation".
Air Current	-	a mechanism caused by a handling regime, for example during the filling of a silo, when a circulating air current from the centre to the walls is induced by the falling particles, which carry fine particles away from the centre, and deposits them at the walls.
Rolling	-	a mechanism by which large or rounded particles roll down the surface of a heap in formation.
Sieving	-	a mechanism by which smaller particles flow downward through a rolling or sliding layer of large particles. Often combined with percolation which, arguably, can be defined as a mechanism in its own right.
Impact	-	a mechanism by which particles with higher coefficients of restitution bounce off a heap surface, and are found further away from the centre of a heap than those with lower coefficients. Could also be called "bouncing".
Embedding	-	an inertia based mechanism which causes larger or denser particles to penetrate a layer on the surface of a heap at its apex, and become locked there.

Angle of Repose	-	a mechanism caused by different angles of repose of different components filled sequentially onto a heap, whereby the component with a lower angle of repose flows over that with a higher angle, towards the edges of a heap.
Push-away	-	a mechanism by which heavier particles falling on the apex of the heap push-away lighter, equally sized, particles towards the edges of the heap. Could equally well be called "displacement segregation".
Displacement	-	a mechanism, originally identified by Williams, by which larger particles rise to the surface of a mixture of large and small particles in stages, as a result of vibrations. The vibrations, which have to have vertical amplitudes, move a large particle upwards. Small particles percolate into the space vacated and prevent its return. The mechanism has been called "floating migration" by Salter [48].
Percolation	-	a mechanism that causes smaller particles to fall through gaps between larger particles as a result of localised shear. Although similar to sieving, this mechanism does not require the larger particles to be in a flowing layer. The mechanism can be activated by vibrations or local shear planes, and may also lead to displacement upwards of larger particles.
Fluidization	-	a mechanism which keeps a fluidizable component of lighter or finer particles on the surface of the contents during, for example, the filling of a container or a silo.
Agglomeration	-	a mechanism which causes very fine particles to form larger aggregates with greater mobility.
Concentration driven displacement	-	a mechanism encountered in rotating systems (e.g. drum mixers) where the mobility of fine particles is much higher than that of large particles, leading them to concentrate in zones.

The different mechanisms are illustrated in Figure 1.

2.1. SEGREGATION TESTERS

2.1.1 Two Dimensional Testers

Right from the start of this work there was a need to determine whether a segregation tester could be two-dimensional or whether it had to be three-dimensional. In the development of these testers, there has been focus on the mechanisms that are

encountered in heap segregation. Understandably, the preference was for a two dimensional tester which requires less material, and from which it is much easier obtain to samples. Several such testers have been made and tested over the years.
These have been described by e.g. Harris and Hildon [50], Shinohara [30], Drahun
and Bridgwater [15] and Bagster [23]. At Telemark College, we further developed a
tester which had been originally conceived of in our earlier existence at Chr.
Michelsen Institute in Bergen [51]. First, we investigated whether we could get
away with half a two dimensional heap. These preliminary tests, which were carried
out in the two geometries shown in Figure 2 by Maltby [52], gave rise to the results
shown in Figure 3a.

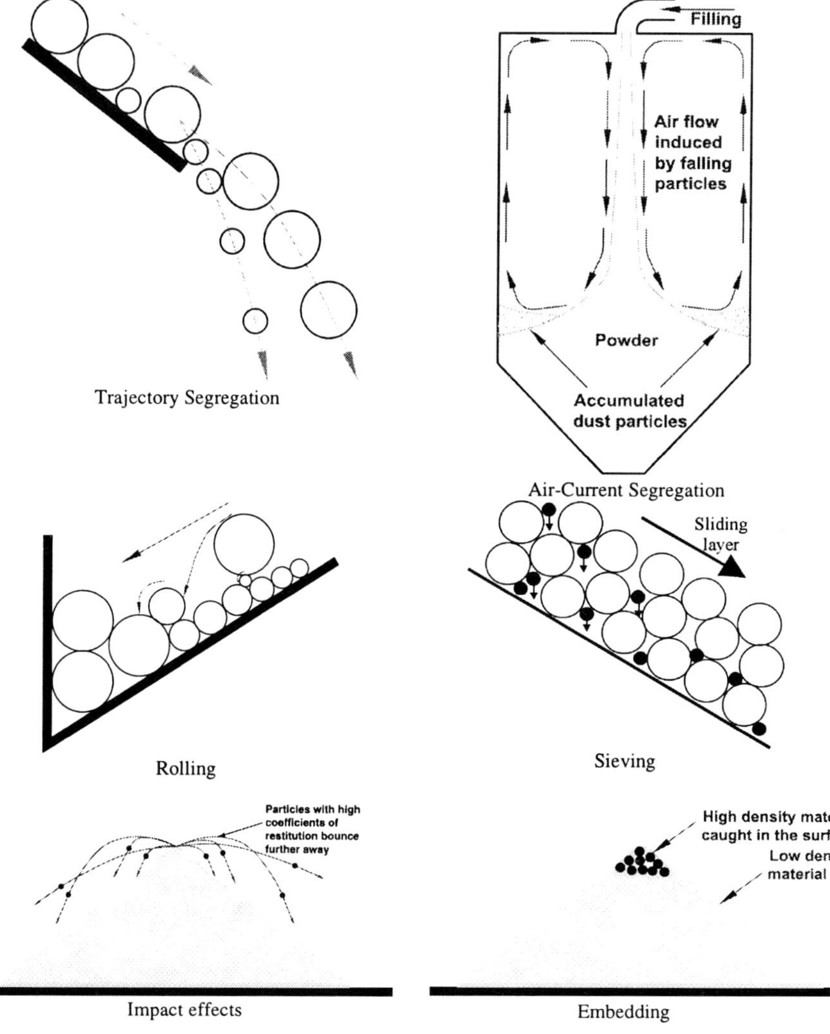

Trajectory Segregation

Air-Current Segregation

Rolling

Sieving

Impact effects

Embedding

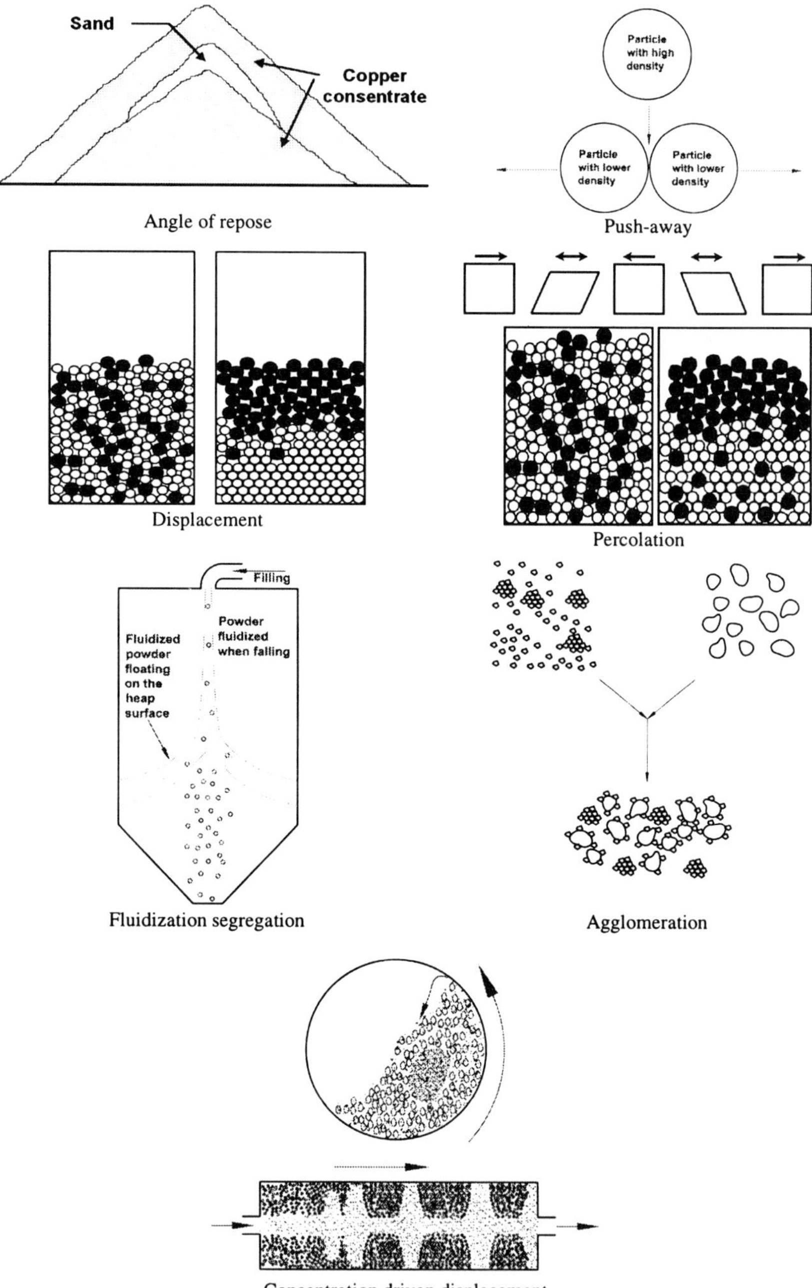

Figure 1.   *Mechanisms of segregation.*

The test material used was a sodium formate with a very wide size distribution ranging from 0.5 to 8 mm. Samples were taken at various points along the surface as shown in Figure 3b, and the content of particles greater than 2 mm in the samples was determined by sieving. The fraction over 2 mm in the feed was 0.22.

Maltby found that the differences were insignificant except for sampling point 1, where the results were found to have significant differences at the 5% confidence level. Maltby concluded that this difference might have been caused by the fact that he used identical feed rates in filling both the full and half heaps, whereas he should, ideally, have halved the rate during the latter tests in comparison with the former. It appears, however, that his successors [46, 48, 53] accepted that any observed differences were not significant enough to worry about, and proceeded with further investigation using the half heap model!

The next investigations were largely carried out by van den Berg [53]. Two free flowing powders were used during the tests: a Leighton Buzzard sand with a size distribution varying from 0.3 to 1.5 mm (median size 0.65 mm) and an alumina with a size distribution of 0.01 to 0.2 mm (median size 0.09 mm). There was no overlap between the sizes. The other pertinent information about the two materials is given in Table 1. The tester geometry used is shown in Figure 4. To ensure a homogenous feed, the components were feed in the desired ratio by two feeders into the static mixer, from where the ready mixture, was discharged into the tester. Previous testing of this system had proved that it produced homogenous mixtures.

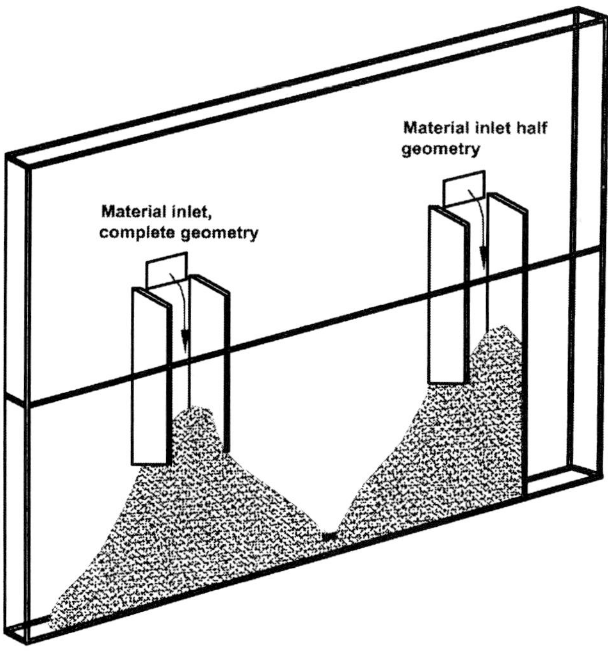

Figure 2.   Tester used in comparing the results of half a heap with those of a full heap [52].

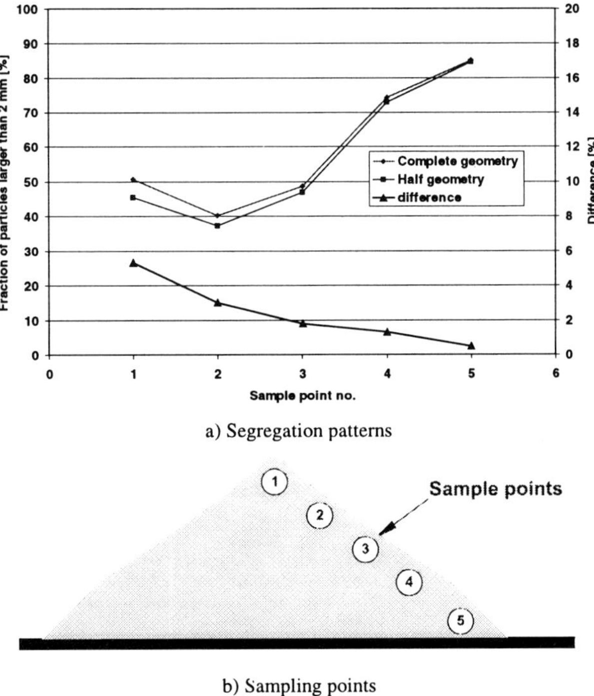

a) Segregation patterns

b) Sampling points

*Figure 3.*Results of tests with half a heap and a full heap [52], showing sampling points.

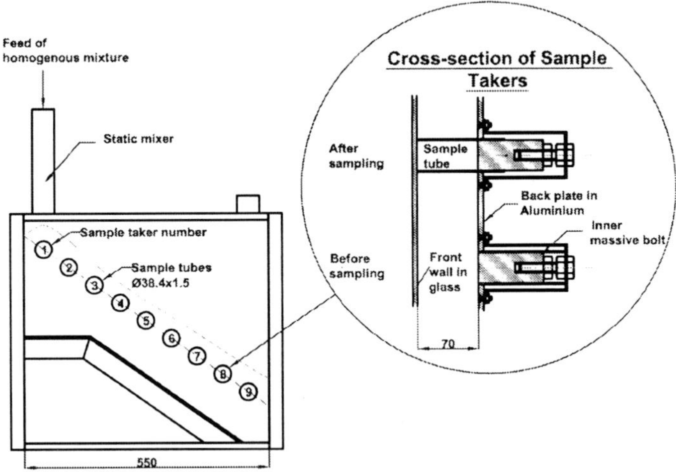

Figure 4. Half-heap tester used by Mosby [46], showing positions and construction of sample takers.

TABLE 1. *Physical characteristics of Leighton Buzzard sand and alumina used in segregation testing at Telemark College*

	Alumina	Sand
Dynamic angle of repose	32°	36°
Static angle of repose	34°	37°
Poured density, g/cm^3	1.00	1.51
Tapped density, g/cm^3	1.12	1.66
Particle density, g/cm^3	3.40	2.65

An investigation was undertaken of the depth (= width) of the tester necessary to eliminate wall effects. Visual observations of the flowing layer indicated little differences in the results between testers of 7 and 11 cm in depth, and the 7 cm unit was used as a standard. Excellent repeatability was found as shown in Figure 5. Similar repeatability was found by Salter [48] from his two dimensional tester. The repeatability is another proof that the mixture that was fed into the tester must have been homogenous.

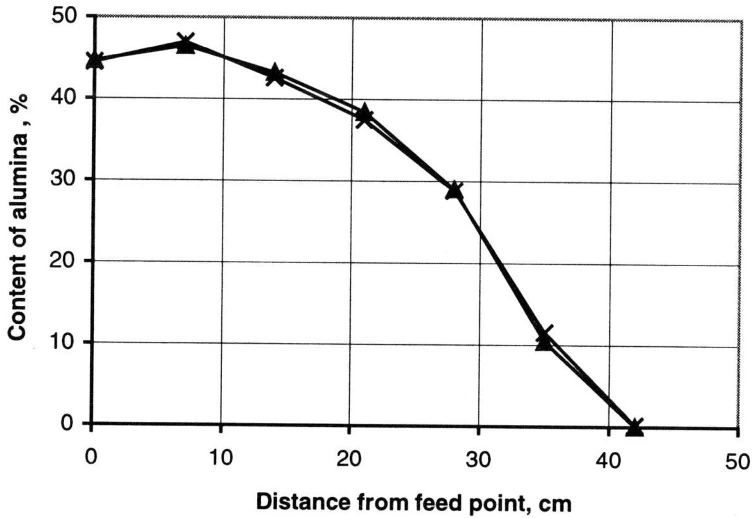

Figure 5.   Results of two duplicate tests with alumina.

The second investigation concerned the variables: mixing ratios (Mi [w%] Figure 6), feed rates (Fr [g/s] Figure 6) and heap length (L [cm] Figure 6). These investigations were the subject of Mosby's Dr.Ing. thesis, and detailed results are found in reference 46. Mosby was able to summarise his results as shown in Figure 6, and his conclusions are given below.

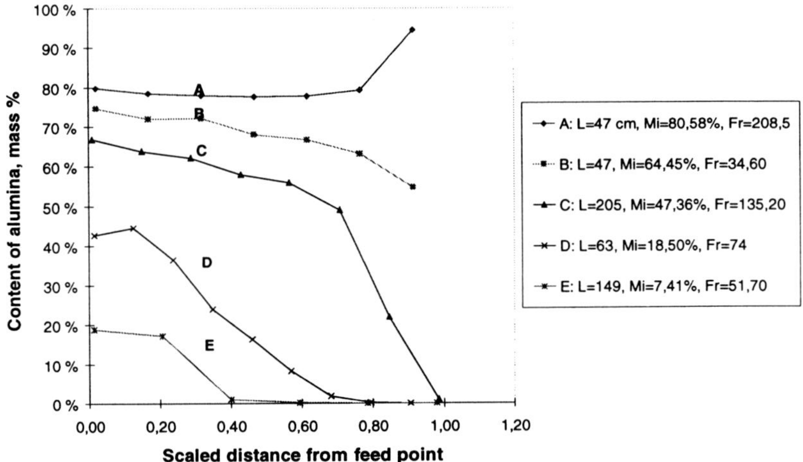

Figure 6.   Characteristic segregation patterns for a mixture of sand and alumina. Mosby [46].

Curve A shows a special form of heap segregation that has not been reported previously. For most of the heap length, the content of alumina is almost constant, but increases near the edge of the heap. The fine particles are clearly over represented here, which is the opposite of what is normally expected. It is thought to be caused by fluidization effects. When relatively fine particles fall onto or slide down a heap surface, spontaneous fluidization will occur. Large particles with densities higher than this fluidized layer will sink to the bottom of the layer and will settle out on the static heap below. In our case, the bulk density of the fluidized alumina layer is less than 1 g/cm^3 while the density of the larger sand particles is 2.65 g/cm^3. In curve A it seems as if the sand has settled out until it is almost depleted at the edge of the heap. Alumina used in these experiments was very easy to fluidize. The experiments with segregation of alumina and sand showed clearly that the fluidization effects only occur if the content of alumina in the feed is high. For all the experiments with a feed content of alumina higher than 70 % fluidization effects were seen.

Curve B has an even, almost linear, decrease in the content of alumina from approximately 75 % near the feed point to about 55 % in the sample taker nearest the end wall of the tester. This is a typical example where the rolling mechanism is the dominating segregation mechanism,. a mechanism that dominates at low feed rates and fines contents more than sufficient to fill to the voids between the particles of the coarser component.

Curve C has the same form as curve B from the feed point till about 75% of the heap length. From 75 % of the heap length until the edge, the content of alumina shows a steep drop. Here two segregation mechanisms have been active. The rolling mechanism has been dominating through the first 3/4 of the heap length, while the sieving mechanism has been dominating in the lowest part of the heap. At the first part there are enough fines to fill all the spaces between the larger particles, and therefore rolling is the dominating effect. When a layer of the mixture flows down the heap, the coarser particles that reach the top of the layer will be accelerated in relation to the flowing layer below. After a certain distance, the content of alumina reaches a value where the alumina is no longer able to fill all the voids. From this point, the content of alumina will drop at a much higher rate.

Curve D has a steep drop all the way from the second sample until the content of alumina reaches 0 at a distance of 80% from the feed point. This is a clear effect of the sieving mechanism where fines are sieved out and settle out near the centre of the heap.

The segregation pattern in curve E is caused by the same mechanism as in curve D, the sieving mechanism. In this experiment the content of alumina reaches 0 after travelling a distance only equivalent to 40% of the heap length. The alumina is then already depleted.

Salter [48], on the other hand, using much coarser particles (all over 1 mm) said that he found little influence of feed rate on the results from his two dimensional tester. In his experiments, the mixing ratio (i.e. fines content in the feed) was the dominant variable, followed by the diameter ratio of the binary mixture used. His results are shown in Figure 7. It is clear that the choice of test materials affects the pattern obtained. Salter's decision to restrict himself to use sizes above 1 mm, effectively eliminated the fluidization mechanism observed by Mosby, and spontaneous percolation was eliminated by restricting himself to diameter ratios less than 4:1.

In Figure 7, one can see an influence of feed rate when comparing tests T53 (100 g/s) with T48 (700 g/s), which appears to indicate that feed rate is important at higher diameter ratios.

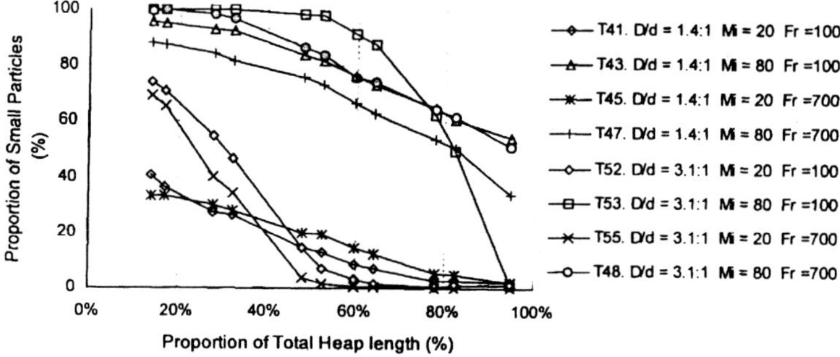

Figure 7.   Patterns of segregation. Effects of mixing ratios, feed rate and diameter ratio. Salter [48].

2.1.2   Three Dimensional Testers

Mosby and Salter both went on to develop three-dimensional testers. Mosby's tester is shown in Figure 8. To reduce the disturbance caused by the sampling device when inserted in the heap, the material inside the sampling device is covered by a plate pressed on to the top of the material. Otherwise the insertion will flatten the heap somewhat, reducing the effect of the segregation that has taken place during filling. Salter's tester was similar to Mosby's, but with a somewhat more sophisticated feeding device. Figures 9 and 10 show the sampling device used.

Figure 8. Three dimensional heap segregation tester developed by Mosby [54].

Figure 9. Insertion of sampling device into three-dimensional tester [54].

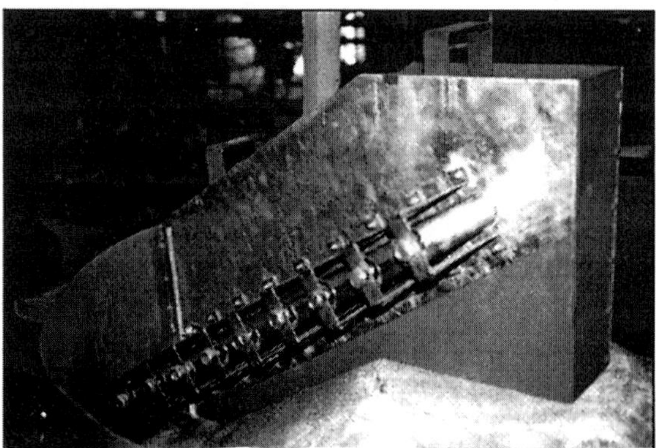

Figure 10.   Sampling tubes fixed to sampling device after superfluous material is drained [54].

The results of parallel tests from the three dimensional testers were found to be less repeatable than those from the half-heap two-dimensional units, but still adequately similar. Mosby's preliminary test results are shown in Figure 11a and Salter's, that were even more reproducible, are shown in Figure 11b.

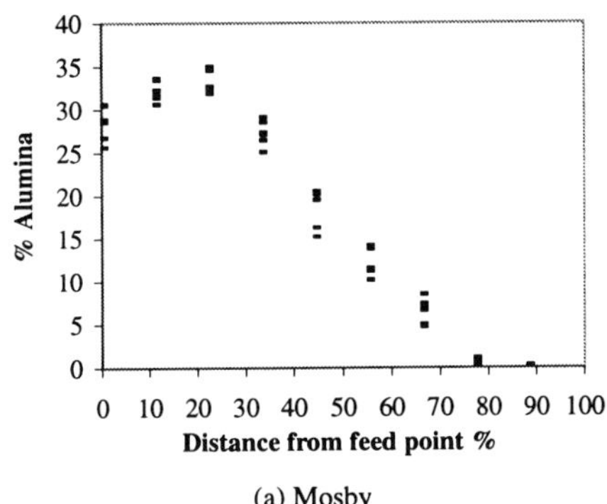

(a) Mosby

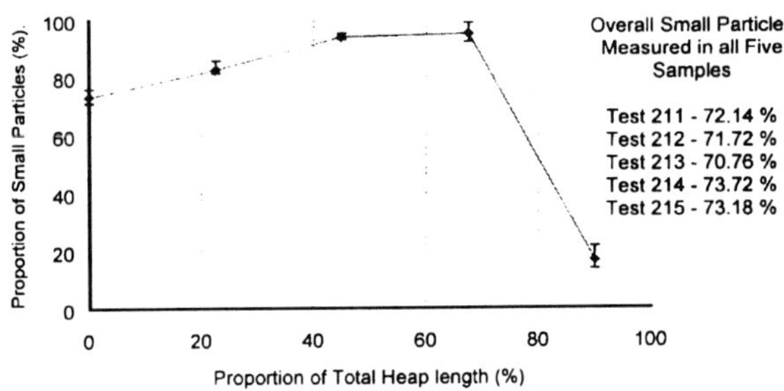

(b) Salter

Figure 11.   Repeatability of results with three-dimensional heap segregation testers.

Both Mosby and Salter found that feed rate had a more noticeable effect with the 3-D testers than with the 2-D half-heap units. Their results are shown in Figure 12.

Finally, the various patterns of segregation observed by Mosby at a feed rate of 260 g/s for various mixing ratios are shown in Figure 13, and for various diameter ratios by Salter in Figure 14. Figure 13 shows that 3-D testers produce segregation patterns qualitatively similar to those obtained in the 2D half-heap units, but there are differences with regard to the distance from the centre at which the various phenomena are observed, as is further discussed in the next section.

S. R. DE SILVA ET AL.

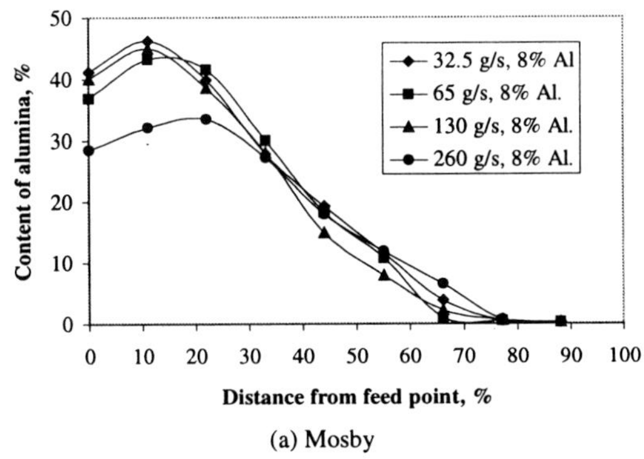

(a) Mosby

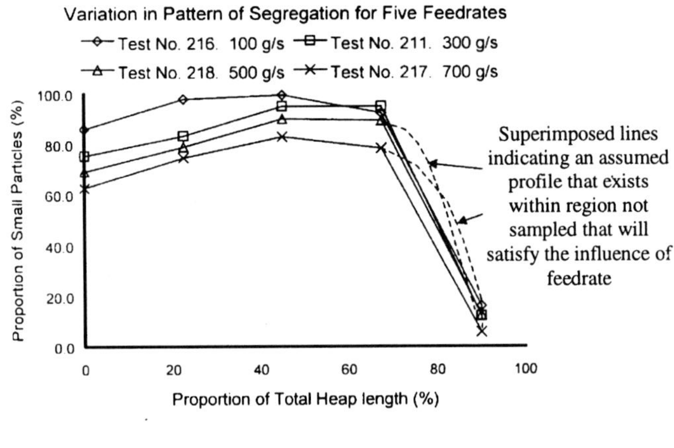

(b) Salter

Figure 12. Effect of feed rate on segregation patterns in a 3-D heap segregation tester.

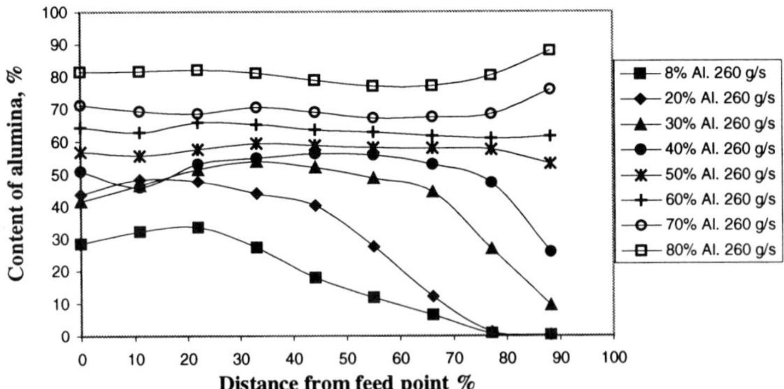

Figure 13. Patterns of segregation in 3-D tester as function of mixing ratio [46].

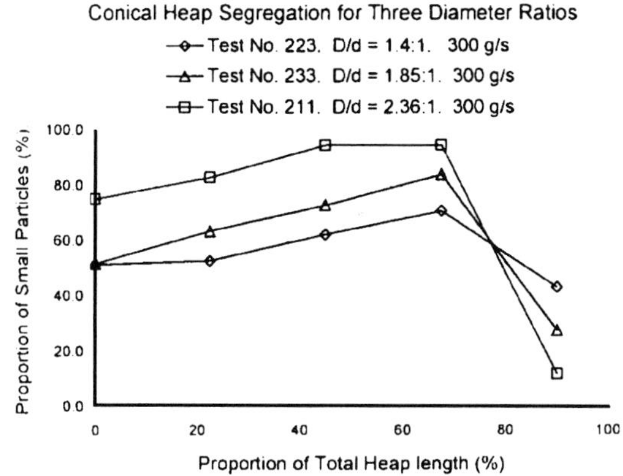

Figure 14.   Patterns of segregation in 3-D tester as a function of diameter ratio [48].

3.   Reconciliation of differences between 2-D and 3-D Testers

As.mentioned before, there are clear differences between the results from the 2-D, half heap, testers and those from 3-D, full heap, testers. Figure 15 shows Mosby's results for mixing ratios of 8%, 20% and 30% (i.e. 8 and 20% alumina in the mixture).

Salter found similar differences. By and large, however, the two types of testers provide the same qualitative information. In observing what happens in the testers, the most significant observation is that layers of material do not flow evenly from the apex to the edges of the heap. The nature of the flow is sporadic and - in both testers – the layers flow in the form of avalanches. This effect is more pronounced in the 3-D units, but was also clearly observed, and discussed in detail, by Salter during tests with his plane flow unit. Both authors conclude that the frequencies of the avalanches are very dependent on feed rate. Mosby concludes that if one were to adjust the feed rate in a 3-D tester to be comparable to that in a 2-D tester by multiplying by the area ratio of the respective surfaces, one should have comparable results. Unfortunately, he does not provide proof of this claim by citing the relevant experiments. Taking the nature of the segregation mechanisms active in a growing heap into consideration, however, it should not be the mass flow rate that one should multiply by the area ratio, but the **volumetric** flow rate, Since it is a given volume that is spreading over an increasing area down the slope of the heap. Mosby does in fact imply this when he says that "a two and a three dimensional heap should grow at the same speed in height". This approach, however, will not account for the nature of the avalanches.

Salter [48] suggests that the equivalent distance from the apex of a 3-D heap to a probe position be calculated on the basis of the formula:

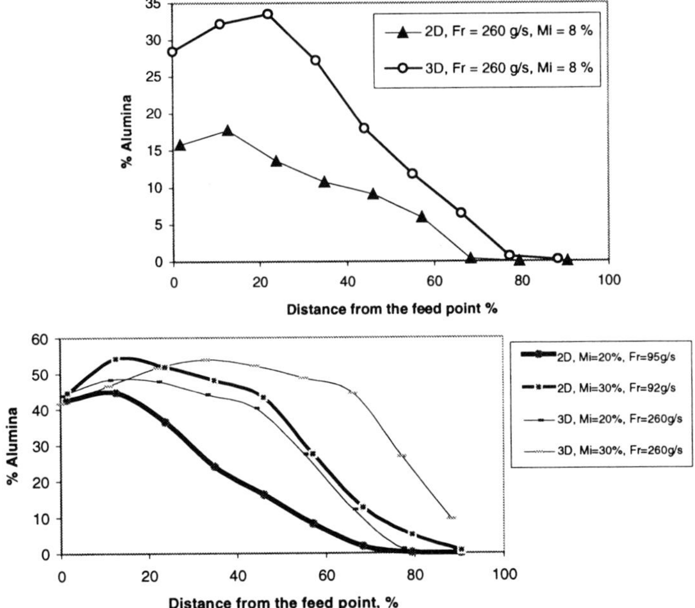

Figure 15. Differences between 2-D and 3-D test results.

$$L = \frac{\sqrt{L_{PF}}}{\sqrt{L_{CH}}} \cdot \frac{100}{1} \%$$

(1)

In equation (1) L_{PF} is the distance of probe from the apex in the 2-D tester and L_{CH} is the distance from the apex to the periphery in a conical, 3-D heap. The reasoning here is the same as Mosby's. Using Equation (1), he predicts the segregation pattern in his 3-D tester from his 2-D results as shown in Figure 16, and noted in the legend as the "scaled profile." He went on to also use the predictive technique on Mosby's results, as shown in Figure 17.The results are encouraging, however, the technique only takes account of an area relationship, and ignores the chaotic nature of the avalanches.

4. Conclusions

From the evidence available so far, it appears that segregation mechanisms that play a role in heap formation can indeed be identified using half-heap two-dimensional testers - at least qualitatively. Quantitative predictions of what exactly happens in a real 3-D situation, based on 2-D tests, are somewhat further away, but may be possible if one can find a method of incorporating the chaotic nature of the avalanches into a physical model. Further work must be directed towards this end.

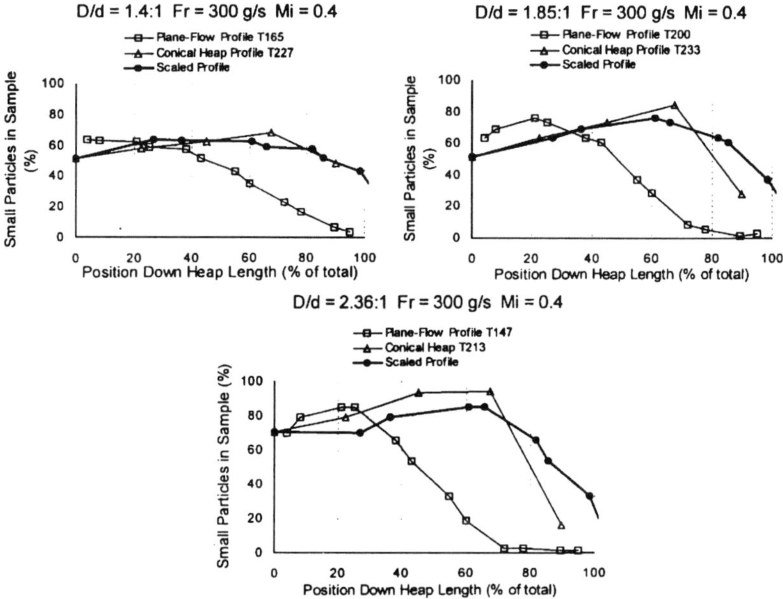

Figure 16. Segregation patterns predicted from 2-D tests, and those measured in a 3-D tester. Salter [48].

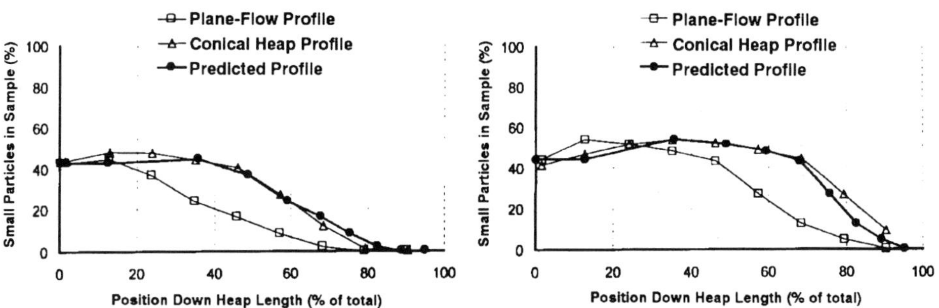

Figure 17. Segregation patterns predicted from 2-D tests, and those measured in a 3-D tester by Mosby [46].

5.   Acknowledgements

The authors wish to acknowledge the support of the Norwegian Research Council, The Swedish Mineral Research Foundation, our POSTEC members, and Hydro Aluminium, in sponsoring our work in segregation over the last several years.

6. References

1. Mitchell, D.R.: Segregation in the Handling of Coal. *American Institution of Mining, Metallurgical and Petroleum Engineers. Trans.* **130** (1938),

2. Sherman, R.A, Kaiser, E.R: Segregation of Coal in an Industrial Steam Plant Bunker. *Combustion,* December 1937, pp. 25-28.

3. Holmes. *Colliery Engr.* (1934) 11.

4. Castanoli: *Amer. Min. Congress Yearbook* (1935), 270.

5. Hebley: *Appalachian Coals, Inc., Fuel Engineers Meeting,* July 1936.

6. Brown, R.L.: The Fundamental Principles of Segregation. *J. Inst. Fuel* **13** (1939), 15.

7. Stock, A.J.: Coal Segregation in Boiler Plants. *Mechanical Engineering* **66** (1944).

8. Garve, T.W.: *J. Amer. Ceramic Soc.* **8** (1925), 668.

9. Williams, J.C.: The Segregation of Particulate Materials. A Review. *Powder Technology,* **15** (1976), 245-251.

10. Williams, J.C., Kahn, M.I.: The Mixing and Segregation of Particulate Solids of Different Particle Size. *Chemical Engineering,* **19**, London, (1973), 169.

11. Williams, J.C.: The Segregation of Powders and Granular Materials. *Fuel Society, Sheffield, Journal Vol. 14.* (1963).

12. Bridgwater, J., Ingram, N.D.: Rate of Spontaneous Interparticle Percolation. *Transactions of the Institution of Chemical Engineers* **49** (1971), 163-169.

13. Stephens, D.J., Bridgwater, J.: The Mixing and Percolation of Cohesionless Particulate Materials, Part I: Failure Zone Formation. *Powder Technology* **21** (1978), 17-28.

14. Stephens, D.J., Bridgwater, J.: The Mixing and Percolation of Cohesionless Particulate Materials, Part II: Microscopic Mechanisms for Particles Differing in Size. *Powder Technology* **21** (1978),.17-28.

15. Drahun, J.A., Bridgwater, J.: The Mechanisms of Free Surface Segregation. *Powder Technology* **36** (1983), 39-53.

16. Foo, W.S., Bridgwater, J.: Particle Migration (short communication). *Powder Technology* **36** (1983), 217-273.

17. Bridgwater, J., Foo, W.S., Stephens, D.J.: Particle Mixing and Segregation in Failure Zones: Theory Theory and Experiment. *Powder Technology* **41** (1985), 147-158.

18. Johanson, J.R.: Solids Segregation - Causes and Solutions. *Powder and Bulk Eng.,* August 1988, 13-19.

19. Johanson, J.R.: Particle Segregation, and What to do About it. *Chemical Engineering,* May 1978.

20. Johanson, J.R.: Solids Segregation Causes and Solutions. *Powder & Bulk Solids, Conf. May 1991, Rosemont, Illinois.*

21. Johanson, J.R.: Solids Segregation - Case Histories and Solutions. *Bulk Solids Handling* **7**[2] (1987), 205-208.

22. Bagster, D.F.: The Effect of Coarse Particle Concentration and of Water Content on the Segregation of Particulate Solids in Bins. *Proc. RELPOWFLO I,* Bergen, Norway, 1985, 289-302.

23. Bagster, D.F.: The Influence of Cohesion on the Segregation Pattern in Bins. Proc. Int. *Conf. on Bulk Materials Storage, Handling and Transportation,* Newcastle, Australia, 1983, 203-206.

24. Bagster, D.F.: Studies of the Effect of Moisture Content and Coarse and Fine Particle Concentration on Segregation in Bins. *KONA* **14** (1996), 138-143.

25. Yu, A. and Standish, N.: Estimation of the Bulk Density of a Coal Stockpile. *Bulk Solids Handling* **11** [3] (1991), 605-612.

26. Standish, N., Yu, A., He, Q.: An Experimental Study of the Packing of a Coal Heap. *Powder Technology* **68** (1991), 187-193.

27. Standish, N.: Studies of Size Segregation in Filling and Emptying a Hopper. *Powder Technology* **45** (1985), 43-56.

28. Shinohara, K., Idemitsu, Y., Gotah, K., Tanaka, T.: Gravity Flow Mechanism of Particles from a Hopper. *Ind. Eng. Chem. Process Design and Development* **7**[3] (1968), 378-383

29. Shinohara, K., Shoji, K., Tanaka, T.: Mechanism of Segregation and Blending of Particles Flowing out of Mass-Flow Hoppers. *Ind. Eng. Chem. Process Design and Development* **9**[2] (1970), 369-376.

30. Shinohara, K., Shoji, K., Tanaka, T.: Mechanism of Size Segregation in Filling a Hopper. *Ind. Eng. Chem. Process Design and Development* **11**[3] (1972), 369-376.

31. Shinohara, K.: Mechanism of Segregation of Differently Shaped Particles in Filling Containers. *Ind. Eng. Chem. Process Design and Development* **18**[2] (1979).

32. Shinohara, K., Miyata, S.: Mechanism of Density Segregation of Particles in Filling Vessels. *Ind. Eng. Chem. Process Design and Development* **23**[3] (1984), 423-428.

33. Shinohara, K.: Some models on particle-segregation in filling hoppers. *Aufbereitungs-Technik, No. 3,* 1985, 116-122.

34. Shinohara, K.: General Mechanism of Particle Segregation during Filling of Hoppers. *Proceedings of 9th CHISA Congress, section H: Particulate Solids, H.3.5,* Prague, (1987).

35. Shinohara, K.: General Segregation Mechanism of Binary Solids Mixtures Filling Two-Dimensional Hoppers. *Aufbereitungs-Technik* **31**[9] (1990), 482-488.

36. Shinohara, K., Enstad, G.G.: Segregation Mechanism of Binary Solids in Filling Axi-Symmetric Hoppers. *Proceedings of 2nd World Congress Particle Technology. September 1990, Kyoto, Japan.*

37. Shinohara, K., Shaitoh, J.: Segregation by Multi-Point Feeding of Binary Solids Mixture onto Two-dimensional Heap. *Intnt. Conf. Bulk Mat. Handl and Transp.; Symposium on Freight Pipelines,* Wollongong, Australia, July 1992.

38. Shinohara, K., Saitoh, J.: Mechanism of Solids Segregation over a Two-Dimensional Dead Man in Blast Furnace. *The Iron and Steel Institute of Japan* **33**[6] *(1993),* 672-680.

39. Savage, S.B., Lun, C.K.K.: Particle Size Segregation in Inclined Chute Flow of Dry Cohesionless *Granular Solids. Fluid Mech.***189** (1988), 311-335.

40. Popplewell, L.M., Campanella, O.H., Sapru and Peleg, M.: Theoretical Comparison of Two Segregation Indices for Binary Powder Mixtures. *Powder Technology* **58** (1989), 55-61.

41. Alonso, M., Satoh, M., Miyanami, K.: Optimum Combination of Size Ratio, Density Ratio and Concentration of Minimise Free Surface Segregation. *Powder Technology* **68** (1991), 145-152.

42. Artega, P. and Tuzün, U.: Flow of Binary Mixtures of Equal Density Gradients in Hoppers – Size Segregation, Flowing Density and Discharge Rates. *Chem. Eng. Sci.* 45[1] (1990), 205-223.

43. Dolgunin, V.N., Ukolov, A.A.: Segregation Modelling of Particle Rapid Gravity Flow. *Powder Technology* **83** (1995), 95-103.

44. Meakin, P. and Jullien, R.: Simple Models for Two- and Three Dimensional Particle Size Segregation. *Physica A* **180**, (1992), 1-18.

45. De Silva, S.R., Enstad G.G.: Bulk Solids Handling in Scandinavia – A Case Study in the Aluminium Industry. *Bulk Solid Handling* **11** (1991), 65-68.

46. Mosby, J.: Investigation of the Segregation of Particulate Solids with Emphasis on the use of Segregation Testers. *Dr.Ing. Thesis, Telemark College/Norwegian University of Science & Technology, 1996.*

47. De Silva, S.R.: Mixing and Segregation in Industrial Processes. IFPRI Report, 1997.

48. Salter, G.F.: Investigations into the Segregation of Heaps of Particulate Materials with Particular Reference to the Effects of Particle Size. *Ph.D Thesis, University of Greenwich, 1998.*

49. Bates, L.: User Guide to Segregation. *1ˢᵗ Ed. British Materials Handling Board/ Bartham Press 1998.*

50. Harris, J.F.G., Hildon, A.M.: Reducing Segregation in Binary Powder Mixtures with Particular Reference to Oxygenated Washing Powders. *Ind. Eng. Chem. Process Des. Develop.* 9[3], (1970).

51. Enstad, G.G.: Segregering av partikulære materialer ved innmating i beholdere (Segregation of Particulate Materials during the Filling of Containers). CMI report 853111-2, 1985.

52. Maltby, L.P.: Segregering ved fylling av haug (Segregation in the Formation of Heaps). *Tel-Tek Report.* May, 1990.

53. Van den Berg, J.G.: Investigations on Segregation Tendencies using Shinohara's Theory. M.Sc. *Dissertation. University of Twente/Tel-Tek,* December, 1992.

54. Bartolome, R.: Three Dimensional Testers for Investigation of Heap Segregation. *Tel-Tek Report,* 1995.

PARTICLE SEGREGATION IN GRANULAR FLOWS DOWN CHUTES

JAMES W. VALLANCE and STUART B. SAVAGE
McGill University
Department of Civil Engineering and Applied Mechanics
817 Sherbrooke Street West
Montréal, Québec, H3A 2K6, Canada

Abstract

Laboratory experiments with dry particles of different sizes and different densities, and different particle-liquid mixtures flowing down rectangular chutes show the relative importance of segregation mechanisms. In uniform steady flow experiments with binary mixtures of small and large particles, the small particles fall downward and the large particles migrate upward. In slow, dry, frictional flows, the downstream segregation is so efficient that zones of 100% small and large particles separated by a concentration jump form. In rapid, dry, collisional flows, segregation is less efficient because of diffusive mixing. Diffusive mixing smoothes vertical concentration profiles so that concentration jumps blur or disappear. The presence of a viscous fluid inhibits size segregation. Little to no segregation occurs when liquid density matches particle density.

In general, segregation patterns and velocity profiles observed in steady, uniform, chute-flow experiments, having solids fractions of 0.4 or greater, result from kinetic sieving, attenuated by diffusive mixing in high energy flows and by viscosity or buoyancy in flows with liquid present. Kinetic sieving comprises two processes. The first is the gravity-driven, size-dependent, void filling mechanism of percolation. In a granular flow subjected to shear strain, small and large particles percolate downward into the void spaces that periodically open beneath them. The small particles percolate downward much more frequently than large particles because the probability of void spaces sufficiently large opening beneath them is greater than that of void spaces large enough opening beneath the large particles. The second mechanism is the expulsion of individual particles out of their layers into adjacent layers owing to contact forces. Expulsion can occur in either direction and does not have to be size preferential. A model based on the above definition of the kinetic sieve gives reasonable agreement with the results of dry granular experiments. A third process, diffusive mixing, becomes progressively more important in increasingly energetic flows.

In steady uniform flows, segregation of particles by density differs fundamentally from segregation of particles by size. In dry granular flows with particles of the same size and differing densities, dense particles efficiently migrate downward in upper parts of the flow but do not efficiently displace light particles in the lower parts of the flow. Concentration jumps observed with particles of contrasting size do not occur because

31

A.D. Rosato and D.L. Blackmore (eds.), IUTAM Symposium on Segregation in Granular Flows, 31–51.
© 2000 *Kluwer Academic Publishers. Printed in the Netherlands.*

neither dense nor light particles percolate preferentially. Instead, imbalances in contact forces, which favor the dense ones to move downward when dense and light particles encounter each other, are the sole cause of segregation.

1. Introduction

Deformation or agitation of mixtures of particles in the presence of gravity commonly results in segregation of the particles if the particles have different physical character-istics. By far the most important of these physical characteristics is contrast in size; however, other parameters such as density, shape, surface roughness, elastic modulus, voidage, and several others may also be important in segregation processes. In mineral processing, gravity concentration devices such as the pinched sluice, spiral, and Reichert cone are used to sort particles according to their size or density as they pass through the device. Particle segregation is vital in mineral processing procedures but is troublesome when uniform mixing is desired as in the manufacture of pharmaceuticals, ceramics and plastics [21]. Despite the importance of particle segregation mechanisms in industrial processes, little experimental evaluation of them appears to exist [13].

This paper discusses particle segregation mechanics in relatively high concentration grain flows moving down slopes, advances an analysis to explain vertical sorting of large and small particles in moving granular flows, and, through experiments, shows the relative importance of the most important segregation processes.

2. Important processes

This section reviews selected literature on the processes that affect the segregation and desegregation of particles in granular flows. In flowing, poorly sorted sediment, these processes include percolation, diffusion, and gravitational settling of particles, in addi-tion to local forces that may push particles into adjacent layers.

2.1. PERCOLATION

Percolation may be viewed as the migration in a gravitational field of small particles through a matrix of larger particles. Percolation may be spontaneous, vibration induc-ed, or shear induced. Spontaneous interparticle percolation is defined as the drainage of small particles through a static arrangement of larger particles [21]. In a gravitational field the fine-grained particles will move downward so that the upper portions of a mate-rial will have relatively small proportions of fine-grained particles and lower part of the material will have large proportions.

The fine-grained particles will be concentrated in the direction of flow if a fluid is forced through the material. Consideration of a triangular arrangement of three large particles having diameter, D, indicates an upper limit of $0.155 D$ as the maximum size of the smaller percolating particles. Percolation of particles in a static array is of less im-portance in granular flows than percolation of shearing or vibrating particles; neverthe-less, comparison of this simple system to more complicated systems is of some value.

Bridgwater and others [4] and Bridgwater and Ingram [3] have shown through dimensional analysis and experiment that the nondimensional percolation velocity,

$$\frac{c}{(2gD)^{\frac{1}{2}}} = fn\left(\frac{d}{D},\ \varepsilon,\ v\right), \tag{1}$$

where c is the percolation velocity of small particles, g is the gravitational constant, d/D is the diameter ratio of the percolating to static particles, ε is the coefficient of restitution, and v is the solids fraction. The density ratio of the percolating and static particles is unimportant since bed particles are not permitted to move.

Thus only the size ratio of percolating to static particles, coefficient of restitution, and solids ratio affect percolation velocities. Generally, c increases as d/D increases; c goes to zero as ε goes to one; and c (in the radial direction) increases as v decreases.

The static beds considered previously permit only small particles ($d/D \leq 0.155$) to percolate. When vibration or shear deform the bed, void size and shape vary with time. As a result larger voids are created so that a wide range of particles can percolate, including the bed particles.

When a dry bed of particles of uniform size containing one large particle vibrates, the large particle tends to rise. In a systematic study of this effect, Williams [28] suggested that the small particles tend to move into spaces that open near the underside of the large particle and that the large particle pivots up over them at the point of contact and thus rises. Williams not only observed that the large particle migrated to the top of the bed but that increasing the density of the particle enhanced this upward migration. This is perhaps surprising since denser particles of the same size as the bed particles are observed to migrate downward on shaking (observation of the authors). In contrast, Rippie and others [17] observed that a threefold contrast in density resulted in no segregation of identically sized particles when vibrated. The explanation of these contradictory observations probably lies in the experimental method of Rippie and his colleagues, which involves vertical agitation in narrow cylinders. Our observations are based on random agitation.

Two closely allied hypotheses purport to explain upward migration of the coarse particle. The first suggests that as the particles are jostled small voids periodically open beneath the large particle and small particles slip into these voids. Because of greater friction between the small particles owing to interlocking and compression of them by the large particle, the small particles are unlikely to move once in place beneath the large particle [28]. The second hypothesis derives from probabilistic arguments of Rosato and others [18]. They suggest that several small particles beneath the large particle would have to be moved simultaneously in order to create a void large enough for it to fit into, but only one particle would have to be moved to create such a void beneath a small particle. Therefore, the probability of opening a void space that is big enough for a large particle to fall into is considerably less than that of opening a void big enough for a small particle to fall into. Thus, large particles will tend to migrate upward because probability favors small particles filling voids beneath them. Both mechanisms probably operate. Vibrational segregation is also commonly observed in mixtures having greater proportions of large particles. Size appears to be the dominant sorting parameter, but density sorting can occur where grains are of nearly subequal size. Rippie and others [17] observed circulation patterns due to wall effects for angular particles. These patterns can influence the distribution of the segregated particles but do not cause segregation of the particles.

Bridgwater and his colleagues [1-2; 5-6, 24) have carried out an extensive set of experiments with a simple rectangular shear cell. They deformed the shear cell in a reciprocating fashion so that the cell alternately took on the shape of a parallelogram slanted to the right and to the left when viewed from the front. Tracer particles could be added at the top, middle, or bottom of the shear-cell apparatus.

Cooke and others [6] and Bridgwater and others [1] give the nondimensional percolation velocity as

$$\frac{c}{YD} = fn\left(\frac{d}{D}, \ Y\left(\frac{D}{g}\right)^{\frac{1}{2}}, \ \frac{\rho_p}{\rho_b}, \ \frac{S_p}{S_b}, \ \frac{\varepsilon}{E_b}, \ \frac{\rho_b g D}{E_b}, \ \frac{f_p}{f_b}\right), \tag{2}$$

where c is the percolation velocity, Y is the bulk shear rate, d and D are the percolating and bulk particle diameters, ρ is particle density, S is a shape factor, E is the elastic modulus, f is the coefficient of friction, and subscripts p and b refer to percolating and bulk particles. YD is interpreted as the relative velocity between shear layers one diameter apart, $Y(D/g)$ as the relative-velocity to free-fall-velocity ratio, $\rho_b g D/E_b$ as the strain associated with the weight of the particles, and ε/E_b as the strain associated with the bulk stress [21]. The ratio of percolating to bed particle diameters, d/D, was determined to be the most important parameter. A tenfold increase in ρ_p/ρ_b resulted in a 24% increase in nondimensional percolation velocity. An increase in ε/E_b resulted in a slight decrease in c/YD. All other parameters from the dimensional analysis had relatively little impact on percolation velocities.

2.2. DIFFUSION OF PARTICLES

In a flowing granular material, diffusion is defined as a tendency of particles to become evenly dispersed. In some degree, diffusion of particles occurs in all granular flows but is attenuated in flows where agitation of particles is limited. In slowly moving grain flows, the particles experience little agitation, and diffusive processes are much weaker than percolation. The strongest diffusion occurs in dry, cohesionless granular material, subjected to energetic flow regimes, in which collisional interactions of the particles are important.

Scott and Bridgwater [25] studied random motion of dry particles in the reciprocating shear cell described above. They placed particles (identical except for differing color) in each half of the cell so that a vertical boundary between them was perpendicular to the horizontal shear plane. They found that diffusion occurs about equally in all three directions despite shearing in only one direction. They also found that the diffusion coefficients were constant.

Hwang and Hogg [14] studied diffusion rates of sand in an inclined chute roughened by sand. They injected tracer particles at the top or base of the flow and then used splitter plates to study particle distribution. The diffusion coefficient depended on the shear rate plus a constant. The shear-rate dependence of Hwang and Hogg's chute-flow experiments was not observed in the oscillatory experiments of Scott and Bridgwater [24] Diffusion processes are more important in fast flow than in slow flow because momentum transfer is dominated by particle collisions rather than by slow frictional rubbing. As the flow rate increases, voidage increases, and particle collisions become more common. In very fast diffuse flow, momentum transfer is by particles migrating from one layer to another, and this too contributes to diffusive dispersion of the particles [20].

2.3. GRAVITATIONAL SETTLING

The most common explanation [e.g., 11] of normal segregation in a granular flow is gravitational settling of large particles in a liquid-particle mixture. The mechanism requires that the concentration of solids in fluid be small enough that the density, viscosity, and strength of the fluid are insufficient to support larger and denser particles. The large, dense particles have greater settling velocities and thus tend to migrate more rapidly toward the bed. Three conditions must be met if this explanation of normal size segregation is to be viable. First, the surrounding fluid must be viscous enough that the particles reach their terminal velocities quickly and thus don't indefinitely accelerate under the influence of gravity. Second, the solids fraction must be small enough that settling is not significantly hindered. If particles are in grain-to-grain contact there is a tendency for the smaller ones to percolate downward through the interstices and the large ones to migrate upward. Third, the flow cannot move as a rigid plug. If the matrix has sufficient strength that plug flow occurs then theoretically neither coarse- nor fine-grained particles should be able to settle in the plug. In some dilute flows, the particles may be randomly dispersed until the waning stages of the flow and only then experience differential settling [11].

2.4. SQUEEZE EXPULSION AND KINETIC SIEVING

Savage and Lun [23] have performed chute-flow experiments using binary mixtures of particles of identical material but different sizes. In their experiment the bed was roughened so that the particles were sheared in the presence of a gravitational field. The small particles percolated to the bottom and the larger ones drifted to the top of the sheared layer. Splitter plates were arranged to measure concentration profiles at various points in the flow. Sorting was so efficient in the streamwise direction that concentration jumps quickly developed and zones of 0% fines, mixed particles, and 100% fines were sharply delineated. Middleton [16] first proposed an explanation for this inverse sorting, which he called kinetic sieving. Dyer [10] observed inverse segregation during hindered settling in mineral processing applications and proposed a similar mechanism to explain it. Savage and Lun [23] refer to this mechanism as the randomly fluctuating sieve. The probability of a particle encountering a big enough void space below to fall into is the basis of the mechanism. For a given solids fraction, the probability of a small particle encountering a big enough opening is greater than that of a big particle encountering a big enough opening.

Kinetic sieving allows particles (both large and small) to percolate only toward the impermeable bed; thus a mechanism for the counter flow of particles is required to give zero net mass flux in the direction perpendicular to the bed. Fluctuating contact forces on an individual particle can cause force imbalances that squeeze the particle out of its own layer into an adjacent one if an opening is available or the force imbalance is sufficiently large. Savage and Lun [23] propose that this squeeze-expulsion mechanism is not gravity driven, is not size preferential, and operates in either direction. It is common to assume that contact forces are more likely to force large particles upward than small ones. However, if small particles percolate downward more often than big ones, size segregation will occur regardless of whether the large particles are preferentially squeezed upward or are equally likely to be pushed either up or down. Through this combination of processes, particles will tend to segregate, with small ones percolating toward the bed and large ones pushed toward the surface. This combination of the randomly fluctuating sieve and squeeze expulsion will be referred to as the kinetic-sieve mechanism.

3. A theoretical framework for size segregation

On the basis of experimental observations, Savage and Lun [23] advanced an analysis of segregation in uniform chute flows of binary mixtures of large and small, dry, cohesionless particles. Using the continuity equation and information entropy concepts, Savage and Lun [23] obtained an analytic solution for the small-to-large-particle frequency ratio, $\eta = \eta(x,y) = f_s / f_l$, in a region of steady uniform flow, in terms of distance downstream, x, distance from the bed, y, the diameters of the large, D, and small, d, particles, and the net nondimensional percolation velocity of the small particles, $\hat{q}_{s\,net}$. Their solution also assumed uniform solids fraction and velocity which increases linearly with distance from the bed. The continuity equation for the small particles is

$$u(y)\frac{\partial \rho_s}{\partial x} + \frac{\partial}{\partial y}\left[\rho_s q_{s\,net}\right] = 0 , \tag{3}$$

where $u(y)$ is the downstream velocity as a function of distance from the bed, $q_{s\,net}$, and ρ_s are the net percolation velocity and mass per unit volume of the small particles. The term ρ_s is given as

$$\rho_s = \frac{\rho_p \eta \left(\dfrac{d}{D}\right)^3}{(1+e)\left[1+\eta\left(\dfrac{d}{D}\right)^3\right]} , \tag{4}$$

where ρ_p is the density of the particles, e is the voids ratio, and d/D is the ratio of the diameters of the small and large particles. The net nondimensional percolation velocity of the small particles is

$$\hat{q}_{s\;net} = \frac{-q_{s\;net}}{D\left(du/dy\right)} \, . \tag{5}$$

The net nondimensional percolation velocity, $\hat{q}_{s\;net}$, can be calculated independently using information entropy concepts [23, p 320]. Substituting (4) and (5) into (3) and rearranging yields

$$u\frac{\partial \eta}{\partial x} - D\hat{q}_{s\;net}\frac{\partial u}{\partial y}\frac{\partial \eta}{\partial y} - D\hat{q}_{s\;net}\eta\left(1+\eta\left(\frac{d}{D}\right)^3\right)\frac{\partial^2 u}{\partial y^2} = 0 \, . \tag{6}$$

We can now repeat the solution of Savage and Lun for a more general power law velocity distribution where

$$u(y) = k\,y^n. \tag{7}$$

Using equation (7) in equation (6) we obtain

$$y^2\frac{\partial \eta}{\partial x} - nD\hat{q}_{s\;net}\,y\frac{\partial \eta}{\partial y} = n(n-1)\,D\hat{q}_{s\;net}\eta\left(1+\eta\left(\frac{d}{D}\right)^3\right). \tag{8}$$

Applying the initial condition $\eta(0, y) = \eta_0(y)$, at $x = 0$ the general solution of equation (8) from the method of characteristics is

$$\eta = \frac{f_s}{f_l} = \frac{\eta_0\left(y^2 + 2nD\hat{q}_{s\;net}x\right)^{\frac{n-1}{2}}}{\left(1+\eta_0\left(\frac{d}{D}\right)^3\right)y^{n-1} - \eta_0\left(\frac{d}{D}\right)^3\left(y^2 + 2nD\hat{q}_{s\;net}x\right)^{\frac{n-1}{2}}} \, . \tag{9}$$

From equation (8), we find that the characteristic surfaces are given by

$$y^2 = -2nD\hat{q}_{s\;net}x + \text{constant}. \tag{10}$$

We can now determine the development of the concentration shock surfaces. In a uniform flow of depth h, and an initially random mixture of large and small particles, $\eta = 0$ at $(x, y) = (0, h)$, and the 0%-fines concentration surface is defined by integrating equation 10 to obtain

$$y(0\%) = \left(h^2 - 2nD\hat{q}_{s\;net}x\right)^{\frac{1}{2}} \, . \tag{11}$$

Similarly, we seek the position of the 100% fines concentration jump, y(100%). Applying the depth-averaged, conservation-of-mass equation for the small particles yields

$$\left[\int_0^h \rho_s u\,dy\right]_{x=0} = \int_0^h \rho_s u\,dy = \int_0^{y(100\%)} \rho_s u\,dy + \int_{y(100\%)}^{y(0\%)} \rho_s u\,dy = \text{constant}. \tag{12}$$

Assuming the solids fraction, $\nu = 1/(1+e)$, is uniform with depth and distance downstream, substituting equations (4), (7), and (9) into equation (12), integrating, and rearranging yields

$$y(100\%) = \left(2nD\hat{q}_{s\ net}\,k_{f\eta_0}\,x\right)^{\frac{1}{2}}, \tag{13}$$

where $$k_{f\eta_0} = \dfrac{1}{\left(k_{\eta_0}^{\left(\frac{-2}{n+1}\right)} - 1\right)}$$ and $$k_{\eta_0} = \dfrac{\eta_0\left(\dfrac{d}{D}\right)^3}{1+\eta_0\left(\dfrac{d}{D}\right)^3}. \tag{14}$$

The distance, x_s, required for complete separation of the small and large particles is obtained by equating (11) and (13) so that $y(0\%) = y(100\%)$ and rearranging as

$$x_s = \frac{h^2}{2nD\hat{q}_{s\ net}\left(1+k_{f\eta_0}\right)}. \tag{15}$$

The distance from the bed at complete separation is

$$y_s = \left(\frac{k_{f\eta_0}}{1+k_{f\eta_0}}\right)^{\frac{1}{2}} h. \tag{16}$$

Equations (9), (11), and (13) together with the relation for $\hat{q}_{s\ net}$ given in Savage and Lun [23] predict the development of the 0% fines concentration line and the 100% fines concentration jump for any n (figure 1). Equation (9) gives the concentration profile between $y(0\%)$ and $y(100\%)$. Concentration jumps develop at the 0%- and 100%-fines concentration surfaces, and these surfaces migrate down and up, respectively, with distance downstream until the concentration jump becomes fully developed where the two surfaces meet. The effect of increasing n, which as we shall see tends to correspond with decreasing slope, improves segregation (figure 1). Figure 1 shows that decreasing n from 1.3 to 0.6 and holding all other conditions constant should increase the distance for complete separation of the large and small particles by a factor of 2.28 (slightly more than $n_1/n_2 = 1.3/0.6 = 2.17$).

In the analysis, the strength of segregation depends on the shear strain rate and hence the velocity gradient. When n is less than 1, the velocity gradient is smaller in the upper

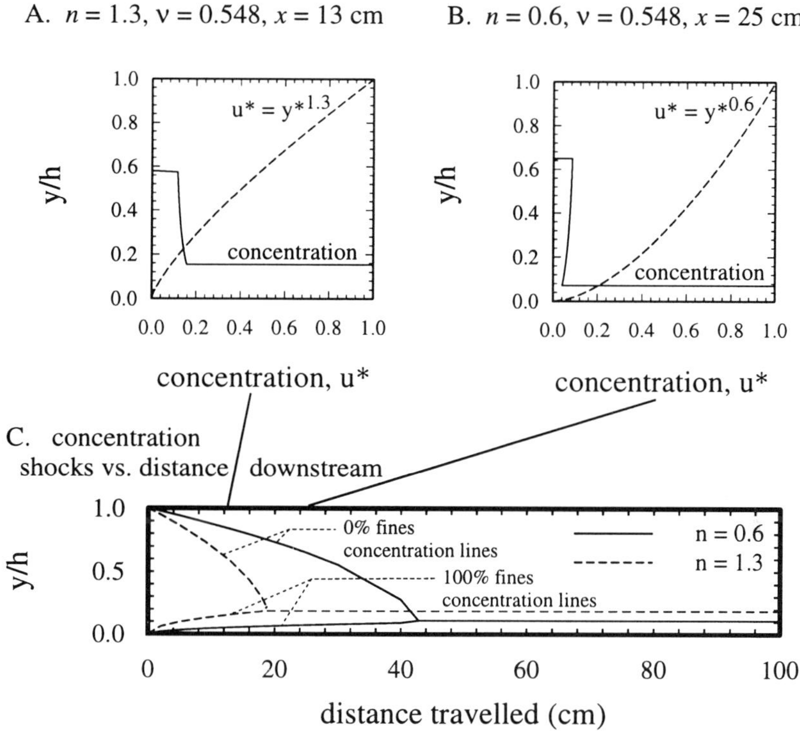

Figure 1. Predicted concentration profiles for small particles in a mixture of large and small particles with a 10% initial concentration of small particles: A.—concentration and $u*$ versus $y*$ with $n = 1.3$ at $x = 13$ cm downstream, and B.— concentration and $u*$ versus $y*$ with $n = 0.6$ at $x = 25$ cm downstream. The solids fraction is 0.548. C. —Position with distance downstream of 0% fall line and 100% accumulation line are shown for the two values of n in which n is the exponent in a power-law velocity profile such that $u* = y*^{n}$ where $y* = y/h$.

half of the flow than it is in the lower half of the flow. When n is greater than 1, however, the strength of the velocity gradients in the upper and lower halves of the flow reverses. Thus, with $n < 1$, the small particles with the greatest distance to move--those that begin near the surface of the flow--percolate more slowly than with $n > 1$, and hence the distance required for full separation is larger (e.g., figure 1 C, $n = 0.6$). In contrast, with $n > 1$, segregation in the upper part of the flow is stronger and the distance required for full separation is smaller (figure 1 C, $n = 1.3$).

Velocity gradients should also control the downstream increase or decrease of small particles within the intermediate zone between the 0% and 100% concentration jumps. Because velocity gradients are stronger near the surface and weaker near the bed where $n > 1$, small particles move into the zone between the 0% and 100% concentration jumps more efficiently from above than move out of it at its base. A net increase of small particles thus occurs in the intermediate zone (figure 1 A). Conversely, where $n <$ 1, vel-ocity gradients are greater near the bed than near the surface of the flow. Small particles

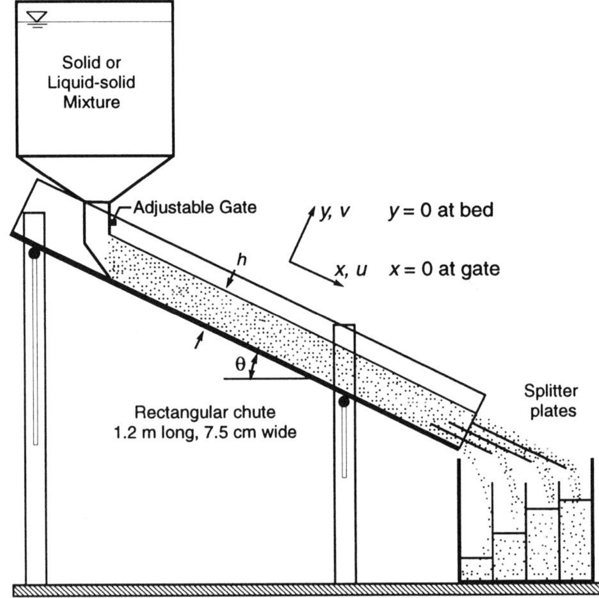

Figure 2. Apparatus used in experiments. The reservoir can be moved to any position along the flume. The flume can be adjusted to any slope.

therefore move out of the intermediate zone more efficiently than come into it from above (figure 1 B). The small particle concentrations thus diminish in the zone between the 0 and 100% concentration jumps where $n < 1$.

4. Experimental results

We performed particle-size segregation experiments for a wide range of bed and slope conditions and simultaneously measured velocity profiles and average velocities for each experiment. These experiments were repeated with mixtures of particles in viscous fluids and with dry mixtures of particles differing in density. On the basis of these results, we infer the relative importance of the various segregation mechanisms discussed in the previous section.

4.1. EXPERIMENTS WITH PARTICLES OF DIFFERING SIZE

The apparatus (figure 2) comprises a flume 7.5 cm wide and 120 cm long, a reservoir that provides a uniform mass discharge of particles, and an array of splitter plates that divides the flow vertically then funnels the vertical splits into 4 separate collector bins. The flume has smooth sidewalls and a bed roughened with particles like those used in the tests. It can be tilted to any angle. The reservoir can be positioned at any point along the flume so that streamwise development of segregation can be measured. By measuring the proportion of fine-grained particles in each of the collector bins, vertical concentration profiles can be measured at any selected distance downstream. In each of the dry size-segregation experiments, the mixture comprises 10% small spherical particles and 90% large spherical particles. The particles, initially, are randomly mixed. Each experiment is allowed to run until a steady state is achieved, then concentrations of small particles are measured from the vertical splits.

The polystyrene particles used in these experiments are nearly round, have an internal friction angle, $\phi = 25°$, and bed friction angle, $\delta = 24°$, on a bed roughened with the same particles. Similarly, the glass particles are round, have an internal friction angle of 23°, and a bed-friction angle of 17.7° on a roughened bed of the same particles. The bed-friction angles reported here are applicable only to the bed roughness, flume conditions, and depths of flow used in these experiments.

Except at a slope of 39° in the rough-bed studies of experiment 1, the experiments are limited to steady uniform flow. Uniform flow on the bed occurs when $\tan\theta > \tan\delta$, $\tan\delta < \tan\phi_d$, and the slope angle, θ, is not so large that accelerating flow occurs. Empirically, uniform flow is possible over the widest range of slopes for depths between 5 and 12 times the dominant particle diameter, D; therefore, the depths in our experiments were in this range.

4.1.1. Size segregation of dry bimodal mixtures of large and small particles

In experiments with dry, bimodal mixtures of large and small particles of equal density ($\rho_s = 1.028$ g/cm^3), a regulated mass flux yielded steady nearly uniform flow with depths between 0.8 and 1.1 cm (5.3D and 7.3D, where D is the diameter of the large particles) and averaging 1.0 cm (6.6D). Most of the flows showed slight downstream variations in depth. Typical flows showed a slight to moderate decrease of depth between the source and 5 cm from the entry. They also showed slight increases of depth within 5 cm of the splitter plates. The mass flux was varied to keep the depth of flow as nearly constant as possible among the experiments. Minor changes in average depth were necessary from experiment to experiment to keep the volume proportions of each of four vertical splits as close to the same as possible so that they would be comparable. The position of the splitter plates was held constant.

Flows of dry bimodal particle mixtures showed that segregation of small particles ($d = 0.97$ mm) to the base of the flow and large particles ($D = 1.51$ mm) to the top of the flow increases dramatically with distance downstream (figures 3 and 4). Figure 4 shows that the segregation is 80% to 90% complete within 20 cm of the entry. This inverse size segregation is strongest at slopes near the bed-friction angle (24°) and becomes weaker at progressively greater slopes (figures 3 and 4). Experimental concentrations for the first split ($0 < y^* < 0.25$, where $y^* = y/h$) are slightly smaller than actual values and those of the second split ($0.25 < y^* < 0.5$) are larger than actual values because some remixing occurs at the first splitter plate.

Downstream segregation strengthens initially and then weakens at a slope of 39°, the largest slope used in the experiments. At that slope, the flow is nonuniform because it accelerates and coincidentally transforms downstream from frictional flow to collisional flow. The gradual desegregation, apparent between 20 to 90 cm of the entry, results from increasing diffusion as the flow undergoes a transition from dense frictional flow to

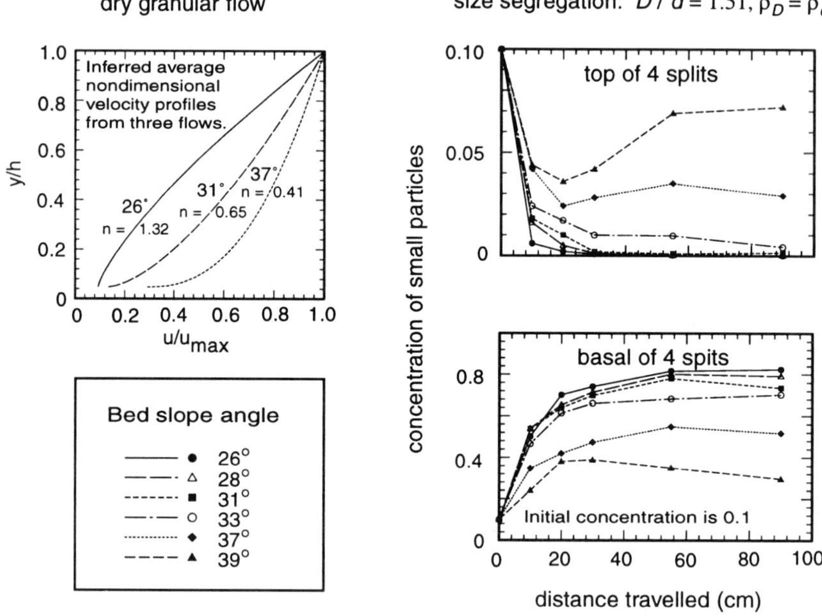

Figure 3. Plots of the concentration of small particles in the uppermost and lowermost of four vertical splits versus distance downstream for bimodal mixtures of large (90%) and small (10%) particles flowing down several slopes as shown. Insets shows the inferred average velocity profiles for slopes of 26°, 31°, and 37° and the line-pattern key for plots to the left.

more diffuse collisional flow, and the particle interaction changes from frictional rubbing to vigorous saltation.

For flows at each slope, we removed the splitter plates and simultaneously measured velocities of the particles at the sidewall, at the roughened bed, and at the surface with video techniques. We also measured mass flux, bulk density, and depth and calculated depth-averaged velocity and bulk solids fraction. A known portion of uniform flow trapped during motion and then weighed allows calculation of bulk density using

$$\bar{\rho} = M / LWh , \tag{17}$$

where M is mass of the trapped particles, L is the length of the trap, W is the width of the flume, and h is the total depth of the flow. The average velocity is calculated as

$$\bar{u} = \dot{m} / \bar{\rho} g h , \tag{18}$$

where $\dot{m}$ is mass flux. Shear strain rate is estimated,

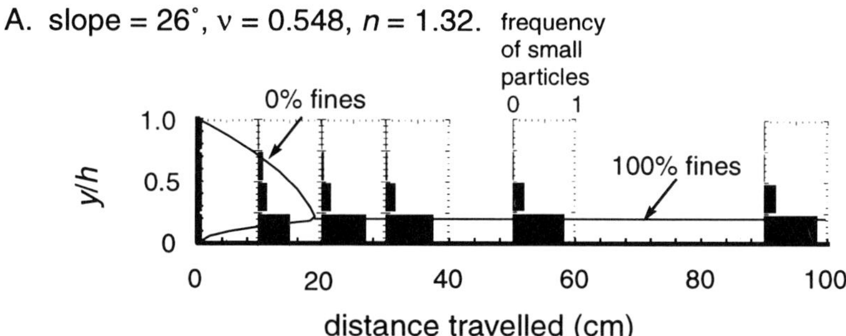

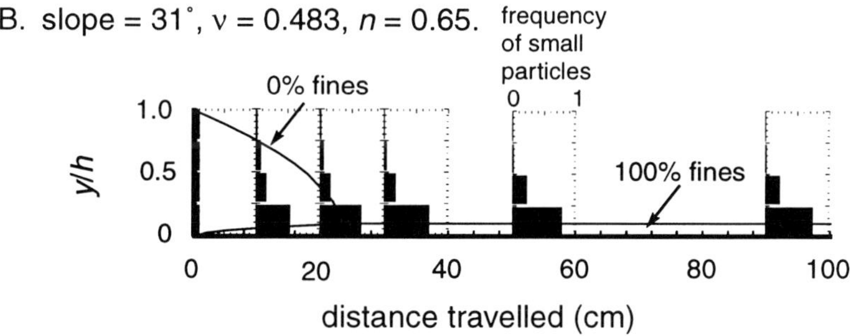

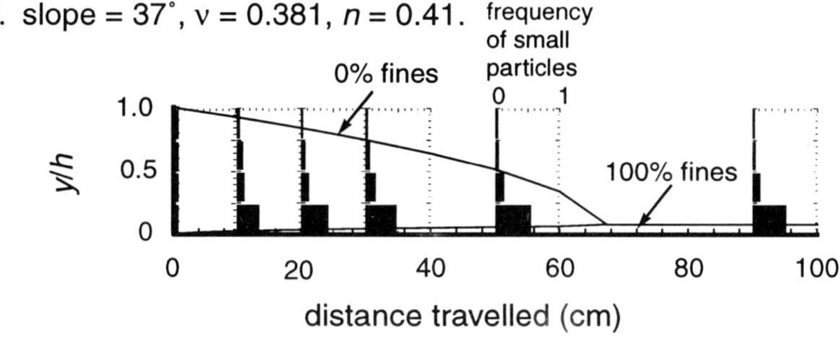

Figure 4. Predicted and experimental concentration profiles of small particles in a mixture of 90% large and 10% small particles. Experimental results are shown in superimposed histograms for the slopes and solids fractions shown in A, B, and C. For comparison, the position of the 0% fall line and the 100% accumulation line are shown for the appropriate values of n in a power law velocity profile such that $u^* = y^{*n}$ where $y^* = y/d$.

$$\dot{\gamma} = \frac{du}{dy} \sim \left(\frac{2\bar{u}}{h} \right). \tag{19}$$

Separate measurements of mass flux and bulk density indicate average velocities for the side and middle parts of the flows. Knowing average and maximum velocities and assuming a power-law fit permit reconstruction of interior velocity profiles. Accounting for slip at the bed and noting that u_{bed} is zero for the no slip condition,

$$u(y)^* - u_{bed}^* = \frac{u(y) - u_{bed}}{u_{max}} = \left(y^*\right)^n, \tag{20}$$

where $y^* = y/h$. The nondimensional average velocity minus average bed velocity is

$$\overline{u}^* - \overline{u}_{bed}^* = \int_0^1 \left(y^*\right)^n dy = \frac{1}{n+1}. \tag{21}$$

Hence, we have

$$n = \frac{1}{\overline{u}^* - \overline{u}_{bed}^*} - 1. \tag{22}$$

For the sidewall velocity measurements, power-law curves give fits within the 95% confidence interval. Malekzadeh [15] measured interior velocity profiles of particle-liquid mixtures with a non-intrusive method and found that basal slip systematically increased with distance away from the sidewall. Power-law curves also provide good fits to Malekzadeh's measured profiles.

In the absence of a non-intrusive method of measuring velocities away from transparent sidewalls, power-law curve fits using equations 20 through 22 give simple interpolations of interior velocities. The method cannot predict velocity profiles with inflection points and would give misleading results for velocity profiles with strong inflections about such points. The method will predict the overall sense of the curvature of the profile--that is whether it is positive and convex up ($d^2u/dy^2 > 0$ where $n > 1$), zero and linear ($d^2u/dy^2 = 0$ where $n = 1$), or negative and concave up ($d^2u/dy^2 < 0$ where $n < 1$). Thus, $n < 1$ indicates that the shear strain rate is largest near the bed, $n = 1$ indicates uniform strain rate, and $n > 1$ indicates that it is largest near the surface.

With measurements of u_{max}, $\overline{u}$, and u_{bed} from experiment 1, we calculate n using equation 22 and reconstruct velocity profiles using equation 20 (figure 3). At the sidewall in dry grain flow experiments, velocity profiles are strongly convex up where the slope relative to the bed friction angle is small ($\theta - \delta$) and become linear or concave up as the slope increases. In the interior of the flow, velocity profiles are linear where $\theta - \delta$ is small and become increasingly concave up as the slope increases (figure 3). Also where $\theta - \delta$ is small, the trend in velocity profile from the sidewall to the middle is from strongly convex up to nearly linear. At greater slopes, the corresponding trend is from slightly convex up to strongly concave up.

Although we designed these experiments to be two dimensional, lateral variations in concentration of the small particles, velocity, and bulk solids fraction are apparent. The concentration of small particles is somewhat larger near the sidewalls than it is in the middle of the flow. This effect is strongest where the slope is near the bed-friction angle and diminishes as the slope increases. The lateral concentrations of small particles are approximately uniform at the largest slopes. At slopes near the bed-friction angle the approximate y-position of the 100%-accumulation surface is greater near the sidewall than it is in the middle. The increased concentration of small particles near the sidewall is associated with diminished depth-averaged velocity and slightly diminished solids fraction

These results show that strain rate is not invariably focused at the bed as is commonly assumed. In the case where $\theta - \delta$ is small, the strain rate is nearly uniform in the middle of the flow, but, near the sidewalls, it is maximum at the surface. Only at greater slopes is the strain rate strongly focused near the bed and, even in these cases near the sidewalls, may be reversed such that it is stronger at the surface.

Small $\theta - \delta$ results in the strongest size segregation. In that case, velocity profiles indicate that strain rate is uniform or greatest near the surface. Increasing $\theta - \delta$ correlates with progressively weaker size segregation, and velocity profiles indicate that the strain rate is greatest near the bed.

4.1.2. *Particle-size segregation in viscous fluids*

In experiments with viscous fluids, the bimodal mixtures that were investigated included both glass particles in water or in a water-ethanol mixture and the polystyrene particles used previously in a neutrally buoyant fluid. The mass flux was regulated to maintain steady uniform flow having depths of between 0.9 and 1.5 cm. Flow-front depths 1.5 to 4 times greater than those of the following (upstream) flows apparently resulted because of higher solids fractions. The following (upstream) flows however had uniform solids fractions and all measurements were made on that part of the flow. It was necessary to use a single splitter plate as multiple arrays of splitters disturbed the flows too much. Splits were taken and measured at three different vertical plate positions at each station.

Flows of liquid-particle mixtures showed increasing segregation of small particles ($d = 0.99$ mm) to the base of the flow and large particles ($D = 1.44$ mm) to the top of the flow with distance downstream (figure 5), but not so dramatically as in dry granular flow (figures 3 and 4). Apparently the presence of a viscous fluid inhibits segregation. Curiously, segregation is slightly weaker in water than is in a fluid 3.7 times more viscous than water. Possibly, a slightly smaller fluid-solid density contrast causes the stronger segregation in the more viscous fluid (compare, $\rho_{\text{particle}} - \rho_{\text{alcohol-water-mixture}} = 2.49 - 0.94 = 1.55$ g/cm^3, $\mu_{\text{alcohol-water-mixture}} = 3.7$ centipoise; and $\rho_{\text{particle}} - \rho_{\text{water}} = 2.49 - 1.0 = 1.49$ g/cm^3, $\mu_{\text{water}} = 1.0$ centipoise). Figure 5 shows that the segregation in these experiments proceeds most rapidly within 25 cm downstream of the source but that segregation continues downstream where $\theta - \delta$ is small. This inverse size segregation of particles in the moving flow is strongest at slopes near $\theta = \delta$ and becomes progressively weaker at greater slopes (figure 5). These experiments generally show that segregation improves with decreasing slope.

4.1.3. *Particle-size segregation in neutrally buoyant fluid*

There is very little evidence of size segregation (d = 0.97 mm; D = 1.51 mm) in flows where the fluid and the particles have exactly the same density ($\rho_p = \rho_f$ =1.028 g/cm^3). Splits from the upper part of the flows at a distance of 95 cm downstream have concentrations of small particles that are the same as the initial concentration within the limits of accuracy of the experiments. In the basal split, the concentration of small particles is

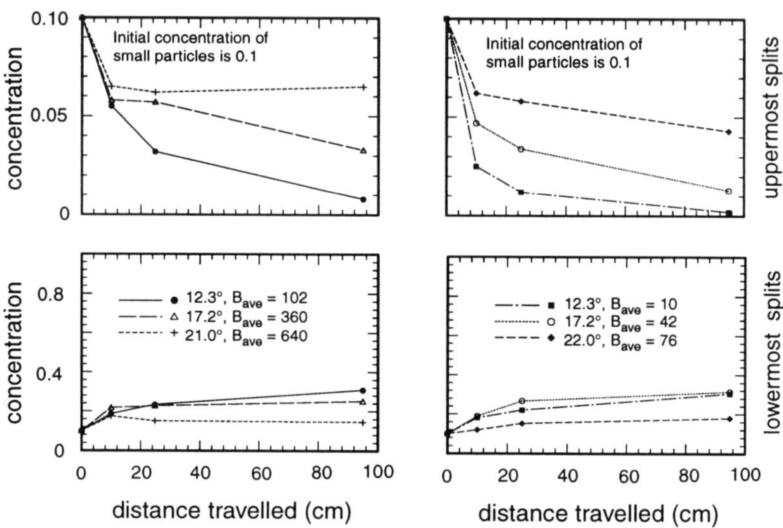

Figure 5. Plots of the concentration of small particles in the uppermost and lowermost splits versus distance downstream for bimodal mixtures of large (90%) and small (10%) particles in liquid and of the velocity profiles at the sidewall for several slopes as shown. The bed friction angle (δ) in these experiments is approximately 11°.

also the same as the initial concentration at slopes where θ - δ is small; however, there is a weak decrease in concentration of the small particles with increase in slope. This normal segregation is perplexing, because the usual hypothesis for explaining normal sorting is that, in a viscous fluid, large particles settle faster than small ones. However, that process cannot occur in a neutrally-buoyant, viscous fluid. Several repetitions of these experiments yielded the same result.

4.2. ASSESSMENT OF PARTICLE-SIZE-SEGREGATION MECHANISMS

4.2.1 Kinetic-sieve hypothesis

Both observations and measurements are compatible with the kinetic-sieve hypothesis of particle-size segregation. In experiments both with and without liquid, particles percolate downward at the transparent sidewall. Percolation events occur when voids sufficiently large open beneath the particles. In slow, frictional, dry-particle, size-sorting experiments, percolation events nearly always succeed before contact forces close the voids. Percolation of the smaller particles is much more common than percolation of large particles. Contact forces may push either large or small particles into adjacent layers. This combination of percolation and squeeze expulsion defines the kinetic-sieve hypothesis.

The results of dry-particle, size-sorting experiments show reasonably good agreement with the modified theory of Savage and Lun [23] (figure 4). The predicted downstream segregation is somewhat stronger than that of the experiments. Some of the divergence between the theory and experiment may be attributed to the assumption inherent in the theory of Savage and Lun that η_0 is relatively close to zero. A second source of divergence is because the theory assumes that when a large enough void opens beneath a particle, the particle falls instantaneously in to it. Observations suggest that in some instances the void is closed again before the particle can enter. The theory correctly predicts that sorting should diminish as slope increases, but it does not predict the degree in which sorting is diminished. As slope increases, flows undergo a gradual transition from a frictional to a collisional flow regime in which the assumptions of the theory are violated. The increasingly large fractions of small particles in the upper two splits are an indication of the increasing importance of diffusive pro-cesses in the flows at the largest slopes (figures 3, 4, and 5).

Observations indicate that the kinetic-sieve mechanism is operable in liquid-particle mixtures. However, because of viscous retardation and reduced gravity in experiments with liquid-particle mixtures, it is more common for voids to close before the particles above can fall into them. For this reason, the inverse size segregation is weakened (figure 5) sufficiently that the theory does not give reliable results.

4.2.2. *Lateral segregation*

The kinetic-sieve process alone cannot explain the lateral variation in concentration of small particles observed in the basal regions of slow frictional flows because the gravity-dependent process of percolation does not operate in the lateral direction. Savage [19] observed a secondary circulation pattern in dry granular chute flows in which particles at the surface migrated toward the center of the flow, down toward the bed, toward the sidewalls, and finally upward near the sidewalls. Such a circulation pattern superimposed on the vertically acting kinetic-sieve process would cause small particles to move toward the corners of flume where they would be trapped because they could not move upward to complete the circulation pattern. Thus, if such a circulation pattern existed, the concentration of the small particles would build up in the corners as the flow moves downstream.

Some slowly moving frictional flows seem to have secondary circulation patterns in which particles at the surface move from the middle of the flow toward the sidewall. This requires an alternate hypothesis that appeals to the probabilistic arguments of Bridgwater and his colleagues [12, 26]. They suggested that small particles will tend to migrate toward the regions of largest solids fraction because the voids are typically smaller in such regions, and big particles cannot enter. The probability for a small particle to encounter a void large enough and enter the region is greater than that for a big particle. Therefore, small particles will tend to move into regions of large solids fraction and big particles will tend to remain in regions of smaller solids fraction. Slow frictional flows have larger average solids fractions near the sidewall than in the middle of the flume. Furthermore, the largest values of v occur with the smallest shear strain rates and hence the smallest du/dy. Generally, in the basal part of the flow, du/dy is smallest near the corners of the flume and largest in the middle part of the flow. The correlation of the largest concentrations of small particles in the region with the smallest velocity gradients supports the probabilistic argument for explaining the lateral distribution of particles.

Density Segregation: $d_1 = d_2$, $\rho_1 / \rho_2 = 2.42$.

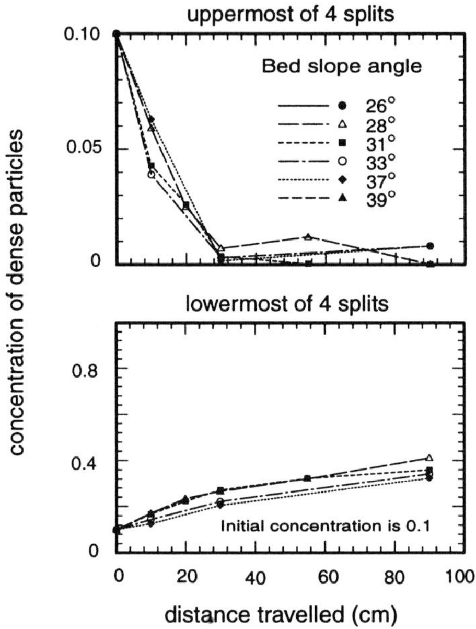

Figure 6. Plots of the concentration of the concentration of dense particles in the uppermost and lowermost of four vertical splits versus distance downstream for bimodal mixtures of light (90%) and dense (10%) particles flowing down several slopes as shown.. Inset in figure 3 shows the inferred average velocity profiles for slopes of 26°, 31°, and 37°.

4.3. DENSITY SEGREGATION IN DRY GRANULAR FLOWS

Chute-flow experiments on particles of identical size ($D = 1.44$ cm) and differing density ($\rho_1/\rho_2 = 2.42$) show segregation of denser particles to the bottom and less dense particles to the top of the flow (figure 6). Experiments were performed on a variety of slopes using the same flume and splitter as before. Because the particles have the same bed and internal friction angles and because experiments were performed with the same depths on the same slopes as in experiment 1, the velocity profiles are the same as those given in figure 3.

Under comparable conditions, density segregation is generally weaker than size segregation (compare figures 3 and 6). The concentration jump that forms with distance in size segregation does not form in density segregation. Only at slopes very much greater than the bed-friction angle is density segregation nearly as strong as size segregation (slope = 37°, figures 3 and 6, compare especially the upper splits).

The randomly fluctuating sieve cannot operate if the particles are the same size. In slow flow, only a mechanism analogous to squeeze expulsion acts in the density segregation process. Density contrast modifies the squeeze-expulsion mechanism because denser particles have a greater tendency to push downward than do lighter ones. Lighter

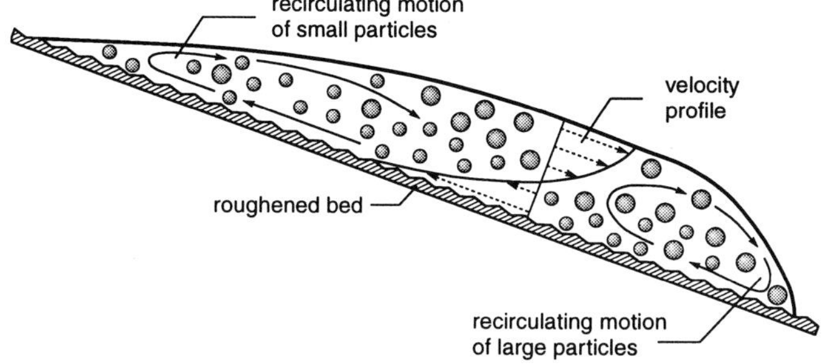

Figure 7. Schematic diagram of stationary granular surge wave in Davies' [7] moving-bed laboratory experiments. Circulatory paths of small and large particles and accumulation of large particles at nose of flow shown.

particles tend to be displaced upward. Thus, enhanced local gravitational forces are added to shear forces acting on individual particles. Zero net mass flux normal to the bed occurs because the tendency of the dense particles to push downward is exactly opposed by the tendency of the low density particles to be displaced upward. The strong concentration gradients and concentration jumps in particle-size sorting occur because diffusion is rather weak in comparison to the random fluctuating sieve in slow dry granular flows. In contrast, diffusion is moderately strong compared to squeeze expulsion processes that are important in slow dry granular flows with particles of contrasting density.

5. Discussion and concluding remarks

Concentrations of coarse particles at the margins of small debris flows and some avalanches have been associated with inverse sorting in the flows [27]. Randomly mixed sand and gravel allowed to slide down a roughened chute having a fifty degree slope connected to a horizontal section by a short curve result in sorting of the gravel to the front and top after the material comes to rest. Similarly when random mixtures of coarse sand and fine sand are allowed to slide down the same chute, the coarse sand tends to be sorted to the top and front of the deposit. Drahun and Bridgwater [8-9] observed that larger particles tended to rise to the top and move toward the front of free surface flows with binary mixtures of dry sand-sized particles. Takahashi [27] suggested that this inverse longitudinal sorting occurs because large particles migrate upward owing to dispersive stress then move to the front because velocities are faster in upper layers. The particle reaching the front tumbles down and is buried in the flow. But if it is bigger than the surrounding particles, it will migrate up to the top again. Savage [22] suggested a similar process in which kinetic sieving rather than dispersive stress serves as the vertical sorting mechanism (figure 7).

Davies [7] has performed experiments where particles flow on a roughened treadmill adjusted so that the flow front remained stationary. At high enough Froude numbers, Davies' flows broke up into individual surges like that in figure 7. When Davies added coarse particles to the mixture he observed that these particles migrated to the front of the flow. Films of Davies' [7] experiments confirm that large particles circulate near the front of the flow, and fine particles migrate toward the tail of the flow as in figure 7. He also observed that when the height of the surge increased longitudinal segregation diminished. In cases where more material was added and the height was greater, shear strain rate in the upper parts of the flow was small. Shear strain is necessary for the kinetic sieve to operate, and, where it is small, longitudinal segregation diminishes.

Experiments show the importance of such processes as percolation, squeeze expulsion, and diffusion in flowing mixtures of particles. For example, change in the energy of a mass flow may affect the relative importance of the sorting processes and hence the effectiveness of sorting. Diffusion, which acts to randomize particle distribution, becomes more important in faster, higher energy flow, whereas percolation is of paramount importance in slow dry frictional flow. Interstitial liquid attenuates particle sorting owing to viscous retardation of percolation. A diminished density contrast between particles and liquid diminishes size segregation markedly.

Generally, difference in size is the most important parameter affecting particle sorting, but differences in density, can also be important. Flowing mixtures of particles varying in density but not size segregate in a fundamentally different way than those varying in size but not density because the latter experience size-preferential percolation and the former do not.

Acknowledgements

James Vallance gratefully acknowledges an International Fellowship from the Natural Science and Engineering Research Council of Canada and a grant from the U.S. National Science Foundation, EAR-9527263. Stuart Savage acknowledges an operating grant from the Natural Science and Engineering Research Council of Canada.

References

1. Bridgwater, J., Cooke, H.H., and Drahun, J.A.: Shear induced percolation, *Institute of Chemical Engineers Symposium Series number 69* , (1985), 171-191.
2. Bridgwater, J., Cooke, H.H., and Scott, A.M.: Inter-particle percolation: Equipment, development, and mean percolation velocities, *Transactions of the Institution of Chemical Engineers* **56** (1978), 157-167.
3. Bridgwater, J. and Ingram, N.D.: Rate of spontaneous interparticle percolation, *Transactions of the Institution of Chemical Engineers* **49** (1971), 163-169.
4. Bridgwater, J., Sharpe, N.W., and Stocker D.C.,: Particle mixing by percolation, *Transactions of the Institution of Chemical Engineers* **47** (1969), T114-T119.
5. Cooke, M.H. and Bridgwater, J.: Interparticle percolation. A statistical mechanical interpretation, *Industrial Engineering and Chemistry Fundamentals* **18** (1979), 25-27.
6. Cooke, M.H., Bridgwater, J., and Scott, A.M.: Interparticle percolation: lateral and axial diffusion coefficients, *Powder Technology* **21** (1978), 183-193.
7. Davies, T.R.H.: Debris flow surges--a laboratory investigation, *Mitteilung No. 96 der Versuchsanstalt fur*

Wasserbau, Hydrologie, und Glaziology an der ETH, (1988), 1-122.

8. Drahun, J. A., and Bridgwater, J.: Free surface segregation, *Institute of Chemical Engineers Symposium Series Number 65* (1981), S4/Q/1-S4/Q/14.

9. Drahun, J. A., and Bridgwater, J.: The mechanism of free surface segregation, *Powder Technology* **36** (1983), 39-53.

10. Dyer, F.: The scope for reverse classification by crowded settling in ore dressing practice. *Engineering and Mining Journal* **127** (1929), 1030-1033.

11. Fisher, R.V. and Schminke, H.U.: *Pyroclastic Rocks*, Springer-Verlag, Berlin, 1984.

12. Foo, W.S. and Bridgwater, J.: Particle migration, *Powder Technology* **36** (1983), 271-273.

13. Holtham, P.N.: Particle transport in gravity concentrators and the Bagnold effect, *Minerals Engineering* **5**, (1992), 205-221.

14. Hwang, C.L., and Hogg, R.: Diffusive mixing in flowing powders, Powder Technology **26** (1980), 93-101.

15. Malekzadeh, M.J.: The flow of liquid solid mixtures down inclined chutes, Unpublished PhD dissertation, McGill University, Montreal, Canada, 1994.

16. Middleton G.V.: Experimental Studies related to problems of flysch sedimentation, in Lajoie, J. (ed.), *Flysch Sedimentology in North America*, Business and Economic Science Ltd., Toronto, 1970, pp. 253-272.

17. Rippie, E.G., Olsen, J.L., and Faiman, M.D.: *Journal of Pharmaceutical Science* **53** (1964), 1360-1363.

18. Rosato, A., Prinz, F., Standburg, K.J., and Swendsen, R.: Monte Carlo simulation of particulate matter segregation, *Powder Technology* **49** (1986), 59-69.

19. Savage, S. B.: Gravity flow of cohesionless granular materials in chutes and channels, *Journal of Fluid Mechanics* **92** (1979), 53-96.

20. Savage, S.B.: The mechanics of rapid granular flows, *Advances in applied mechanics* **24** (1984), 289-376.

21. Savage, S.B.: Interparticle percolation and segregation in granular materials: A review, *Developments in Engineering Mechanics*, Elsevier, 1987, pp. 347-363.

22. Savage, S.B.: Flow of granular materials, in Germain, .P., Piau, M. & Caillerie (eds.), *Theoretical and Applied Mechanics*, IUTAM, 1989, 241-266.

23. Savage, S.B. and Lun, C.K.K.: Particle size segregation in inclined chute flow of dry cohesionless granular solids, *Journal of Fluid Mechanics* **189** (1988), 311-335.

24. Scott, A.M., and Bridgwater, J.: Interparticle percolation: a fundamental solids mixing mechanism, *Industrial Engineering & Chemistry Fundamentals* **14** (1975), 22-27.

25. Scott, A.M., and Bridgwater, J.: Self-diffusion of spherical particles in a simple shear apparatus, *Powder Technology*, **14** (1976), 177-183.

26. Stevens, D.J, and Bridgwater, J.: The mixing and segregation of cohesionless particulate materials part II: Microscopic mechanisms for particles differing in size, *Powder Technology* **21** (1978), 29-36.

27. Takahashi, T.: Debris flow on prismatic channel., *Journal of the Hydraulics Division, ASCE* **106** (1980), 381-395.

28. Williams, J.C.: *Journal of the Fuel Society* **14** (1963), 29-34.

VISUALIZATION OF SEGREGATION IN GRANULAR FLOWS INSIDE SILOS

A. SAMADANI, A. PRADHAN and A. KUDROLLI
Department of Physics, Clark University,
Worcester, MA 01610, USA.

Abstract
We study the segregation of multi-sized granular matter inside a quasi-two dimensional
silo discharging from an orifice at the bottom. We visualize the flow and show that poly-
disperse and bi-disperse beads segregate with larger particles found at the center of the
silo surrounded by smaller particles. In case of bi-disperse particles we measure the relative
concentrations of the particles as a function of position and time, and obtain quantitative
information of the evolution and extent of segregation as a function of size ratio and the
weight fraction of the particles.

1. Introduction

Segregation is one of the most intriguing properties displayed by granular material under
excitation. This property is often exploited in industrial processes such as mining, but
is often an undesirable feature when constituent matter needs to be thoroughly mixed.
A number of mechanisms have been identified which can lead to segregation in granular
matter consisting of two or more types of particles [1-4]. The presence of velocity gradients in the flow or gravitational field can lead to segregation. Interaction of the granular
matter with the boundary usually has to be taken into account to explain the observed
segregation [5].

The flow of granular matter in most industrial situations is very complicated and more
than one mechanism may be present that may interact with each other. Therefore, it is
often necessary to have a quantitative measure of the role of each mechanism. However,
these calculations are very difficult due to the multitude of parameters that are involved.
The difficulties include a lack of a rigorous theory that describes granular flow from the
very rapid to the quasi-static regime [6]. Furthermore, few experiments have been conducted in which quantitative data has been obtained to characterize the development of
segregation. Therefore, there is a great need for experiments that give quantitative results
of the evolution of segregation.

One of the important categories of flows are those that are driven by gravity. Examples
include flows down pipes and hoppers during transportation and storage of granular matter. Among the few quantitative studies of development of segregation in such situations is
the work of Savage and Lun [2]. They developed a microscopic random void filling model
to describe the concentration profiles of bi-disperse particles flowing down an inclined

A.D. Rosato and D.L. Blackmore (eds.), IUTAM Symposium on Segregation in Granular Flows, 53–60.

plane. Particles from different heights in the layer were channeled into various bins to test their predictions. Quantitative studies of segregation using bi-disperse particles have been done in discharging silos [7], where the weight ratio of the two particles was obtained as a function of time. Fluctuations were observed in the ratio toward the end of the discharge. With the availability of high resolution digital imaging and processing, it is now possible to obtain quantitative data for the development of segregation as a function of space and time.

The flow inside silos as they are drained has been studied in some detail [8-11]. The flow is continuous for an orifice size that is much greater than the diameter of the grains. The top surface of the granular matter is flat as it discharges if the height to which the grains are filled in the silo is much greater than the width. For smaller ratios, the surface is inclined, which occurs toward the end of the draining. For a wide rectangular silo, the flow is confined to the central region where the material flows very rapidly. Significant regions with no flow are found towards the sides. The flow deep inside the silo is described by the kinematic model [8]. This model gives velocity profiles which are Gaussian centered at the symmetry axis of the silo. Therefore we see that the flow is complicated, but some work on its characterization exists which can be used to quantitatively study segregation in such a system.

In this paper we report a preliminary experimental study of segregation inside a silo that is discharging granular matter from an orifice at the bottom. A quasi-two dimensional silo geometry with glass walls is used to visualize the granular matter inside and quantitative aspects of the development of segregation are measured. Using imaging we show, that the segregation occurs at the surface and larger particles are carried along the stream lines to the center of the silo where the flow velocity is highest. We extract the relative concentration profiles of the particles inside the silo as a function of time. We report the data as a function of weight fraction of the initial mixture. A more detailed report that also includes the discussion of the flow will be presented elsewhere [12].

2. Experimental system

The experiments are conducted in a quasi-two dimensional silo, the dimensions of which are given in Fig. 1. The orifice has a rectangular shape with dimensions 0.25″ by 0.5″ and its linear dimension is at least 10 times the diameter of the beads used in the experiments. The flow inside the silo is observed to be two-dimensional and no significant change in the results were found by doubling the width of the silo. The granular matter is visualized through the transparent glass walls using a Kodak ES 1.0 Digital CCD camera with a resolution of 1000 pixels by 1000 pixels. The silo is illuminated uniformly from the front. The images are directly captured by a frame grabber on a computer and then analyzed. The experiments are conducted with non-cohesive glass beads and the humidity is controlled to be less than 15%. By using glass side walls and glass beads in the experiment we minimizes static charge build-up.

3. Experiments with poly-disperse particles

Poly-disperse clear glass beads are used and the segregation occurs inside the silo. The development of the flow and the velocity fields are similar to the case of mono-disperse particles which were used to characterize the flow inside the silo and to check the two

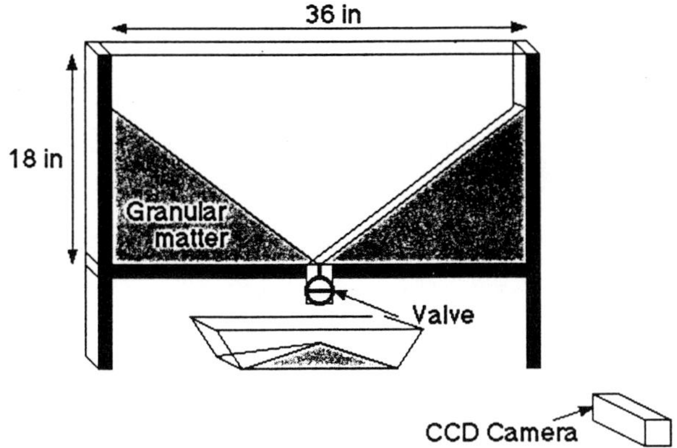

Figure 1. Schematic diagram of the silo. The side walls are transparent glass to allow imaging. The width of the silo is 0.5″ and the dimensions of the rectangular orifice at the bottom is 0.25″ by 0.5″. A valve is also used to control the flow rate Q.

dimensionality of the system [8,12]. Figure 2 shows an image of a portion of the silo at a time $t = 200\,$s after flow is started. The surface is observed to be inclined and is close to the angle of repose of the grains. The smaller beads scatter more light and therefore regions with smaller beads appear bright compared to regions with the larger beads. One, observes that the central region appears darker corresponding to the larger particles and therefore segregation is observed in the flow. However the light intensity cannot be easily correlated to size concentration and the results are qualitative in this case.

We can also observe horizontal layering at the sides of the silo. These layers occur during the pouring of the granular matter when the silo is initially filled before being drained. Therefore the horizontal layering indicates the static regions. It can be also noted from Fig. 2 that the flow at the surface is confined to a few layers towards the sides of the silo.

At the surface a kink is observed above which the material is stationary. The kink moves periodically from the bottom of the inclined surface to the top. This kink motion results in periodic avalanching of material at the surface. This periodic avalanching is shown in Fig. 3 by monitoring the change in height of the surface at the center of the silo in 4 s time intervals. Because periodic avalanching is not observed in mono-disperse particles under similar conditions of orifice size and flow rate, the periodic motion is related to the multi-sized nature of the grains and segregation. We also note that similar kinks have been observed in situations where rough and smooth particles are used to fill silos. In that case, stratification and segregation has been observed [13].

4. Experiments with bi-disperse particles

To obtain quantitative data for the evolution of segregation as a function of position inside the silo, particles of two sizes with different colors were used. The size ratio of the beads is denoted by r, and the ratio by weight of the two particles in the mixture is denoted by P_W. The mixture is poured as uniformly as possible into the silo. After the orifice is opened, the particles are observed to segregate as they flow down the surface. A region

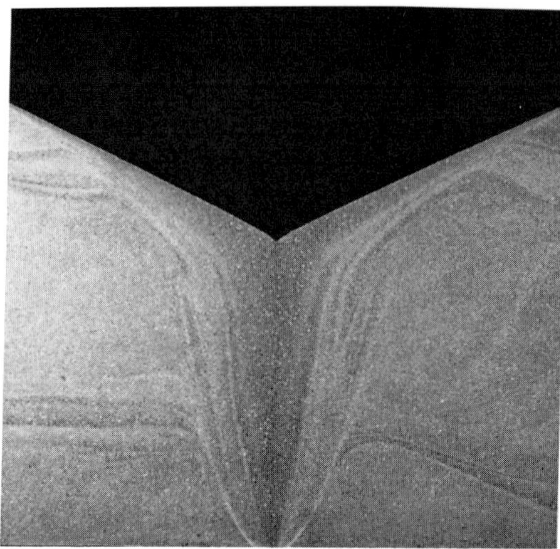

Figure 2. Size segregation inside the silo is observed when poly-disperse particles are drained from an orifice. Glass beads with diameters distributed between 50 μm and 500 μm where used in this case. The larger particles scatter less light and appear dark, whereas the regions with smaller particles appear brighter. The horizontal layering develops due to segregation during the initial pouring of the mixture to' fill the silo. These layers indicate the static regions in the bottom left and right of the silo. The dimensions of the image are $12'' \times 12''$.

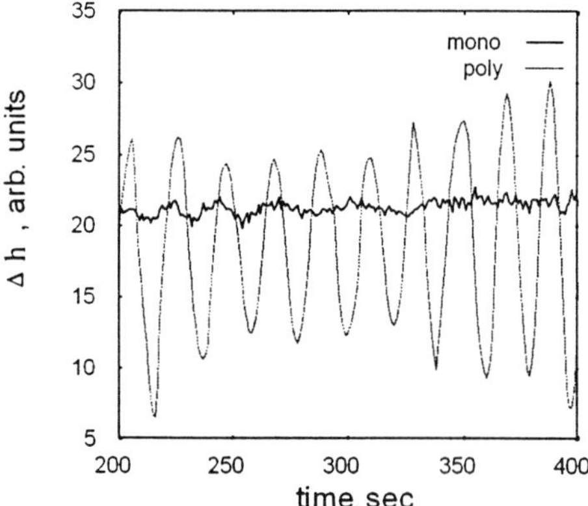

Figure 3. Periodic avalanching of particles at the surface is observed in poly-disperse beads ($Q = 15$ g/s). This phenomena is illustrated by plotting the change in height Δh of the surface near the bottom of the incline as a function of time. These oscillations are not observed in mono-disperse particles for similar flow rates.

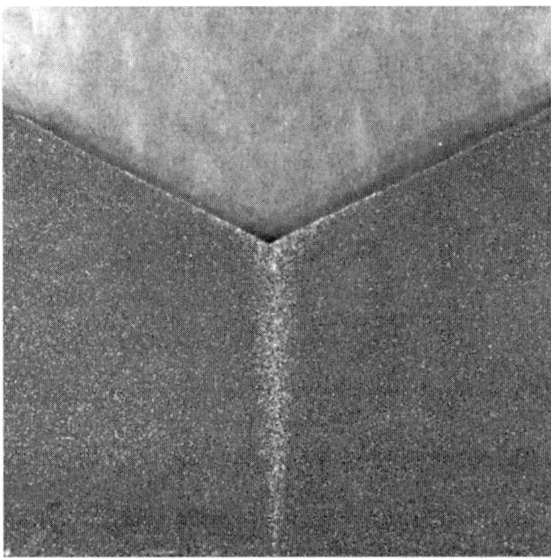

Figure 4.   Image of segregated particles inside a discharging silo. The regions with large particles appear white and the small particles appear dark ($r = 1.2$, $P_W = 15\,\%$, $Q = 15$ g/s.)

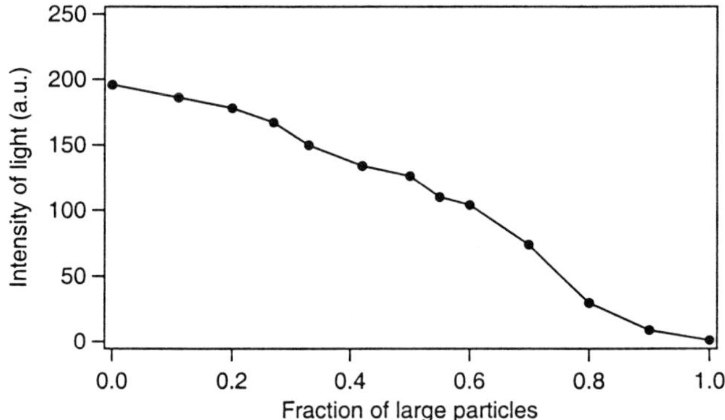

Figure 5.   Calibration curve corresponding to the light intensity recorded by the CCD camera versus the fraction of large particles. The data corresponds to particles used in Fig. 4. ($r = 1.2$).

of large particles first develops at the minima of the inclined surface. They then move toward the orifice in a narrow region surrounded by the smaller particles. Figure 4 shows an image of the resulting segregation 200 s after the development of the flow for $r = 1.2$. Note the similarities between Figs. 2 and 4 which show that the darker regions observed in Fig. 2 correspond to larger particles. It is remarkable that segregation is observed for such small r.

To quantitatively study the development of segregation, we first calibrate the light intensity as a function of the fraction of large particles in the mixture (see Fig. 5.) This

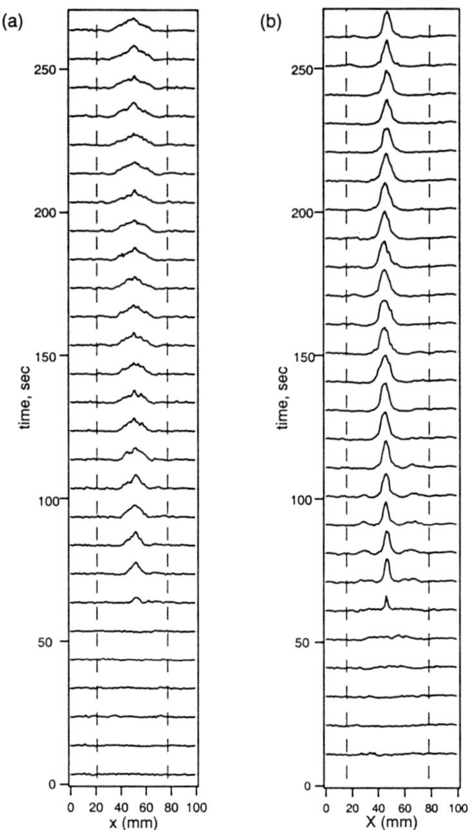

Figure 6. The fraction of large particles by weight in a horizontal region 5 cm above the orifice as a function of time. (a) $r = 1.2$, $P_W = 30\%$. (b) $r = 2$, $P_W = 10\%$. The development of the peak indicates segregation. Note that the peak develops only after about 60 s in both cases, indicating that the segregation occurs at the surface. The region inside the dashed lines are in motion.

is accomplished in a separate series of experiments in which the average light intensity is measured in a small area corresponding to known ratio of particles. Using this calibration, we plot the ratio of the particles (by weight) as a function of time in a region 5 cm above the orifice (see Fig. 6a.) The regions inside the two parallel lines in Fig. 6 denote the region in motion at that depth inside the silo. The data in Fig. 6a indicates that there is no segregation in the area under observation for 60 s after the flow starts. In this time, the velocity gradients deep inside the silo are well developed and therefore indicates that velocity gradients in the bulk do not lead to segregation. The observed segregation occurs at the surface, and the streamlines of the flow are such that the large particles are carried to the center where particles have higher velocities. The time 60 s corresponds to the time which the particles at the surface take to reach the region under observation near the

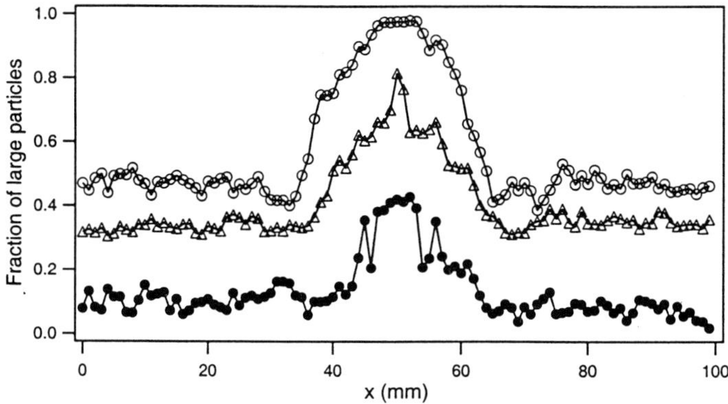

Figure 7. The fraction of particles as a function of horizontal distance as in Fig. 5 for $P_W = 15\%$ ($\bullet$), $P_W = 30\%$ ($\triangle$), $P_W = 50\%$ ($\circ$) ($r = 1.2$). The plots correspond to $t > 250\,\mathrm{s}$ and are time independent. Note that the maximum fraction of large particles is less than 1 for higher values of P_W indicating saturation of segregation.

orifice.

We studied the dependence of the extent of segregation on the weight fraction P_W of the large particles in the initial mixture. The results for three different P_W are plotted in Fig. 7. The plots indicate that the segregation is not 100% for small P_W, but increases and saturates as the fraction of large particles is increased. The boundary between the large and the small particles is also diffused in this case.

Experiments with higher size ratio ($r = 2$) were also conducted. In this case complete segregation was observed at smaller fractions, $P_W \sim 10\%$ and the boundary separating the large and small particles was sharper (see Fig. 6b). Thus as one expects, the segregation is stronger for larger values of r. Even in this case, no segregation is observed in the bulk flow of the silo for the first 60 s because the fraction of large particles is constant across the container in the region under observation. Apparently, the velocity gradients are not strong enough inside the silo to cause gradients in the concentration profiles.

5. Summary

Segregation is observed in the flow inside the silo when multi-sized particles are drained. By visualizing the flow we are able to show that the segregation occurs at the surface as the material avalanches down. By using bi-disperse particles we are able to obtain quantitative data on the development of segregation. The results of the flow inside the silo and the dependence of the segregation on the flow rate and the size of the particles will be presented elsewhere [12]. These results are useful for making quantitative comparisons to predictions of models that need to be developed to explain the segregation.

This work was supported partially by the donors of Petroleum Research Fund under grant ACS-PRF # 33185-G9, and one of the authors (A.K.) was also supported by the Alfred P. Sloan Foundation.

References

1. Rosato, A. D., Strandburg, K. J., Prinz, F., and Swendsen, R. H.: Why the brazil nuts are on top: size segregation of particulate matter by shaking, *Phys. Rev. Lett.* **58** (1987), 1038.
2. Savage, S. B., and Lun, C. K. K.: Particle size segregation in inclined chute flow of dry cohesionless granular solids, *J. Fluid Mech.* **189** (1988), 311.
3. Bridgewater, J.: Mixing and Segregation Mechanisms in Particle Flow, in *Granular Matter* A. Mehta, Ed., Springer-Verlag, 1993, 161.
4. Jenkins, J. T.: Particle Segregation in collisional flows of inelastic spheres, in *Physics of Dry Granular Media*, Kluwer Academic Publishers, 1998, 645.
5. Knight, J. B., et al.: Experimental study of granular convection, *Phys. Rev. E* **54** (1996), 5726; Knight, J. B.: External boundaries and internal shear bands in granular convection, *Phys. Rev. E* **55** (1997), 6016.
6. Campbell, C. S.: Rapid Granular Flows, *Annu. Rev. Fluid Mech.* **22** (1990), 57.
7. Arteaga, P., and Tuzun, U.: Flow of binary mixtures of equal-density granules in hoppers-size segregation, flowing density and discharge rate, *Chem. Engg. Sci.* **45** (1990), 205; Tuzun, U., and Artega, P.: A microstructural model of flowing ternary mixtures of equal-density granules in hoppers, *Chem. Engg. Sci.* **47** (1992), 1619.
8. Nedderman, R. M., and Tuzun, U.: A Kinematic Model for the Flow of Granular Materials, *Powder Tech.* **22** (1979), 243.
9. Tuzun, U., and Nedderman, R. M.: Experimental evidence supporting kinematic modelling of flow of granular media in the absence of air drag, *Powder Tech.* **24** (1979), 257.
10. Mullins, W. W.: Experimental evidence for the stochastic theory of particle flow under gravity, *Powder Technol.* **9** (1974), 29.
11. Baxter, G. W., Behringer, R. P., Fagert, T., and Johnson, G. A.: Pattern Formation in Flowing Sand, *Phys. Rev. Lett.* **62** (1989), 2825.
12. Samadani, A., Pradhan, A., and Kudrolli, A.: Size segregation of granular matter in silo discharges, *Phys. Rev. E* **60** (1999) December issue.
13. Cizeau, P., Makse, H., Stanley, H. E.: Mechhanism of granular spontaneous stratification and segregation in two dimentional silos, *Phys. Rev. E* **59** (1999), 4408.

SEGREGATION DURING GRAVITY FILLING OF STORAGE BINS

D.B. HASTIE and P.W. WYPYCH
Centre for Bulk Solids and Particulate Technologies
University of Wollongong, Wollongong NSW 2522, Australia

Abstract

Binary mixtures of particles have been prepared in varying proportions and charged into a rectangular two-dimensional perspex bin and hopper. To record the degree of segregation, various methods of mechanical sampling have been investigated and found unsuitable. A new non-intrusive sampling method, which incorporates a personal computer, a video camera and special software to capture and process the relevant images, had to be developed to record the extent of segregation. This technique has been found accurate, convenient and cost effective.

From the experimental results, a relationship between degree of mixing and product proportion has been developed and presented using previously developed mixing indices. Some unexpected trends at certain locations in the bin and hopper also have been included and discussed.

1. Introduction

The segregation of particles during the handling of bulk solids can have a detrimental effect on the quality of products in industrial processes. Some typical examples include: food stuffs, such as muesli and mixed cereals, where strict ratios of the constituents are required; poly granules, where homogeneity is required for packing.

Many different types of segregation have been reported in the literature, such as: heap segregation, although other mechanisms may be involved (eg sifting, fluidisation, dynamic effects), Mosby et al. [1]; sifting, Carson and Marinelli [2], Williams [3]; trajectory effects, Carson et al. [4]; fluidisation effects, Carson and Marinelli [2], Johanson [5]; dynamic effects, Johanson [5].

Another interesting type of segregation has been observed when materials fall over great heights into storage facilities. In such situations air is entrained into the material, Cooper et al. [6], and then released violently at the point of impact. Circulating air currents are produced in the bin and fine material is deposited towards the walls. The resulting build-up of fines can cause flow problems and reduced discharge capacities.

It is believed that the extent of segregation caused by many of the above mechanisms can be reduced by modifying the handling procedures, as well as the design and operation of the process equipment. The various issues that need to be considered here include: the identification of the relevant segregation mechanisms; a reliable

A.D. Rosato and D.L. Blackmore (eds.), IUTAM Symposium on Segregation in Granular Flows, 61–72.
© 2000 *Kluwer Academic Publishers. Printed in the Netherlands.*

sampling technique to extract the representative portions of the mixture; an efficient method to quantify the degree of segregation present in the given mixture.

This paper reviews existing methods that have been developed for the measurement, quantification and assessment of segregation (eg mechanical probes, mixing indices). A new non-intrusive method of sampling, that had to be developed specifically for this project, also is described. Together with suitable mixing indices, this new method is used to evaluate the extent of segregation caused by the gravity filling of various binary particle mixtures into a two-dimensional bin. Although the data presented in this paper are applicable only to the two-dimensional test bin, the observed trends and segregation mechanisms are expected to be indicative of three-dimensional systems.

2. Sampling Methods

Sampling of a mixture to determine the degree of segregation present has been performed in numerous ways in the past, by using sampling probes, coloured dyes and computer imaging, the technique used being determined by the mixture content. For example, Carley-Macauly and Donald [7] used the sampling probes shown in Figure 1 to take samples of fine products, each sample containing a maximum of 60 particles. Rochowiecki [8] used the sampling probe shown in Figure 2 to take a maximum of 200 particles in each sample. Rumpf and Mueller [9] used an organic dye to coat a portion of fine calcite powder. Samples were taken then washed in water to remove the pigment. Using a photometer, the intensity of the pigment solution was then determined.

Broyles et al. [10] used an image analyser constructed around a personal computer to perform shape analysis on materials. By hand, analysing 100 particles was very time consuming, but the image analyser cut down this time to as little as 10 minutes.

Scott and Bridgwater [11] believed that approximately 40 samples had to be taken from a mixture to obtain decent results. However, this made the analysis monotonous.

Rogers and Clements [12] listed two limits to the amount of sampling which could be conducted: (i) the sampling probes being placed too close together, the insertion of one interfering with the product being collected by another, and (ii) analysing large numbers of samples being too time consuming and expensive.

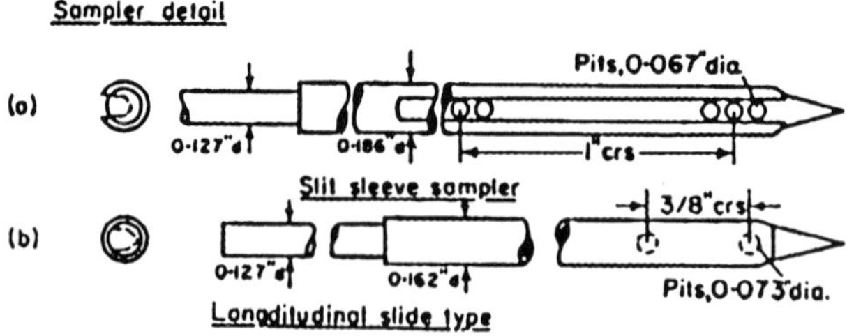

Figure 1. Sampling probes used by Carley-Macauly and Donald [7].

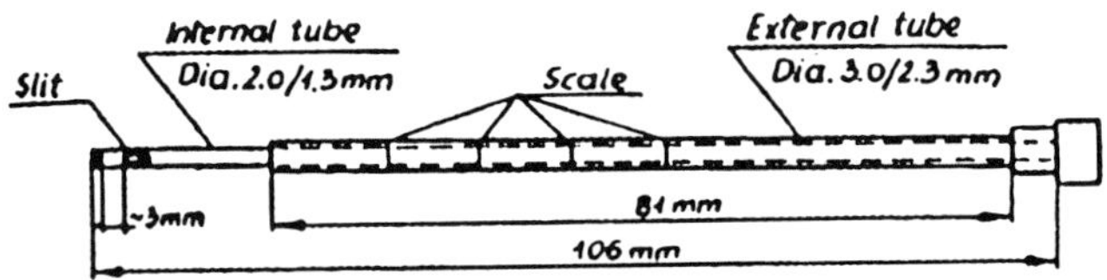

Figure 2. Sampling probe used by Rochowiecki [8].

3. Mixing Indices

Mixing indices are used to evaluate the degree of mixing, or mixedness, in a mixture of particles.

Numerous indices have been developed by a number of researchers over the years and involve statistical parameters such as variance or standard deviation [13]. A short list of mixing indices is provided in Table 1.

TABLE 1. Various mixing indices.

Equation	Mixing Indices	Author(s)	Range of Values (Segregation to Mixed)
(1)	$M = \sigma/\sigma_o$	Fan et al. [14]	1 to 0
(2)	$M = \sigma_o/\sigma$	Smith [15]	1 to $\gg$1
(3)	$M = 1 - \sigma/\sigma_o$	Fuerstenau and Fouladi [13]	0 to 1
(4)	$M = \sigma^2/\sigma_o^2$	Fan et al. [14]	1 to 0
(5)	$M = 1 - \sigma^2/\sigma_o^2$	Hogg et al. [16]	0 to 1
(6)	$M = \sigma^2/\sigma_o^2 - 1$	Fan et al. [14]	0 to -1
(7)	$M = (\sigma_o^2 - \sigma_o^2) \times (\sigma_o^2 - \sigma_r^2)^{-1}$	Lacey [17]	0 to 1
(8)	$M = (log\sigma_o - log\sigma) \times (log\sigma_o - log\sigma_r)^{-1}$	Williams [18]	0 to 1

Note: $\sigma_o = [p(1-p)]^{0.5}$, $\sigma_o^2 = p(1-p)$, $\sigma = [(\Sigma(p_i-p))/(n-1)]^{0.5}$ and $\sigma^2 = (\Sigma(p_i-p))/(n-1)$.

Generally the mixing indices range in value from 0 to 1, 0 being for full segregation and 1 for complete mixing, or vice versa depending on the representation of the variance or the standard deviation in the formula. There are others however that do not follow this trend and can range from 0 up to extremely large numbers.

Many of the mixing indices have been derived by modifying existing indices. For example, Williams [18] stated that Ashton and Valentin modified the index derived by Lacey, shown as Equation (7), due to there being a discrimination present as the mixing process was coming to completion. The modified equation became that shown as Equation (8) and although there was still some discrimination, it was an improvement.

4.   Sampling Probes

Five new sampling probes were designed and built, as shown in Figures 3 and 4: a flat tip, two conical and chisel tips (25 and 50 mm long). The sixth tube shown in Figure 3 is inserted inside the other five to collect the samples. Using a sampling box representing a small portion of the full-scale rig, sampling tests were performed on each design. The outer sampling probe tubes had an outside diameter of 19.05 mm and the inner tube, an outside diameter of 15.95 mm.

While testing, observations immediately showed that for the flat-ended probe, a substantial quantity of product was being displaced in front of the sampling probe while being inserted. This was also the case for the other four tips but to a reduced degree. In a full-scale situation this could have a detrimental effect on the product in adjacent sampling areas giving rise to inaccurate results. This resulted in the use of mechanical sampling probes being abandoned.

5.   Test Products

Rape seed (1.0 to 2.0 mm, ρ_s = 1165 kg m^{-3}, ρ_{bl} = 660 kg m^{-3}) and white plastic pellets (3.8 to 5.6 mm, ρ_s = 910 kg m^{-3}, ρ_{bl} = 535 kg m^{-3}) were selected for the following reasons: comparable loose poured bulk densities (to minimise density effects); approximately spherical particle shape; two distinctly different coloured products (for visual observations); distinct size difference (ease of separation); non-degradable products (to eliminate/minimise replacement of sample); low static (for perspex bin).

6.   Test Rig

The rig used for the full-scale testing consisted of: a 20 litre Forberg mixer [19] and feed bin, Figure 5(a), where binary mixtures were prepared for testing; the observation rig, Figure 5(b), where the feed bin was positioned above the perspex bin and hopper. Flow from the feed bin was controlled via a full-bore ball valve and discharge from the perspex bin was achieved by a slide valve, both pneumatically controlled. The perspex observation bin had internal dimensions 1500 × 1000 × 150 mm and the hopper had a hopper half angle of 30° to the vertical and an outlet size of 150 mm. A grid of 110 × 110 mm was drawn on the bin and hopper, giving 134 sampling squares.

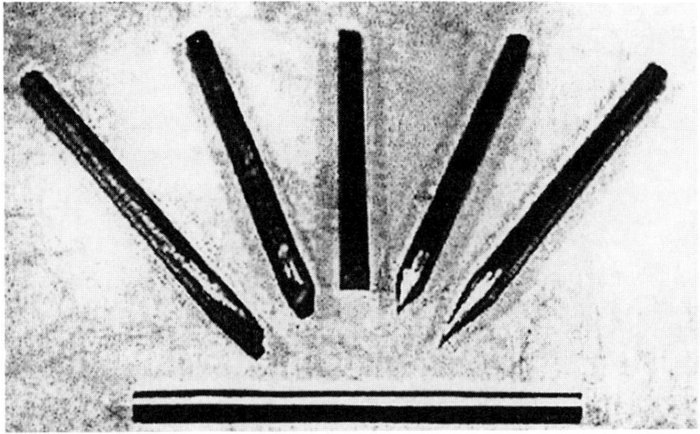

Figure 3. Sampling probe designs.

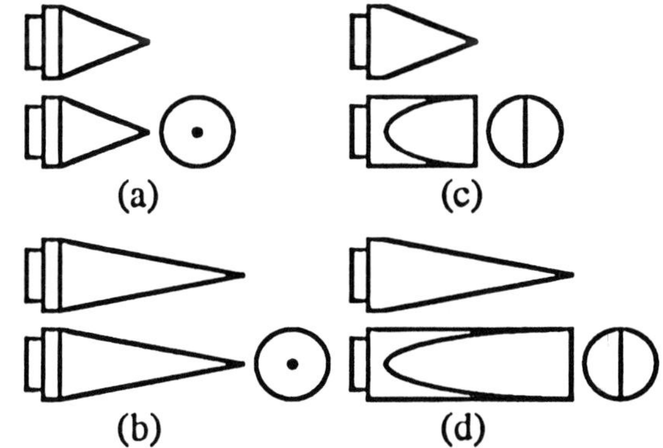

Figure 4. Conical and chisel sampling probe tip designs (25 or 50 mm long).

7. Computer Image Sampling and Analysis

For the reasons stated in Section 3, a computer-based method was sought where non-intrusive sampling could be employed. With the aid of a video camera, a piece of computer hardware called "Snappy" was obtained which captured high resolution digital images and placed them onto a computer. Once saved, the images were analysed to extract the required data to determine the degree of segregation. The software package "Optimas" was used for this purpose.

By analysing the images in black and white, two threshold regions could be selected, the dark region for rape seed and the light section for white plastic pellets. Once these thresholds were selected, the percentage areas of each could be calculated easily by "Optimas". These percentages were then recorded for later use in calculating the mixing indices.

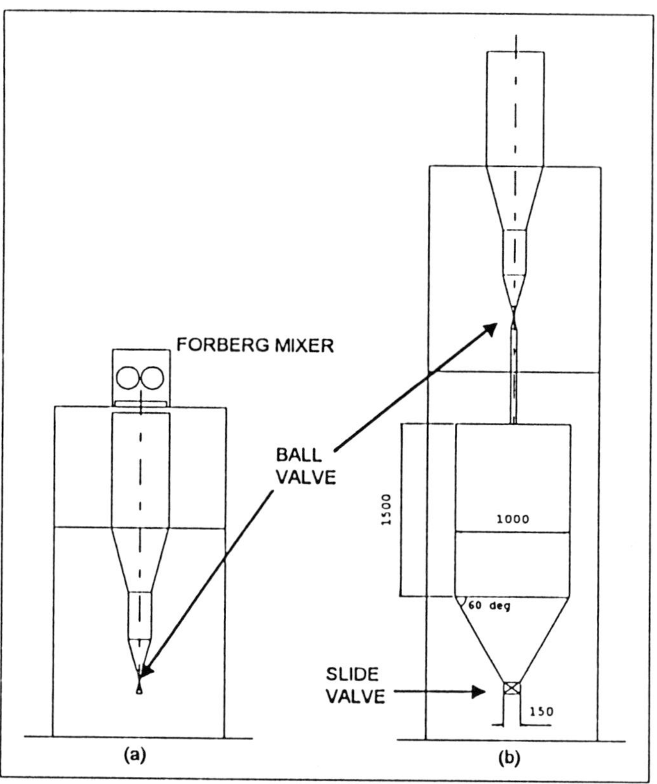

Figure 5. (a) Mixer and feed bin arrangement;
(b) Feed bin, perspex bin and hopper arrangement.

8.   Testing

8.1   QUALITY ASSURANCE OF MIXTURE

Before full-scale testing could begin, it had to be determined whether any segregation was occurring in the feed bin. Three 20 litre mixtures were prepared, see Table 2, and each mixture was dropped from the feed bin onto a long horizontal conveyor belt. Using a specially designed 100 mm wide parallel-plate cutter, a sample was taken from near the front, centre and rear of the belt and then analysed.

To rule out any significant segregation of the mixtures before they entered the perspex bin, three 240 litre batches of product were prepared, as would be used in a full-scale test. Ten cup samples were taken at regular intervals from the stream as it entered the top of the perspex bin and then analysed.

The maximum variation in the conveyor belt and stream sampling tests was 1.8% and indicated that the segregation that was occurring to the mixture before entering the perspex bin was minimal.

TABLE 2. Results from conveyor belt and stream sampling tests.

Initial product mixture	Results from Conveyor Belt Sampling Tests		Results from Stream Sampling Tests	
	Rape seed (% vol)	White plastic pellets (% vol)	Rape seed (% vol)	White plastic pellets (% vol)
25-75 % vol	24.6	75.4	23.2	76.8
50-50 % vol	49.1	50.9	49.5	50.5
75-25 % vol	75.5	24.5	74.1	25.9

8.2   FULL-SCALE TESTING

Full-scale testing was performed on binary mixtures containing 10 to 75% rape seed (by volume), each test comprising 240 litres of mixed product. All tests were recorded on video for later analysis. Once a mixture was prepared for testing, the ball valve at the bottom of the feed bin was opened and the product allowed to fall into the perspex bin.

Various segregation mechanisms were observed in the perspex bin as it was filling. The sifting mechanism was the most prominent, the smaller rape seed particles concentrating at the centre of the bin while the larger white plastic pellet particles concentrating at the sides of the bin. Another segregation mechanism present was that due to dynamic effects. From the apex of the heap, rape seed could clearly be seen bouncing outward and a small crater present where the stream made contact with the heap. This effect would have been continually changing due to decreasing drop height as the perspex bin filled with product.

After a test had been completed, each sample point was captured using "Snappy" and a video camera. Figure 6 depicts the results of the 50% rape seed and 50% white plastic pellets (by volume) test, the segregation, as explained above, can be seen clearly.

9.   Mixing Index Calculations and Graphs

The maximum number of sampling squares used in any one test was 107, due to the heap slope at the top of the pile causing some squares to be empty or partially full. To allow comparative results from one test to the next, the same number of sample squares was analysed for each test, reducing the number of sampling squares for analysis to 89. Figure 7 shows a representation of the sampling bin with all the sampling squares.

Once "Optimas" had been used to analyse each of the sample squares, graphs were produced, as shown in Figure 8, showing the segregation in each horizontal section of the perspex bin and hopper for the 50% rape seed and 50% white plastic pellets (by volume) test. The vertical axis represents the recorded volumetric percentage of rape seed in a given sample square and the horizontal axis the sample square in question.

Two mixing indices were chosen to quantify the degree of segregation in each test: Equations (3) and (5). These two were chosen mainly for their range of values, 0 being for segregation and 1 for fully mixed. Three sets of values were produced for each test, one for the 26 sample squares in the hopper, one for the 63 sample squares in the bin and one for the entire 89 sample squares. The results are shown in Figures 9 to 11.

Figure 6. Segregation test, 50% rape seed
and 50% white plastic pellets (by volume).

Figure 7. Sample square locations.

The graphs show that the segregation occurring in the hopper is very sporadic with
no visible trend, but has a slight overall increase in mixture quality as the volumetric
percentage of rape seed is reduced. This is most probably due to the increased influence
of the dynamic effects due to the narrower angled walls of the hopper.

The curves representing the degree of segregation in both the bin and the entire bin
and hopper showed very similar characteristics. As with the segregation in the hopper,
there was an increasing mixture quality as the volumetric percentage of rape seed was
reduced, however, at the 25% mark, a pronounced dip was observed in both curves.

This was unexpected, possible reasons for this occurrence are: a characteristic of this
combination of particles; and/or a result of using this particular geometry of bin and
hopper.

Further investigations along these lines were not pursued due to being beyond the
scope of the current research project.

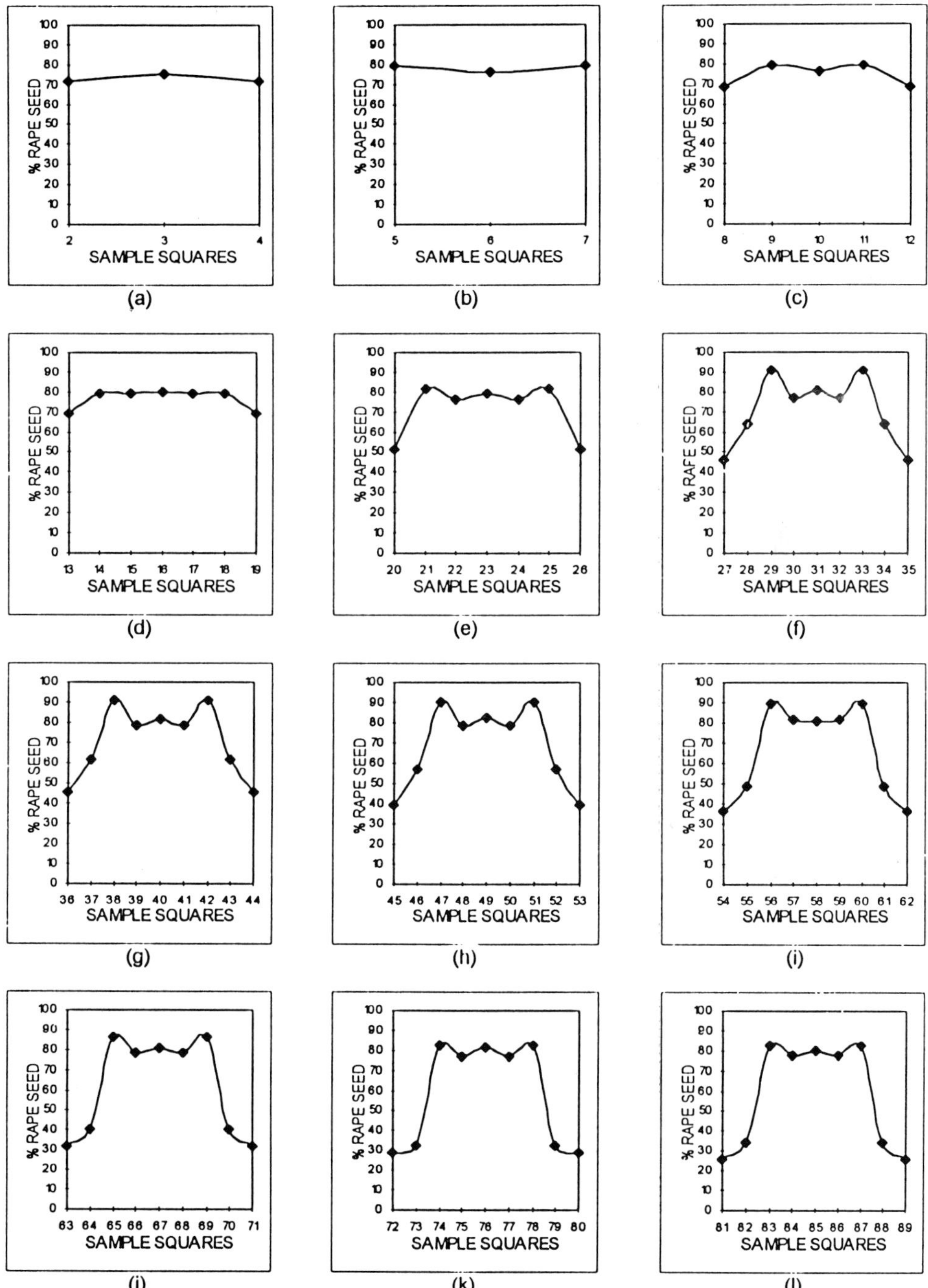

Figure 8. Distribution of segregation in horizontal rows in the hopper and bin
for 50% rape seed and 50% white plastic pellets (by volume) test.

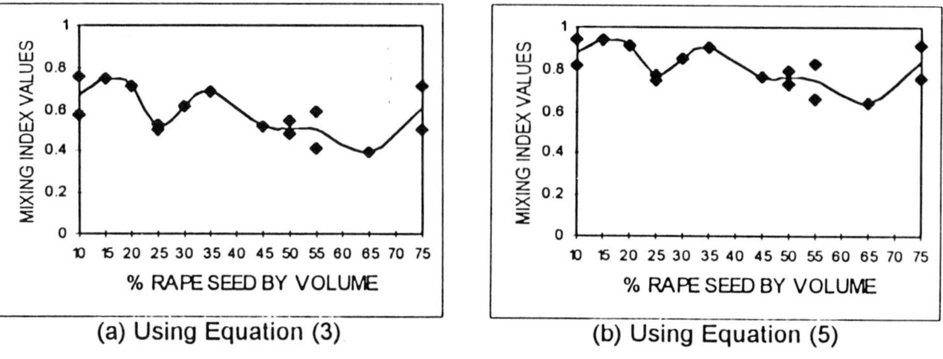

Figure 9. Mixing index values vs vol % of rape seed in hopper section.

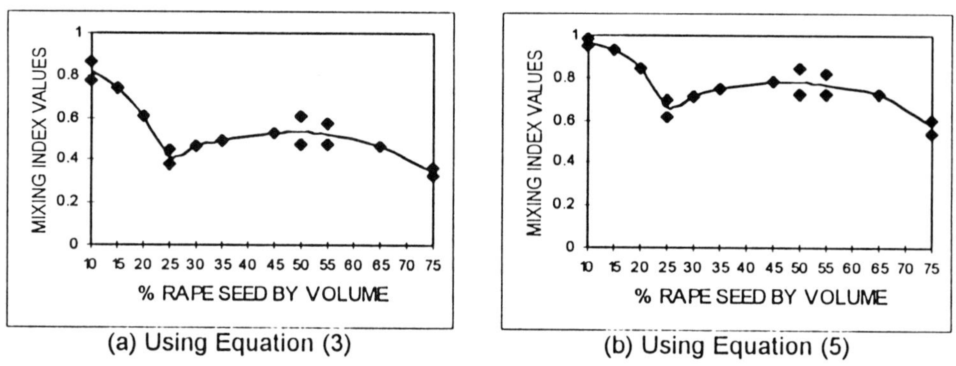

Figure 10. Mixing index values vs vol % of rape seed in bin section.

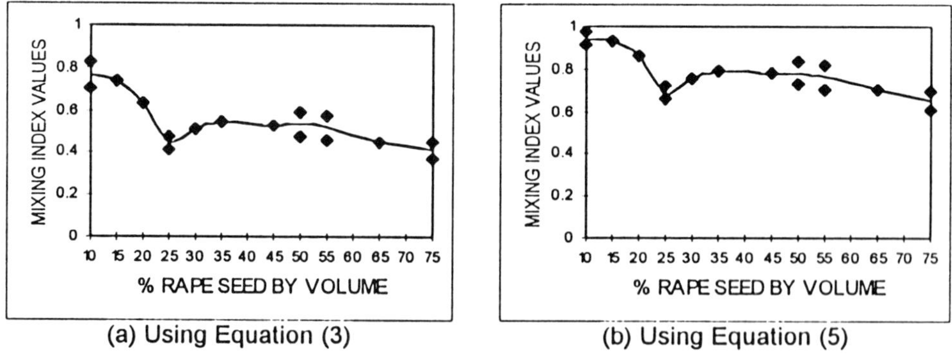

Figure 11. Mixing index values vs vol % of rape seed in combined hopper/bin sections.

10. Conclusions

Existing sampling methods were found unsuitable for the proposed work in this research resulting in several new sampling probes being designed and tested. However, intrusive sampling was discarded and a non-intrusive computer-based sampling method was developed in its place where a two dimensional sample of product was analysed.

Test involving the conveyor belt and in-line stream sampling proved that very minimal segregation occurred in the mixture between leaving the Forberg mixer and entering the Perspex observation bin, showing that the segregation that did occur was as a direct result of the filling operation.

As the testing program was being performed, segregation could clearly be seem. Segregation mechanisms such as sifting and dynamic effects were quite visible. The impact of dynamic effects decreased as each test progressed, as the drop height was continually decreasing.

The mixing indices used clearly represent the segregation present in each test. By combining the results from all tests performed, graphs for the hopper section, bin section and the combined hopper and bin section were produced, showing the variations from different mixtures.

11. Nomenclature

M	Mixing index (-)
n	Number of sample points (-)
p	Numerical proportion of particle with highest quantity in mixture (-)
p_i	Percentage of product present in a sample square (-)
ρ_{bl}	Loose poured bulk density (kg m^{-3})
ρ_s	Solids density (kg m^{-3})
σ	Standard deviation among samples (-)
σ_o	Standard deviation before mixing (-)
σ_r	Standard deviation of the complete random mixture(-)
σ^2	Variance among samples (-)
σ_o^2	Variance before mixing (-)
σ_r^2	Variance of the complete random mixture (-)

12. References

1.   Mosby, J., de Silva, S. R. and Enstad, G. G.: Segregation of particulate materials - mechanisms and testers, *Kona* **14** (1996), 31-43.
2.   Carson, J.W. and Marinelli, J.: Characterize bulk solids to ensure smooth flow, *Chemical Engineering,* April 1994, 78-90.
3.   Williams, J.C.: The segregation of particulate materials – a review, *Powder Technol.* **15** (1976), 245-251.
4.   Carson, J., Royal, T.A., and Goodwill, D.J.: Understanding and eliminating particle segregation problems, *Bulk Solids Handling* **6** (1986), 139-144.
5.   Johanson, J.R.: Particle segregation... and what to do about it, *Chemical Engineering*, May 1978, 183-188.

6. Cooper, P., Smithers, T. and Wypych, P.W.: Air entrainment and dust generation in free falling streams of bulk material, *Bulk Handling Asia '95 Conf.*, Singapore, Sept. 19-21, 1995, Turret Group plc, pp. 238-246.
7. Carley-Macauly, K.W. and Donald, M.B.: The mixing of solids in tumbling mixers, Part I, *Chemical Engineering Science* **17** (1962), 493-506.
8. Rochowiecki, A.: Segregation of particulate solids in horizontal drum mixer, *Bulk Solids Handling* **1** (1981), 501-506.
9. Rumpf, H. and Mueller, W.: An investigation into the mixing of powders in centrifugal mixers, *Transactions of the Institution of Chemical Engineers* **40** (1962), 272-280.
10. Broyles, D.A., Rimmer, H.W., and Adel, G.T.: Rapid shape analysis of crushed stone using image analysis, *Kona* **14** (1996), 167-178.
11. Scott, A.M. and Bridgwater, J.: Interparticle percolation – a fundamental solids mixing mechanism, *Industrial and Engineering Chemistry Fundamentals* **14** (1975), 22-27.
12. Rogers, A.R. and Clements, J.A.: The examination of granular materials in a tumbling mixer, *Powder Technol.* **5** (1971/72), 167-178.
13. Fuerstenau, D.W. and Fouladi, J.: Degree of mixedness and bulk density of packed particles, *American Ceramic Society Bulletin* **46** (1967), 821-823.
14. Fan, L.T., Chen, S.J., and Watson, C.A.: Solids mixing – annual review, *Industrial and Engineering Chemistry* **62** (1970), 53-69.
15. Smith, J.C.: Mixing chemicals with soil, *Industrial and Engineering Chemistry* **47** (1955), 2240-2244.
16. Hogg;, R., Cahn, D.S., Healy, T.W., and Fuerstenau, D.W.: Diffusional mixing in an ideal system, *Chemical Engineering Science* **21** (1966), 1025-1038.
17. Lacey, P.M.C.: Developments in theory of particle mixing, *Journal of Applied Chemistry* **4** (1954), 256-268.
18. Williams, J.C.: The properties of non-random mixtures of solid particles, *Powder Technol.* **3** (1969/70), 189-194.
19. Forberg, H.: Modern mixing, theory and praxis, *Powder Handling and Processing* **4** (1992), 318-321.

STUDY ON THE MIXING OF DISCS IN A GALTON'S DEVICE

L. BRUNO and A. CALVO
Grupo de Medios Porosos.
Facultad de Ingeniería. Universidad de Buenos Aires.
Paseo Colón 850 (1063) Buenos Aires. Argentina

I. IPPOLITO, S.BOURLES, A. VALANCE and D.BIDEAU
Groupe Matière Condensée et Matériaux.
Bât 11A. Université de Rennes 1. 35042. Rennes Cedex. France

Abstract

Mixing of granular materials is difficult because commonly used mixing methods can lead to undesired segregation. We are concerned here with diffusion as a mechanism allowing mixing.

A Galton's device consists essentially in a regular array of obstacles that can modify the direction of movement and velocity of a falling particle (bead or disc), each time the particle collides with an obstacle. In this kind of system the particle flies freely until it collides inelastically with an obstacle. If the energy lost in the collision balances the kinetic energy gained during the fly, the particle can eventually achieve a diffusive motion where velocities before and after the collision are not correlated.

Now, what happens if we let many particles go, at the same time, through a Galton's board? Can this device mix, for example, particles of two quite different sizes?

To answer these questions, we built a 2D Galton's board by sticking obstacles between a wooden wall and a glass, separated by a small gap ε. An initial well characterized distribution of discs of two different diameters (thickness slightly lower than ε) is dropped into the gap at t=0. Afterwards, the distribution of discs at the exit is analyzed in order to study the role of obstacles and collective effects in the mixing mechanism.

1. Introduction

Mixing operations are encountered widely throughout productive industry in processes involving physical and chemical change. Although much of our knowledge on mixing has developed within the chemical industry, many other sectors carry out mixing operations on a large scale. Thus, mixing is a central feature of many processes in the food, pharmaceutical, paper, plastics, and rubber industries.

A.D. Rosato and D.L. Blackmore (eds.), IUTAM Symposium on Segregation in Granular Flows, 73–80.

It is therefore unfortunate that very few scientists and engineers receive a sufficiently thorough grounding in the fundamentals of mixing processes. In addition, there are no widely accepted design codes associated with mixing

Diffusion is the mixing mechanism that takes place when two different liquids or gases come into contact. In this case, the random movement of molecules due to collisions causes the mixing. Can we understand the mixing of granular matter as analogous to diffusion in liquids and gasses?

The words *diffusion* and *diffusion coefficient* are commonly used in granular matter literature [1-3]. The diffusion coefficient is generally computed as the growing rate of the mixing zone of identical beads that only differ in color, when put in a vibrating box or other mixing mechanism. It must be underlined that in all these cases the mixing due to self-diffusion is not the *real mixture* required in practical uses.

Harnby *et al.* [4] explain that the mixing of a particulate system differs from that of liquid systems in three important respects: (1) There is no particulate motion equivalent to the molecular diffusion of gases and liquids. (2) There is no relative movement of the particles without an energy input to the mixture. (3) Particulate and granular components do not usually have the constant properties of molecular species and can differ widely in physical characteristics, so grains can segregate during the mixing process.

The ultimate element of a particulate mixture is several degrees of magnitude larger than the ultimate molecular element of the liquid mixture. This implies that care must be taken when defining mixing. If we can design a mixer that gives the grains the necessary energy to flow and, at the same time, changed randomly the grains velocities, we could, in principle simulate the diffusive movements of molecules in a liquid or gas. In theory, this kind of system will not segregate grains with different physical properties.

Kaye [5] in his recent and comprehensive book on powder mixing describes a mixing device he developed that mixes the ingredients of a free flowing powder by allowing the particles to undertake a randomwalk with large Levy flights. The random motion is achieved by means of angled holed plates that are interposed in the way of the flowing particles.

Based on these ideas, we study experimentally the 2D flow of granular material through a dispersive device: a Galton board. It consists of a regular array of obstacles that deflect the grains in random directions. The flow is gravity driven.

2. Experimental Set-Up

The Galton device consists of a smooth wooden plate and a glass plate (120 cm wide and 80 cm height) separated by a small gap in which disc particles are allowed to fall.

In our experiment we use plastic discs of different diameters (*species*) (0.4, 0.6 and 1 cm) and 0.12 cm thick as flowing particles. The obstacles, plastic discs, (0.4 cm diameter, 0.12cm thick) are fastened to the wooden wall forming a hexagonal network, with a lattice spacing of 5cm (in order to have a well defined characteristic length). This distance was chosen to be 5 times larger than the diameter of the bigger flowing discs (1cm) in order to avoid the formation of arcs [6]. The glass prevents the discs from flying out of the device when flowing; at the same time, it displays the experiment's development.

Figure 1 shows the experimental device, with a typical initial and final distribution. A well-characterized distribution of discs is obtained by allowing the particles to fall through the array of obstacles. More precisely, we fill two boxes (located just at the entrance of the Galton's device) with two different disc species.

Then we release the two particle species at the same time. While the discs are falling they collide inelastically with the obstacles and the other falling discs, loosing energy and changing their velocity direction. They fly freely between collisions. The fall takes around one second. As they exit from the obstacles zone, the discs accumulate orderly (heap up) into different bins (1.5 cm wide).

The final spatial distribution is filmed using a CCD camera. It is then analyzed using an image processing software that permits to obtain the final position (x and y coordinates) and area of each particle in the final distribution. Such procedure allows us to distinguish particles of different sizes. Subsequently the histogram of each species in the x and y direction is built.

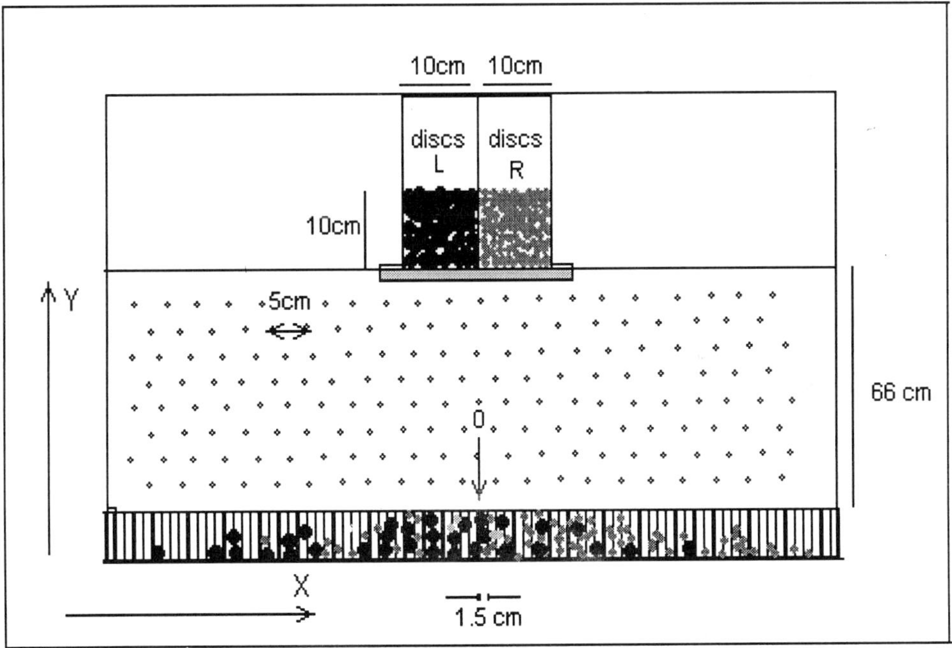

Figure 1. Experimental Set-Up.

3. Analysis and Results

3.1 TEST EXPERIMENTS

We start a set of experiments with only one disc species in order to test our Galton's device. In these experiments only one box was filled out. A typical final spatial distribution along the x direction for three different sizes of discs is presented in Figure 2. The last graph in Figure 2 shows the final distribution for 1 cm discs dropped in the absence of obstacles.

The spreading of the particles clearly depends on the disc size. In the no-obstacles case the discs do not spread as much as they do when obstacles are present. Note that the smaller the discs are, the narrower is the final distribution. We think that this is due to the fact that the effective distance between obstacles *seen* by the falling particles depends on the disc size, and then, on the mean number of collisions. As a consequence, a smaller particle will travel longer distances without colliding, i.e. no changing its direction of movement. This could give rise to a narrower final spatial distribution. Will this affect the mixing?

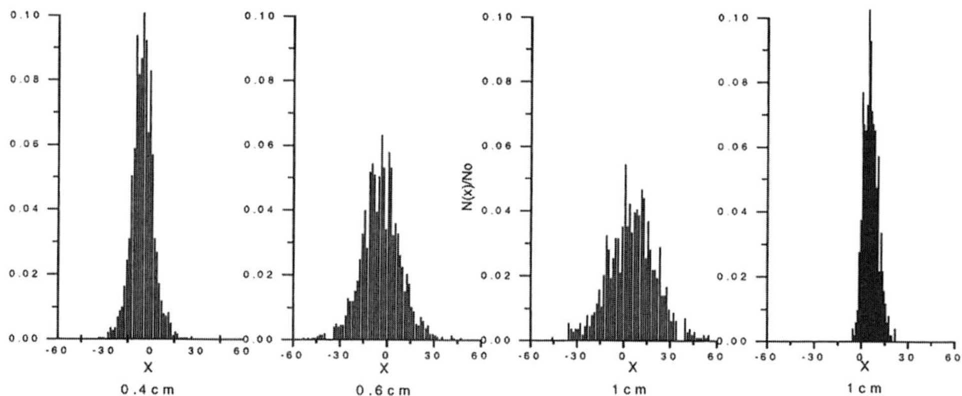

Figure 2. Test Experiments: Final distribution along the x-direction for disc diameter 0.4, 0.6 and 1cm in the presence (gray histograms) and absence (black histogram) of obstacles.

3.2 MIXING EXPERIMENTS

The main difficulty encountered when describing mixing processes is the determination of a control parameter that gives a quantitative description of the degree of mixing.

In our experiments we define a mixing parameter in x direction $M(x)$ as:

$$M(x) = \frac{P_R(x)}{P_L(x)} \qquad\qquad if\ x<0$$

$$M(x) = \frac{P_L(x)}{P_R(x)} \qquad\qquad if\ x>0 \qquad\qquad (1)$$

where $P_L(x)$ (P_R) is the spatial distribution along the x direction of the species L (species R) which is dropped from the left box (right box). Thus:

$$P_L(x) = \frac{N_L(x)}{N_{OL}}$$

$$P_R(x) = \frac{N_R(x)}{N_{OR}} \qquad\qquad (2)$$

$N_L(x)$ ($N_R(x)$) is the number of disc species L (R) at position x in the final distribution while N_{OL} (N_{OR}) is the total number of disc species R (L). We can rewrite (1) as:

$$M(x) = \frac{N_R(x)}{N_L(x)}\, a \qquad\qquad if\ x<0$$

$$M(x) = \frac{N_L(x)}{N_R(x)}\frac{1}{a} \qquad\qquad if\ x>0 \qquad\qquad (3)$$

$a = \dfrac{N_{OL}}{N_{OR}}$ being the composition of the ideal mixture. With these definitions of the mixing parameter the ideal mixture will be achieved when $M(x)=1$. Note that the mixing parameter defined in this way will only make sense when statistical fluctuations can be neglected, i.e. when the number of particles in each bin is big enough. In our experiments we will only calculate the mixing parameter for bins containing more than 10 particles.

The left hand side of Figure 3 shows the histograms of the final distribution in the x direction for three different mixing experiments. Each histogram shows the cumulative results of 15 runs. For each run the right box contained approximately 75 discs of 1cm diameter, while the left box was filled successively with approximately 490, 220 and 75 discs of 0.4, 0.6 and 1cm diameter, respectively. Actually, the two boxes were filled up to the same level (the areas occupied by the two species were identical but due to their different size, they differed in number). The last experiment corresponds to the case where the two species are of the same size. So in order to distinguish the discs coming from the left box from the ones coming from the right box, one species was painted. On the right hand side of Figure 3 the corresponding mixing parameter, as defined in equation (3), is displayed. As expected from the distributions shown in Figure 2, the mixing spreads more when trying to mix bigger particles.

In Figure 4, we have overlapped the mixing parameter for the three different cases shown in Figure 3. We have also added the mixing parameter for the case when no obstacles are present.

                          L.BRUNO ET AL.

This experiment was done with discs of 0.6 cm diameter for species R and 1 cm diameter for species L. Note that the mixing parameter seems to be not very dependent on the two species size ratio while it is sensitive on whether there are obstacles or not.

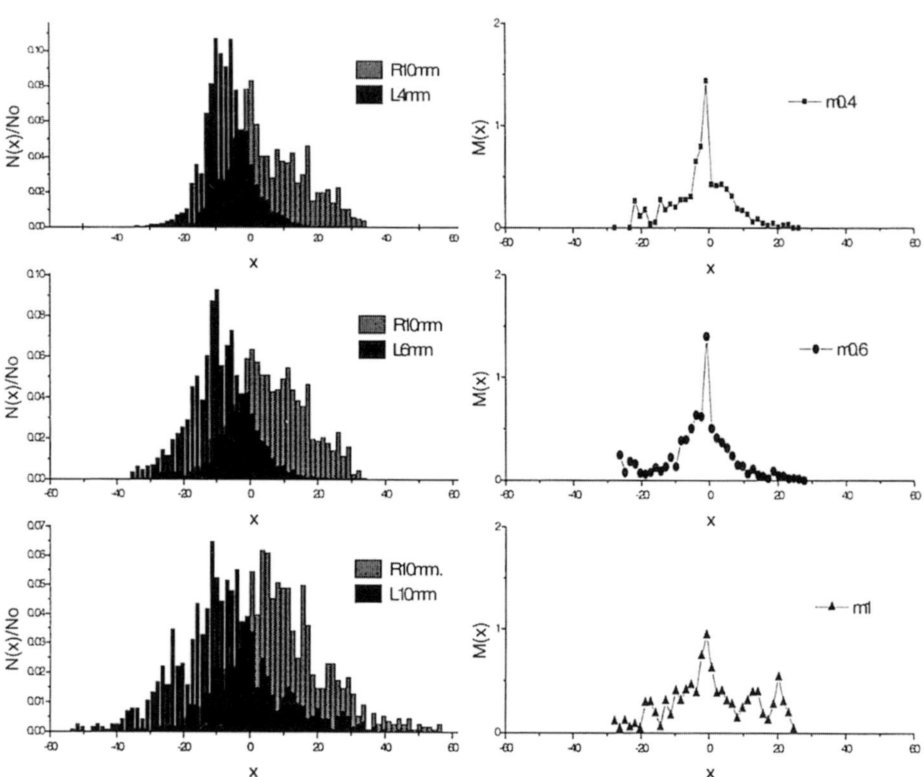

Figure 3. The final distributions for three mixing experiments are displayed on the left. The letter L and R followed by a disc size in the legend refers to whether these discs were dropped from the left or right box, respectively. On the right, the corresponding mixing parameter (c.f. Eq(3)) is shown

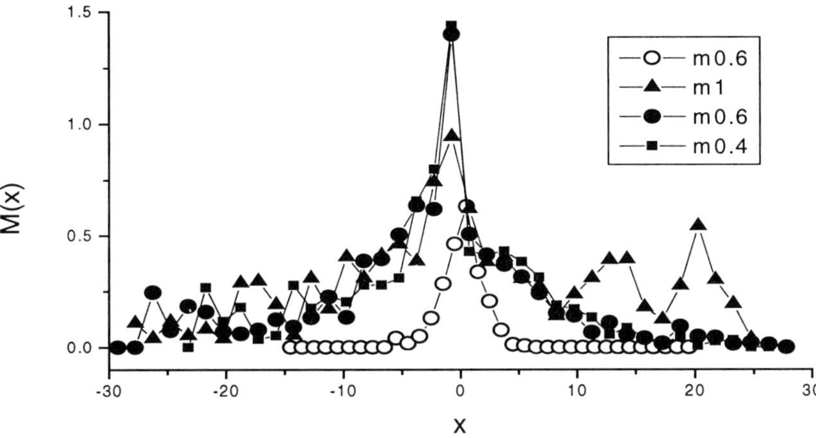

Figure 4. Mixing parameter *M(x)*, for different mixing experiments in the presence (solid symbols) and in the absence (open symbols) of obstacles.

One can also analyze the segregation along the *y* direction. In order to do this, for each bin we calculate the probability of finding a disc of species *L* (or *R*) at a position *y*, i.e. along the height of the bin column. The *y* spatial direction is in fact a measure of the temporal evolution of the mixing process. If the mean transit time that particles spend inside the Galton's board depends, for example, on their size, one would expect that the probability to find the smaller particles at the bottom of the final distribution will differ from that of the bigger ones. This effect will be enhanced if the total height of the Galton's board is increased.

In Figure 5 we display the results for one representative bin (in this case the first bin next to zero in the positive *x*-direction) of 0.6 mixing experiment. The horizontal axis in Figure 5 corresponds to a *y*-distance, which is related to the arrival order of a small or big disc. The probability of species *L* (or *R*) in a bin was obtained normalizing the number of discs *L* (or *R*), i.e. N(Y), by the total number of discs present in the bin considered, i.e. Nt. In this way, one would expect to get a constant value for the probability in the *y* direction.

Figure 5 shows that the probability for both species is approximately the same in our system. We can conclude that there is no significant temporal segregation for the range of the disc sizes used in the experiments.

Finally, we have compared the distributions of the species *L* found in the mixing experiments (*c.f.* Figure 3) to those obtained in the experiments where only one species is dropped (*c.f.* paragraph 4.1). This comparison allows us to analyze the influence of the other species on the final distribution. The distributions looked essentially the same, but their centers seemed to be displaced. These displacements were quantified by means of the first order moment of the distribution, and turned out not to be significant.

                    L.BRUNO ET AL.

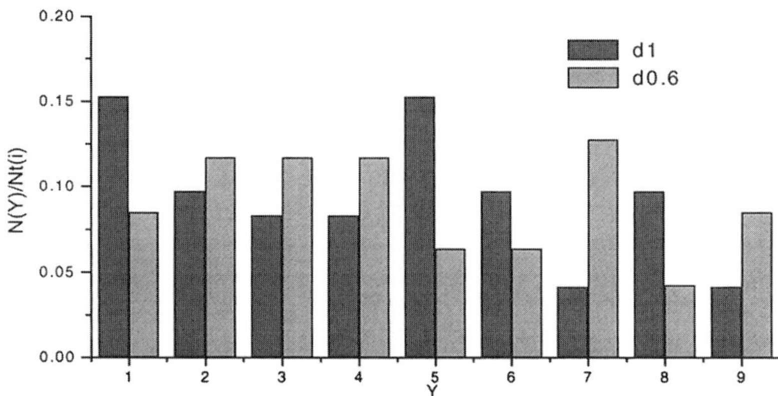

Figure 5. Representative probability distribution in *y* spatial direction for discs of 1cm and 0.6 cm mixing experiment contained in one bin (the first positive from zero).

4. Discussions

These very preliminary experiments seem to show that a Galton board could be effective in mixing granular matter. They also show that the mixing does not depend significantly on the size ratio -within the studied range- of the ingredients one want to mix, but on the presence of the obstacles lattice. The distance between obstacles appears to be the key parameter, that should be taken into account in an improved definition of the mixing parameter.

Changing the distance between obstacles and developing non regular lattices are the next step for future experiments.

5. Acknowledgements

We would like to thank N. Nerone, G. Drazer, M. A. Aguirre, M. Piva, R. Chertcoff, J. E. Wesfreid and R. Delannay for very fruitful discussions and suggestions This work was supported by the Program PICS CNRS CONICET 561. L.B. has also been supported by a UBA grant

6. References

1.   Zik, O. and Stavans, J.: Self-diffusion in granular flows, *Europhysics Letters,* **16** [3] (1991), 255-258..
2.   Natarajan, V. V. R., Hunt, M. L. and Taylor, E.. D.: Local measurements of velocity fluctuations and diffusion coefficients for granular material flow, *J. Fluid Mech.,* **304** (1995), 1-25.
3.   Scott, A. and Bridgewater, J.: Self-Diffusion of spherical particles in a simple shear rate apparatus, *Powder Technology,* **14** (1976), 177-183.
4.   Harnby, N., Edwards, M. F. and Nienow, A.W.: *Mixing in Process Industries*, Butterworths, 1992.
5.   Kaye, B.H.: *Powder Mixing,* Powder Technology Series. Ed. Chapman & Hall, 1997.
6.   Guyon, E. and Troadec, J.P.: *Du sac de billes au tas de sable.* Editions Odile Jacob, 1994.

SIZE SEGREGATIONS IN SNOW AVALANCHES
Observations and Experiments

J. MCELWAINE and K. NISHIMURA
Institute of Low Temperature Science, Hokkaido University
Kita-19, Nishi-8, Kita-ku, Sapporo, JAPAN 060

1. Introduction

In general, well-developed dry snow avalanches consist of at least two stratified layers: snow cloud at the top and dense flowing snow at the bottom [1]. It is well known that the snow cloud sometimes travels faster and farther than the flowing part and may cause serious damage in the runout area. However, since the dense flow often involves most of the mass of the avalanche and is very destructive, understanding its characteristics is of practical significance.

In order to increase our knowledge of avalanche dynamics and to contribute to avalanche zoning and to the design of protection structures, we have carried out natural snow avalanche observations (Shiai-valley, Japan and Ryggfonn, Norway) and also experiments at a ski-jump. In this paper we briefly introduce both approaches and also preliminary results of size segregation experiments.

2. Measurements

In the Shiai-valley, a systematic investigation of natural powder snow avalanche has been under way since 1989. In winter, snow usually accumulates to more than 20 m in the valley and the air temperature falls to below −15 °C [2]. Shiai-valley runs from an elevation of 1600 m (a.s.l.) to the Kurobe River at an elevation of 600m; its length is about 2000 m, the vertical drop is 1000 m, and average angle of inclination is 33 deg.. At the main observation site, at the midpoint of the avalanche path, instruments were set to measure avalanche impact pressures, wind velocity, wind pressure, atmospheric pressure, temperature and ground vibration. Avalanche movement was recorded with three video cameras. Most of the equipment was installed on two steel mounds of cylinders 0.3 m in diameter and 5 m in height. Data were recorded at a rate of 1 kHz by the data acquisition system in an underground room. Detailed information on the measurement system can be found in Kawada *et al.*[2] and Nishimura *et al.*[3].

Figure 1 shows the air velocities calculated from the static pressure difference between in an avalanche cloud and in the underground room during the passage of a dry snow avalanche on January 29, 1996. Since the snow cover was observed to be

A.D. Rosato and D.L. Blackmore (eds.), IUTAM Symposium on Segregation in Granular Flows, 81–88.

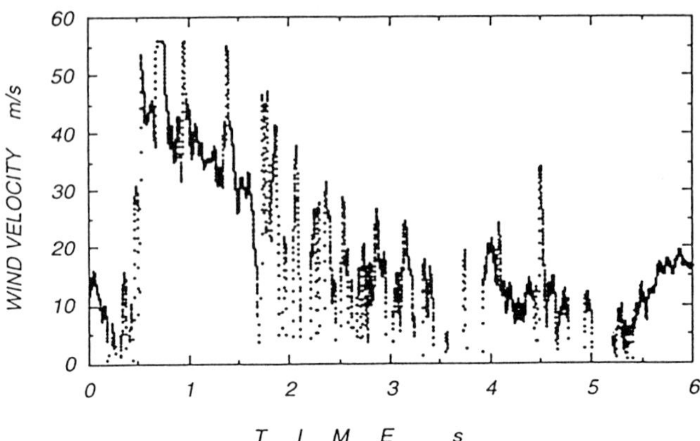

Figure 1.   Air velocities in a snow cloud calculated with the recordings of the
static pressure depression at the Shiai-valley on January 29 1996.

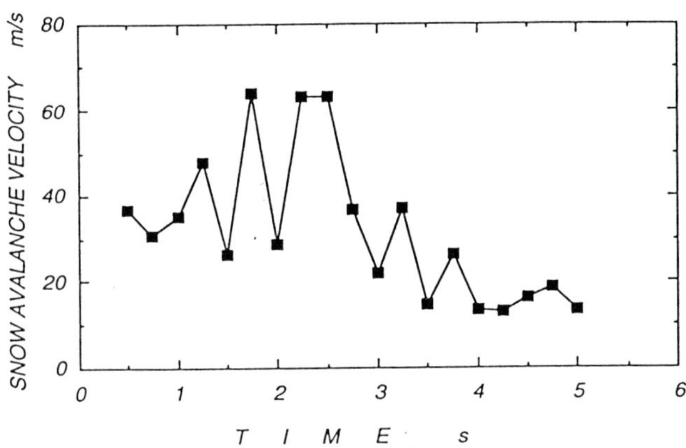

Figure 2. Velocities of the lower flowing layer of the avalanche calculated by
correlating the data from the impact pressure sensors at the Shiai-valley
on January 29 1996.

about 2 m deep, it presents the recording at around 3 m above the snow surface. It was
the largest avalanche in seven years, and was strong enough to damage the observation
tower and to destroy some instruments. The velocity of the snow cloud showed a rapid
increase to more than 56 m/s, the limit of measurement with this system. The velocity
then declined gradually with periodic fluctuations.

Velocities of the lower flowing layer were also calculated by correlating the data
from the impact pressure sensors. The cross-correlation function was calculated at 0.25
s intervals from a time series of impact data to find the average internal snow flow
velocity. The velocity was obtained every 0.25 s from a combination of the lag time that

gave the highest correlation and the distance between the two measuring points. Further details of the method have been given in Nishimura et al. [3]. Calculated velocities are presented in Figure 2 which shows that the magnitude and the variation of the velocity in the snow cloud are in approximate agreement with the velocity of the dense layer, which suggests a close interaction between the two. Since the interval of calculation here is much coarser than the sampling rate of static pressure depression (10^3 s^{-1}), we cannot compare the data directly.   However, it is noteworthy, that the velocities of the snow flow also show a periodic change with a dominant frequency of around 1 Hz [5]. Such longitudinal wave-like characteristics were also found by McClung and Schaerer (1985) for both wet and dry avalanches.

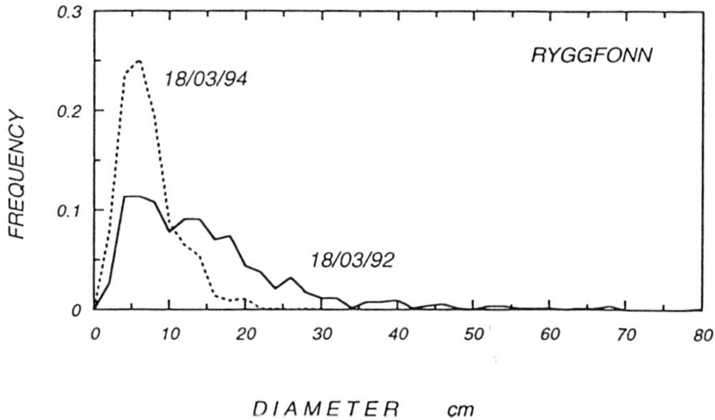

Figure 3.   Size distribution of snow blocks observed in avalanche debris in Ryggfonn, Norway. One on March 18, 1994 was a dry snow avalanche and the other one, released on March 18, 1992, was a wet one.

Although precise data are very difficult to obtain, it is known that the flowing layers are usually composed of fluidized snow and a number of snow blocks. Figure 3 shows the size distribution of snow blocks observed in avalanche debris in Ryggfonn, Norway, both of which were released with an explosion at 1530 m (a.s.l.) and ran down 1600m on avalanche path of 28° mean slope[6-7]. One on March 18, 1994 was a dry snow avalanche and the other one, released on March 18, 1992, was a wet one. In Figure 3 we see the wet one has a wide size distribution and is larger in average (16.0 cm) than the dry one (8.3 cm). In the latter case we also measured the size distribution of snow blocks along the avalanche path every 20m (see Figure 4); zero on the horizontal axis corresponds to the front of the avalanche debris. Although it has been reported in the debris flow that the bigger rocks tend to appear near the front due to segregation processes [8], the snow avalanche in Figure 4 shows a rather uniform distribution.

However, Issler et al. [9] observed that the larger snow blocks existed nearer the surface than the smaller ones by the pit observations of snow avalanche debris. Thus size segregations seem to happen in snow avalanches as well.

In addition it should be also noted that the avalanche balloons have been developed as a safety device that can reduce the risk of avalanche burial and are already practically in use. In case of an avalanche, the skier triggers the balloon, which is

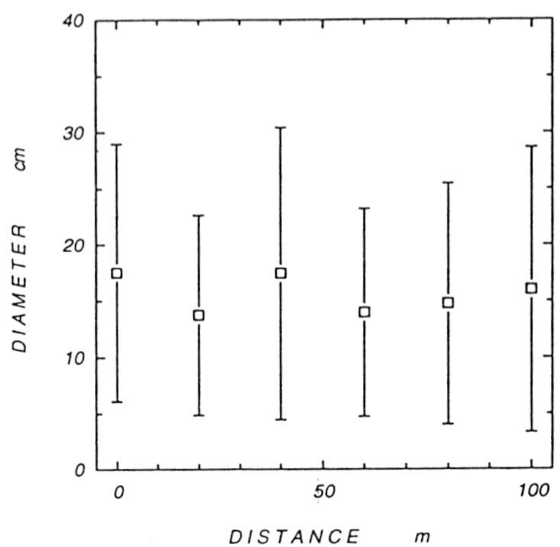

Figure 4 Size distribution of snow blocks along the avalanche path measured in Ryggfonn, Norway on March 18, 1992. Zero on the horizontal axis corresponds to the front of the avalanche debris.

folded and carried in a specially designed backpack, by releasing pressurized gas from a cartridge. It helps to prevent the victims being buried by increasing his volume. The victim rises through the surrounding smaller snow blocks and particles to the surface of the layer, where, due to the higher velocity at the surface, they move on to the front of the avalanche. Static buoyancy alone cannot explain the effect of the avalanche balloon, since flow densities are typically less than 400 kg/m^3 which is about the density of the victim with the inflated balloon. Tschirky and Schweizer[10] carried out the field test and proved its efficiency.

3. Experiments at a ski jump

Perhaps it is right to say that snow avalanches are made up of granular materials. After a dry snow avalanche starts, the snow blocks are broken into smaller lumps or even ice particles. On the other hand, after a wet snow avalanche stops we find a number of snow balls in the debris as shown in Figure 3. Hence, some of the results from studying granular flows can be applied to snow avalanche modeling [11], but unfortunately most of the theories and numerical simulations developed so far appear too simplified to realistically describe the snow avalanches. To investigate granular flows, we carried out inclined chute experiments with snow and ice spheres in a cold laboratory and obtained the profiles of density and velocity as functions of inclination and temperature [12]. However, it was unclear whether the flow reached the steady-state in the 5.4 m long chute. Thus, as a next step, we have started avalanche experiments on the Miyanomori ski jump in Sapporo, because it offers the longest inclined plane under controlled conditions. In winter, natural snow (300 kg maximum) was used. In summer, on the other hand, we have released up to 500,000 ping-pong balls to perform three dimensional granular experiments [13].

The ping-pong balls used in this study were 37.7 mm in diameter and weighted 2.48 g. Since the effect of the air drag acting on such a light ball is fairly large, the flow velocities were expected to arrive at steady state within a short distance. In fact, Nohguchi et al. [14] found in their 22 m long chute experiments that the front velocity

Figure 5.   250,000 ping-pong ball flow along the ski jump.

of ping-pong ball flow became nearly constant at 10m downstream of the starting point. Furthermore, they concluded with their similarity analysis that the ping-pong flow on the 100 m long slope corresponded to the natural powder snow avalanches which run down for a few kilometer distance[13,14].

In the experiments, up to 500,000 ping-pong balls were stored in a large container set on top of the landing slope. They were released simultaneously by opening the gate of the container. The flow accelerated down along the 150m long and 30 m wide slope, the floor of which was made with an artificial grass and its inclination amounted to 36 deg. from K to P point. The individual movements of the balls and the behavior of the flow were recorded with several video cameras (Figure 5).

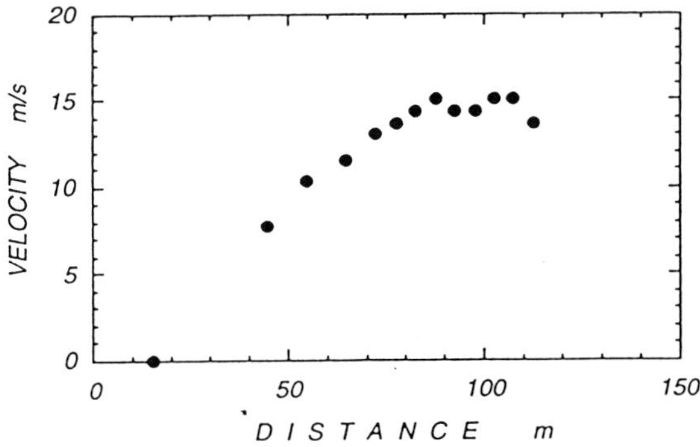

Figure 6.   Leading edge velocity of the ping-pong ball flow.

Figure 6 shows the leading edge velocity as a function of runout distance when 250,000 ping-pong balls were released from 15 m down the top. The flow accelerated linearly with the distance down the inclined artificial grass floor and its velocity eventually amounted to 15 m/s 65 m down from the starting point. Then the flow kept

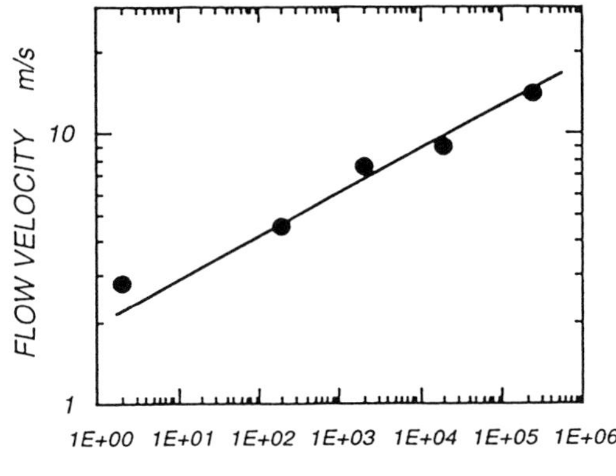

Figure 7.　Leading edge velocities from K to P point as a function of released ball numbers. The line was derived theoretically by Nohguchi *et al.* (1996).

the velocity almost constant for 30m until the inclination started to decrease; that is a steady granular flow moving at its inherent terminal velocity was obtained. The flow spread out laterally and longitudinally as it moved down the slope (Figure 5) and, after passing the steepest part, the flow came to a stop on the braking track.

The flow velocities and run out distance strongly depended on the number of released balls. The leading edge velocities measured from K to P point are given in Figure 7 as a function of the number of released balls. The velocity showed a remarkable growth from 2.8 to 15 m/s as the number of ping-pong balls increased from 2 to 300,000. Generally not only air drag but also particle-particle and particle-floor collision act to reduce the velocity. In fact, when two balls were released the velocity was only 2.8 m/s; each ball ran down individually without interaction. This velocity is

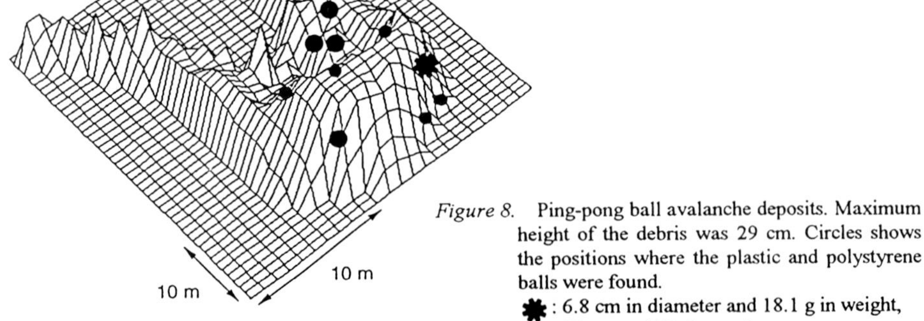

Figure 8.　Ping-pong ball avalanche deposits. Maximum height of the debris was 29 cm. Circles shows the positions where the plastic and polystyrene balls were found.
✳ : 6.8 cm in diameter and 18.1 g in weight,
● : 15 cm and 30.5 g, and
• : 10 cm and 10.6 g .

much less than the free fall velocity of a ping-pong ball, which is $U_t = 9.4$ m/s. However, with an increasing numbers of balls the free fall velocity was reached and surpassed. In fact the largest velocity was 15 m/s which is 1.5 times larger than the free fall velocity.

In the experiments, as the number of balls increased, the head and tail structure became clearer and clearer. When 250,000 balls were released, the thickness of the head was higher than 60 cm which corresponds to about 16 particles diameters. Although the individual balls changed positions rapidly, the leading edge flowed like a consolidated body. Hence, it is reasonable to say that the size of the head gives a strong effect on the flow velocity change listed in Figure 7.

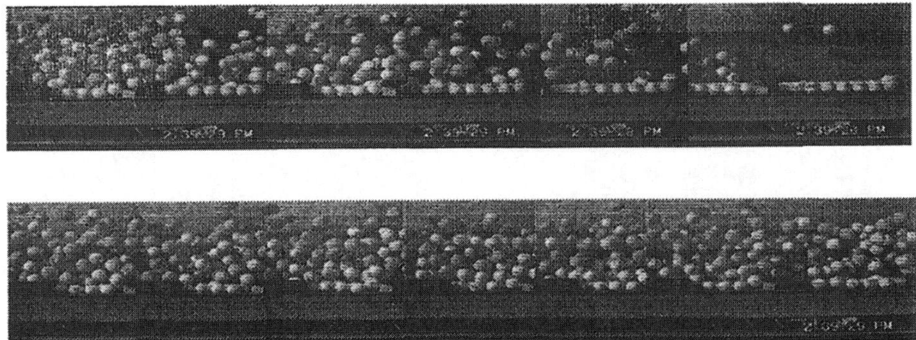

Figure 9.   Ping-pong flow along the chute of the Shinjo Branch of Snow and Ice Studies, NIED. The flow consists of larger balls (44 mm in diameter and 2.2 g in weight) and ordinally ones (38 mm in diameter and 2.5 g in weight).

In addition to the velocity and particle concentration measurements described above, we have put some particles (balls) with different size and densities from the ping-pong balls in the container in order to look at the segregation process in the flow. Although more experiments and careful analysis will be necessary, it might be noted that a plastic ball, the diameter of which is about twice that of ping-pong ball (6.8 cm in diameter) and the density of which is comparable (18.1 g in weight), appeared at the front part of the debris (Figure 8). Further, aiming at investigating segregation process in granular flows along the chute, we have also carried out the ping-pong flow experiments in the chute of the Shinjo Branch of Snow and Ice Studies, NIED, Japan. The chute has a slope angle of 30° and a width of 1 m. One side of the wall consists of glass plate, while the other side is made of wood. Larger ping-pong balls, 44 mm in diameter and 2.2 g in weight are used as well as ordinally ones. All the balls were kept in a container on top of the chute and released simultaneously. Before an each run the ping-pong balls are located all over the floor to make a rough and uniform surface. Figure 9 shows the flow situation. At this stage no clear findings have been obtained, but we expect that the combination of the above experiments and the simulation of 3-dimensional, inhomogeneous two-phase flows , will make it possible to reveal the segregation process in avalanche flows.

4. Acknowledgements

The authors gratefully acknowledge all our colleagues who joined our measurements at Shiai-dani and experiments at the ski jump. It is certain the work could not have been completed without their cooperation. These projects were partly supported by the Grant-in-Aid for Co-operative Research and Science Research of the Ministry of Education, Science and Culture, Japan.

5. References

1. Schaerer, P.A.: Friction coefficient and speed of flowing avalanches, International Association of Hydrological Sciences Publication 114 (Symposium at Grindewald 1974 – Snow Mechanics), 1975, pp. 425-432.
2. Kawada, K., Nishimura, K., and Maeno, N.: Experimental studies on a powder snow avalanche, *Annals of Glaciology* **13** (1989), 129-134.
3. Nishimura, K., Narita, H., Maeno, N., and Kawada, K.: The internal structure of powder-snow avalanches, *Annals of Glaciology* **13** (1989), 207-210.
4. Nishimura, K. and Ito, Y.: Velocity distribution in snow avalanches, *Journal of Geophysical Research* **102** [B12] (1997), 27297-27303.
5. McClung, D.M. and Schaerer, P.A.: Characteristics of flowing snow and avalanches impact pressures, *Annals of Glaciology* **6** (1985), 9-14.
6. Nishimura, K., Maeno, N., Sanderson, F., Kristensen, K., Norem, H., and Lied, K.: Observations of the dynamic structure of snow avalanches, *Annals of Glaciology* **18** (1993), 313-316.
7. Nishimura, K., Sandersen, F., Kristensen, K., and Lied, K.: Measurements of powder snow avalanches, - Nature -, *Surveys in Geophysics* **16** (1995), 649-660.
8. Curry, R.R.: Observation of Alpine mudflows in the Tenmile range, Central Colorado, *Bull. Geol. Soc. Am.* **77** (1966), 771-776.
9. Issler, D., Gauer, P., Schaer, M., and Keller, S.: Staublawinenereignisse im winter 1995, seewis, adelboden und col du pillon, *Interner Bericht* **694**.
10. Tschirky, F. and Schweizer, J.: Avalanche baloons – preliminary test results, *Proceedings of the International Snow Science Workshop, 1996 in Banff*, 1997, pp. 309-312.
11. Savage, S.B.: The flow of granular materials, *Chemical Engineering Science* **38** [2] (1983), 189-195.
12. Nishimura, K., Kosugi, K., Nakagawa, M.: Experiments on ice-sphere flows along an inclined chute, *Mechanics of Materials* **16** (1993), 205-209.
13. Nishimura, K., Keller, S., McElwaine, J., and Nohguchi, Y.: Ping-pong ball avalanche at a ski jump, *Granular Matter* **1** (1998), 51-56.
14. Nohguchi, Y., Nishimura, K., Kobayashi, T., Iwanami, K., Kawashima, K., Yamada, Y., Nakamura, H., Kosugi, K., Abe, O., Sato, A., Endo, Y., Kominami, Y., and Izumi, K.: Similarity of avalanche experiments by light particles, *Proceedings of INTERPRAVENT 1996* – Garmich-Partenkirchen, Tagungspublikation, Band 2, pp. 147-156.

ANALYZING AND OVERCOMING INDUSTRIAL BLENDING AND SEGREGATION PROBLEMS

JAMES K. PRESCOTT and JOHN W. CARSON
Jenike & Johanson, Inc.
One Technology Park Drive
Westford, MA 01886 USA

1. Introduction

Obtaining a uniform blend of dry bulk solids, and maintaining that blend through downstream equipment, is a problem faced daily by engineers and operators in industries as varied as pharmaceuticals, foods, plastics, fiberglass, and battery production. Ineffective blending, or the inability to control particle segregation, is always costly in terms of rejected materials, extra blending time, and defective end products.

The most important requirement in any industrial blending application is to have properly mixed material at the point in the process where it is needed. This is not necessarily the same as requiring that material in a blender be properly mixed, since subsequent handling of a blended material often results in significant de-blending due to segregation. In general, materials that mix easily also separate easily by segregation.

If one is to be effective in addressing blending and segregation problems, a holistic approach must be taken. This involves incorporating an understanding of sampling, blending behavior, segregation mechanisms, and material flow behavior. If the focus is only on one aspect (*e.g.*, blending), the problem may not be solved and, in fact, may be made worse.

As an example, consider if one is experiencing quality control problems with a tableting press. Assays of the tablets are showing unacceptable variations in one or more essential parameters. Rather than jump to conclusions based upon poor or inadequate sampling data, or focus only on a single piece of equipment (*e.g.*, the tableting press itself), a holistic approach is required. This would include:

- conducting a proper sampling program,
- investigating the blender to determine how good a job it is doing in blending the material,
- determining the segregation potential of the material by various segregation mechanisms, and
- determining the material's flow properties to be able to understand how it flows through bins, hoppers, feeders, chutes, etc.

A.D. Rosato and D.L. Blackmore (eds.), IUTAM Symposium on Segregation in Granular Flows, 89–101.

2. Segregation Mechanisms

There are five industrially important mechanisms by which a blend of particles separate from each other. These are:

Sifting/percolation. Under appropriate conditions, fine particles tend to sift or percolate through coarse particles. For segregation to occur by this mechanism there must be a range of particle sizes. (A minimum of 2:1 is often more than sufficient.) In addition, the mean particle size of the mixture must, in general, be greater than about 100 microns, the mixture must be relatively free flowing, and there must be relative motion between particles. This last requirement is very important, since without it even highly segregating blends of ingredients that meet the first three tests, will not segregate.

Relative motion can be induced, for example, as a pile is being formed, as material moves rhythmically over the idlers of a belt conveyor, as particles tumble and slide down a chute, or within a converging flow channel where there is either stagnant material outside of the flow channel or a significant velocity profile within it.

The result of sifting/percolation segregation is usually a side-to-side variation of particles. In the case of a bin or stockpile, the smaller particles will generally be concentrated under the fill point, with the coarse particles concentrated at the outside of the pile. Within a converging flow channel, larger particles can be more easily forced to change direction to coincide with the streamlines of the flow channel than smaller particles. The latter therefore tend to move vertically downward into a slower moving or stagnant region.

Air entrainment (fluidization). Variations in particle size or density of fluidizable powders often result in a vertical striation pattern, with the finer/lighter particles concentrated above larger/denser ones. This can occur, for example, during the filling of a bin or railcar. Whether or not the powder is pneumatically conveyed into the container, or simply drops free-fall through an air stream, it may remain fluidized for an extended period of time after filling. In this fluidized state, larger and/or more dense particles tend to settle to the bottom.

Particle entrainment. Similar to the air entrainment mechanism, particle entrainment occurs primarily with fine powders which vary in particle size or density. As a result of these variations, the finer/lighter particles remain suspended in air longer than larger/denser ones. The result can be unusual patterns of segregation. For example, a dilute-phase pneumatic conveying line, which terminates at the sidewall of a receiver, will induce various patterns of air current within the receiver depending upon the incoming velocity, the solids loading ratio, the amount of material already in the receiver, the location and type of dust collection system within the receiver, etc.

Another example involves material dropping free-fall into a container. The larger/denser particles will tend to remain concentrated in an area near the incoming stream, whereas smaller/lighter particles will be transported into slower moving or even stagnant air.

Trajectory mechanism. Particles exiting from a chute or mechanical conveyor may land at varying distances from their exit point, resulting in a side-to-side variation in the receiving container or pile. This is generally the result of particles having different size, shape, or density, which results in varying momentum, retarding effect of the air, etc.

Dynamic effects. This is an unusual mechanism, but when it occurs can be very difficult to control. It occurs when particles have different resiliency, such that when a pile is formed some particles bounce while others do not.

3. Flow properties

A material's flow properties influence the type of flow pattern which will develop, as well as whether or not segregation will be a concern and which mechanism is likely to dominate.

The most important flow characteristics of a bulk solid are its flow function (strength as a function of consolidating pressure), angles of internal friction, angles of wall friction, bulk and particle densities, and permeability [1]. Most of these properties are typically measured using a Jenike Shear Tester [2].

Most highly segregating bulk solids have little cohesive strength. Hence their flow functions are generally flat and indicate very small minimum outlet dimensions are required to prevent arching.

4. Flow patterns

The type of flow pattern which develops in a bin or hopper often has a greater influence on the degree of segregation which develops upon discharge than any other single factor, including how the material is charged into the container.

Two basic flow patterns have been identified: mass flow and funnel flow [3,4]. Generally, mass flow is preferred when dealing with segregating materials, but this is by no means an automatic assurance that segregation will be eliminated. For example, if the segregation mechanism is that of air entrainment, a vertical striation pattern is likely to develop as the container is filled. The first-in/first-out flow pattern associated with mass flow will maintain this vertical striation upon discharge.

Even for materials that segregate in a side-to-side fashion, mass flow can still present a problem due to the velocity profile that develops within the hopper section. Depending upon how close the hopper angle is to the mass flow limit (which is a function of the flow properties of the bulk material, type of wall material and its surface finish, hopper geometry, and other factors) will have a strong influence on the ratio of particle velocity at the hopper wall to particle velocity along the center line of the flow channel [5].

5. Sampling

Sampling is essential to identify a blending or segregation problem. Sampling is done one of two ways: either by collecting material from a stationary bed of material (*e.g.,* from a container), or by collecting material from a moving stream.

5.1. SAMPLING THIEF LIMITATIONS

Sampling from a stationary bed is often accomplished with a sampling thief. This probe is inserted into the material, and then a small cavity is uncovered to allow collection of a small sample. The cavity is then covered and removed from the material. This practice is very common in the pharmaceutical industry, particularly for sampling blends while still in a blender. Unfortunately, this approach has severe limitations, many of which are cited elsewhere. Collectively, these limitations lead to sampling bias. Some of the more pronounced sources of bias are because:

- The insertion of a thief disturbs and smears the bed, resulting in a sample that does not represent the material that was there prior to the thief being inserted [6].
- Results from a thief are highly operator-dependent; changes in the angle of insertion, insertion rate, twisting or rocking during insertion of the thief or collection of the sample, all contribute to variability that can yield significantly different findings [7].
- Thief results are a function of its depth of penetration, such that even if the blend were uniform, results would be different top to bottom [8].
- The lack of a standard thief design adds further uncertainty; it has been shown that merely changing the design of a thief can change blend results from unacceptable to passing; without changing how the blend was achieved. In a real application, it would be difficult to know which thief was giving the "correct" results.

It has been said that, as the name implies, a thief is not to be trusted. In recognizing these limitations, industry has been asking for a "perfect" thief, so attempts have been made to improve upon the design of the thief to give results that are more accurate. The focus has been how to collect a sample without disturbing it. Certainly, thief design is improving. However, even the use of a "perfect" or trustworthy thief will not provide all necessary information to ensure an acceptable product is produced.

In evaluating a thief, one must not confuse acceptable results of the samples collected with acceptable performance of the thief. A thief that smears the bed will produce samples that more closely resemble each other, in comparison to a thief that does not smear the bed. On the other hand, a thief may induce significant errors, demonstrating variability that does not exist. One thief manufacturer made a claim that in comparing their thief to one produced by their competitors, their sample thief consistently gave results indicating the customer's blend was acceptable, while their competitor's thief consistently yielded results indicative of a failing blend. While this may make the customer happy when their blend passes with the newly acquired thief, one should question whether it should have passed.

5.2. THE THIEF AS A COMPARATIVE TOOL

Of more importance than sampling bias is what the thief results are used for. Thief data from within a blender is useful for determining when a blend is achieved. As blending progresses, the variance of samples collected from within the blender will decrease over time. After a while, this variance will level off, with some "noise" due to sampling error. When the variance has flattened out, no further blending will be achieved with additional blending time. Note that if blending is too long, the material may segregate within the blender, resulting in over-blending and an increase in the variance again. By comparing the results over time, blending time can be optimized. This can be used as an indication of the endpoint, in other words how long the blender should be operated before the best possible blend is achieved.

Thief data can also be used - with caution - for identifying gross differences between locations. For example, if one were concerned about a possible dead spot within a blender, one could compare thief data from that location in comparison to other locations. Though the sample results from that location may be acceptable by itself, a consistently different value may be cause for concern. Likewise, if thief samples are on average higher from the center of a bin in comparison to the perimeter of the bin, segregation is likely occurring within the bin. However, sampling bias may still come into play, as the results may be a function of depth or penetration into the bed.

5.3. THE USE OF THIEF DATA AS AN ABSOLUTE

While thief data can be used to compare one sample to another, the absolute results of the samples (the highest or lowest single sample value, or the actual variance of the samples) are meaningless, even for a "perfect" thief. Since there will be further transfer or handling steps prior to the completion of manufacturing the product, additional interparticle motion causing segregation or blending will occur, changing both the peak values and variance of the samples taken. One must remember that when taking a single sample from a bed of material, this sample does not represent a collection of particles that would have ended up in one single package (or tablet).

However, the FDA has issued a guide [9] to the pharmaceutical industry that thief samples from within the blender must be held to specific requirements for the actual assay of individual samples and standard deviation of the collection of samples. These values are in fact more stringent than the requirements for the final product. The argument is that since segregation may occur after the blender, the tighter requirements are required to ensure uniformity of the final product. However, there are some faults with this logic.

First, achieving an adequate or even perfect blend (however proven) is academic if segregation takes place at any point after the blender. A perfect blend will not ensure that the final product will even be acceptable. Holding the standard for blending tighter than that of the end product, while well intended, is meaningless without considering the extent of segregation (or for that matter, re-blending) that takes place through subsequent handling. In essence, a holistic approach must be taken.

Second, the intent of sampling *within* a blender is to determine one of two things: that the desired endpoint of the blending cycle has been achieved, or that the blending process is repeatable or sufficiently robust from batch-to-batch. In considering this, Carstensen et. al. correctly asserts that "blending validation is a validation of an operation, not a product, and the final analysis rests with consistent good performance of content uniformity" [10]. In other words, one can sample from a blender to show repeatable results. However, the absolute value of those results should not be compared to any standards for the product, since these should be compared to the product alone.

Instead of setting arbitrary acceptance criteria for powder samples, "The in-process powder blend uniformity data should be generated to develop a data base that could be used in future processing as a guide or tool to discover uniformity problems, and pinpoint the part of the process where content uniformity problems arise" [11]. The absolute results from sampling within a blender (or other point mid-process) should not be used as an indication of the quality of the final product, unless compared to prior history for that product in that process.

5.4. A RATIONAL THIEF SAMPLING PROGRAM

If a thief is used, the sampling program (location and number of samples) should be executed in a rational, not random, manner. If the true goal is to identify and avoid problems, samples should be collected from those areas that are cause for concern. These areas are identified based on an understanding of the material behavior and the system under consideration. For instance, in evaluating a tumble blender, the potential dead spots near the trunnions should be sampled. Likewise, if sampling is conducted in a bin that has been quickly filled with a fine, aeratable powder, the focus should be at the walls of the bin and top surface of the material. It is here where fines-rich zones are likely to be found. On the other hand, armed with this knowledge, plant operators have been known to deliberately avoid these areas so as to keep the production lines running, and to keep their daily routine smooth, without regard for the quality of the end product. This behavior has been observed at several different companies.

The number of samples required cannot be determined ahead of time. If the goal is to locate a problem area (poor blend), one must know *a priori* the size and location of the potential problem area to determine the number of samples. In addition, the problem area is only a concern if it affects the product, so one must know how the size, location, and composition of the problem area will transfer from that point in the process through all downstream equipment. This can only be determined with prior history and a holistic analytical approach to the problem. By the time this could be completed, it may be easier to address the blending and segregation problems by themselves.

It is important to remember that, when segregation is a concern, the goal of sampling is not to show that the collection of samples represents the average "statistically representative" blend, but rather to try to find problems. On the other hand, a customer receiving a railcar of material would want to sample it to determine that they have, on average, received what they expected. In this case, reblending is usually performed, so "statistically representative" samples are desired.

5.5. SAMPLING MATERIAL IN MOTION

Since sampling a stationary powder bed has limitations, a better approach is to obey the two "Golden Rules" of sampling, which are: collect the sample while it is in motion, and collect a full stream sample. This eliminates the bias of using a thief. If the sample is collected from the process itself (e.g. upon discharge from the blender while filling a bin), this represents the true state of the material at that point in the process.

In considering the performance of a blender, one must consider not only achieving an adequate blend within the blender, but also whether or not that blend can be discharged without inducing segregation. For example, two different double cone tumble blenders may yield blends with the same variability within the blender. Thief data from each blender would in principle be identical. However, if the material is highly prone to sifting segregation, the flow pattern within the blender during material discharge is critical. A typical double cone blender handling a "normal" industrial product will discharge in a funnel flow pattern. As a result of material flowing against stagnant material, fines tend to percolate out of the flow channel into the stagnant region. The result may be an unacceptable discharge stream from the blender. On the other hand, if designed correctly, the cone can be configured for mass flow such that segregation does not occur, resulting in a uniform material stream upon discharge.

5.6. REDUCING SAMPLE SIZE

If a full stream sample is collected, the resulting sample is often too large for analysis. In order to reduce it to a smaller size, a sub-sample must be collected from the larger one. This process, if done incorrectly, can itself induce significant errors. The larger sample collected is immediately prone to segregation once it is in a container. Sub-sampling with a scoop or spoon, or cone-and-quartering techniques, will result in further confusion as to the actual results. Instead, a non-biased splitting technique is needed. This can be accomplished using a sample splitter, which will uniformly divide a sample into two smaller ones, or a spinning riffler, which can separate a sample into many smaller ones (six to sixteen is common). This process can be repeated to get to a sufficiently small size. In this manner, each smaller sample represents the larger one.

Note that there is less sampling bias with a larger sample size. However, a larger sample size does not show the level of blend intimacy on a smaller scale. In this respect, a larger sample tends to show gross differences in blend uniformity. In sampling the discharge from a blender to determine blend uniformity, this is ususally more critical than small-scale differences if there is further material handling (storage in a bin, transfer through a chute).

5.7. SAMPLE THE PRODUCT

A good way to take full-stream samples, while the product is in motion, is to sample the product itself, once it is packaged or processed to its final form. For instance, as cans of coffee are filled, each can represents a full stream sample of material. Analysis of cans taken from the production line can give insight into what is occurring, so long as other variables are documented at the time. For example, if there is a surge bin prior to the packaging equipment, it is important to note the level in the bin. For simplicity, it may be best to initially fill the bin from empty, then let it discharge completely, while cans are collected at regular intervals. As with collecting samples using a thief, additional samples should be taken from potential "hot spots", such as at the beginning or end of the run. In this example, a plot of particle size vs. time would likely reveal the trend of segregation induced by the bin and other upstream equipment.

In the pharmaceutical industry, the ideal full-stream, unit-dose sampler is the tablet press itself. A plot of assay over time for individual samples (tablets) collected at regular intervals throughout the run will demonstrate the true behavior of the process as a whole. By taking a holistic approach, one can focus the most intensive sampling on the end product, perhaps with a few additional samples to support theories as to what is observed. With an understanding of the flow pattern and segregation mechanisms which may be occurring upstream of the press, one can usually deduce why variations are being observed. Instead, the common practice is to simply collect blender samples and tablets from the beginning, middle and end of a run. Unfortunately, if these are deemed unacceptable, the batch must be re-run with a more intensive sampling plan to correctly diagnose the problem.

5.8. STATISTICAL CONSIDERATIONS

Taking multiple thief samples at many locations, and applying statistics to separate out variations due to sampling bias alone can identify the extent of sampling bias. One will not be able to take a single sample and adjust or prorate it in such a way to subtract out the bias that was induced. Instead, this bias can be compared to the variance of multiple, comparative samples that are determined, to understand what is statistically significant when comparing thief samples.

In general, however, statistics are misused. The biggest fault is the assumptions made about powder blending, segregation, and flow behavior that simply are not valid. For example, the number of samples required to determine blend uniformity often assumes a normal distribution of the quality of the material. Likewise, it is sometimes assumed that mid-process samples will only get better (or worse) with subsequent handling, depending on the needs of the study at the time. Another misapplication of statistical reasoning is that if the average of a collection of samples is not equal to the known average of the blend, then bias of the sampler (thief) is to blame (*e.g.*, the active concentration within a drug product is known to be 100% of label claim, but the sample average is 95%). While this certainly may be due to bias in the thief, it may also be due to biased sampling locations (or limitations of where samples can be taken from).

Statistics cannot be used without a full understanding as to the physics (segregation mechanisms, flow patterns) behind the problem. Even a full understanding of the state of a blend is academic if segregation occurs before the end product is produced. Likewise, understanding as to flow patterns and segregation mechanisms must be supported by rational sampling and proper statistical analysis.

6. Case Studies

6.1. PHARMACEUTICAL TABLETS

Pharmaceutical solid dosage forms (e.g. tablets) are sometimes produced by way of direct compression. In this process, the individual ingredients are blended together without any steps to chemically or mechanically bind them to each other. Typically, the blend is then discharged from a blender to one or more portable bins or drums for transport. These containers then feed a press hopper at the press, and the blended powder is compressed into its final form. Often, the blend is made up of a fine active drug substance (perhaps with a mean particle size of 5 to 20 microns), along with one or more excipients (a typical mean particle size may be 100 microns). This vast difference in particle size makes the blend prone to segregation.

One manufacturer was experiencing high tablet assays at the end of a batch. For their product, they discharged a batch of material from the blender into several drums. After filling, a conical hopper was placed on top of the drum; the drum was then inverted and set above the press. Once the drum was empty, another was set in its place.

Multiple thief samples taken from both the blender and drums did not show any high assays. Therefore, it was strongly suspected that material was segregating during compression.

To diagnose this problem, a holistic approach was required. First, it was determined that the high assays were occurring at the end of each drum, not just at the end of the batch. Therefore, this problem was not likely a result of a non-uniform blend within the blender. Also, there was a concern that the problem could be coming from the last material out of the drum, or the last material out of the press hopper.

To evaluate this, mid-way through the discharge of the drum, the valve on the drum cone was closed, and the remaining material in the press was discharged. No high assays were observed, an indication the press hopper was not to blame.

The original thief samples taken from the drums were not taken from primary areas of concern; namely, the very top surface. Additional powder samples taken from the top surface of the material showed assays well above 120%.

Flowability tests of the powder showed that the drum hopper was funnel flow. Further, segregation tests revealed the material was highly prone to segregation by fluidization and less prone to sifting segregation.

With all of this information, the problem was easily identified. Upon filling of the drums from the blender, the material segregated by fluidization and dusting. The stream of material impacted the pile and became aerated. The fluidized material caused air to move upward, carrying fines to the top surface. Dust was generated, which was rich in active drug, which settled on the top surface. After the drum cone was attached to the drum, it was inverted. This resulted in the layer of fines sliding or avalanching to the side of the drum; an active-rich pocket then remained at the side of the drum. Upon discharge of the drum, the funnel flow pattern resulted in this pocket of material discharging at the very end of the drum.

To correct this problem in the short term, a dust sock was attached to the bottom of the blender. The drum was then elevated (via a fork truck) so that the bottom of the drum was close to the blender valve. Once the valve was opened, the drum was slowly lowered to the floor, resulting in slow, controlled fill of the bin.

Unfortunately, this was a slow process, which also required both shorter drums (to fit up close to the valve) and an operator to estimate an appropriate fill time (rate of lowering the drum). However, it did prove the concept was valid. A long-term solution is to leave the drum on the floor and to use a sock that can be retracted automatically.

6.2. PACKAGED FOOD PRODUCT

Manufacturers who sell bulk material in packages have minimum weight requirements that must be met. Since some variability in the weights is inevitable, packages must be overfilled in order to ensure the minimum is achieved. However, this excess product is essentially given away which, although a benefit to the customer, results in lost profit to the manufacturer. The cost can run in hundreds of thousands of dollars each year per production line.

A manufacturer of a pre-cooked food product realized that their weight variations were high, and tied to the product flow through upstream equipment. In this process, there are several similar production lines. In each line, material is accumulated in a surge bin prior to packaging. Although a certain weight was required in each box, they are filled on a volumetric basis. Therefore, a consistent bulk density is critical to achieving consistent weight.

It was observed that when the material level in the surge bin was high, boxes were heavy. When the level was lower, boxes were light. Initially, plant personnel were suspicious that the higher level compacted the material, resulting in a higher density product; as the level was lowered, less compaction resulted in a less dense product. However, this hypothesis had two flaws. First, the surge bin was located about 10 ft. above the packaging equipment, to allow room for other plant equipment. A long, narrow sloping chute brought product from the bin to the packaging equipment. Unlike a liquid head, any changes in head pressure within the bin would not be transmitted to the packaging equipment, as wall friction within this chute would quickly take up any additional load. The second flaw is that the material was nearly incompressible. Though the bin was only eight feet tall, it would have taken pressure differences beyond that seen at a 100 ft. depth to produce sufficient variations in density to give the weight variations that were observed.

Upon realizing head pressure could not be to blame, segregation was suspected. Though the material was closely sized, attrition in the upstream processes resulted in broken particles. Since the material was relatively coarse (a mean particle size between 8 and 10 mesh) and free flowing, sifting segregation was suspected. To confirm this, boxes were collected throughout the run, which were then weighed and screened. The results confirmed segregation was occurring. Table 1 follows showing the weight of some boxes along with the particle size determined by sieving. The acceptable weight was 794 ± 20 g.

TABLE 1: Measurement of particle size for each box of a given weight

Size, cumulative	Box weight					
	823 g	809 g	804 g	800 g	785g	769 g
+6 mesh	2.8%	2.5	2.8	3.2	3.5	7.1
+8 mesh	20.2	24.1	39.2	42.5	34.4	56.4
+10 mesh	74.6	72.8	83.2	82.5	85.7	89.2
+12 mesh	95.0	95.5	95.9	94.0	97.8	97.9

Flowability tests of the material showed that the surge hopper was funnel flow. Furthermore, the bulk density of the +8 mesh material was significantly lower than that of the −10 mesh material, easily accounting for the variations that were observed.

With all of this information, the problem was easily identified. The inlet and outlet of the funnel flow surge bin are on the centerline of the bin. Upon filling the surge bin, material segregates by sifting. This results in a concentration of fines in the center of the bin, and a concentration of coarse material at the wall. When the level was high, the fines were discharged through the flow channel that formed above the outlet; as the level dropped, the predominantly coarse product from the wall discharged. Without a full understanding of what really was occurring, the symptoms could have been misinterpreted as being caused by the pressure in the bin, rather than the segregation problem that it was.

To correct the problem, one bin was selected from the worst-acting production line, and it was retrofitted with an insert to convert it to mass flow. In mass flow, all of the material is in motion, so that the coarse and fines discharge together. Once implemented, the weight variations were significantly reduced, and this line, which had been giving the worst variations, became the "god child" of the lines. Subsequently, other bins on the other lines were also retrofitted with such an insert.

6.3. ALUMINA PRODUCTION

Alumina is typically calcined through either a rotary kiln or Lurgi-type fluid bed calciner in the final processing stage. Given the nature of the complete process, the alumina produced contains a wide particle size range. This wide size range incorporates "superfine" material (–325 mesh), which is detrimental to aluminum production if segregated from the bulk and found in high localized concentrations. The segregated superfine alumina may lead to increased anode effects, which results in lower efficiency and quality of aluminum produced via smelting operations. Segregated alumina with high concentrations of –325 mesh material may also cause problems with the operation and control of pot feeders, and the density and solubility of alumina in the pots, etc.

An alumina producer was experiencing periods of high concentrations of superfine material being loaded into ships. In this particular case, the calcined alumina was loaded into a product silo, then discharged onto a belt conveyor which transported the material to a rail car loadout silo, the rail cars being utilized to load a large, squat shiploading silo. All silos were flat bottomed, with the product silo discharging at its periphery via a series of open air slides. Similarly, the rail car loadout silo discharged via five gravity flow ports positioned asymmetrically on the silo base, with the shiploading silo discharging via twelve covered air slides positioned midway between the symmetry axis and the silo periphery at 30° radial increments.

Due to the number of handling steps in this process, and silos that were not designed from a material flow properties standpoint, the alumina had many avenues to segregate. Hence, given the multi-staged production process, a holistic approach was required to minimize the segregation problem.

Often, the best way to minimize segregation is to enforce a mass flow pattern in a silo rather than a funnel flow pattern. Other methods of minimizing the segregation leading to high superfines is to alter the operation of the process, or modify the inlet chute distribution and/or discharge arrangement.

In this case, and in the majority of alumina segregation problems of this nature, the material segregates in a slightly different manner to that which was described in the previous examples. Specifically, the primary mechanism of segregation is associated with filling alumina into silos, where air currents developed by the filling process deposit the superfines in regions of stagnant material on the top free surface.

Given the squat nature of the three silos, the associated flow pattern is funnel flow. In this type of flow pattern, the material flows through a more or less vertical channel directly above the silo discharge ports. Material from the top surface forms an angle of drawdown into a flow channel when the material is discharged. As more material is discharged, the drawdown angle is maintained, and the funnel gets larger as material from the top surface moves down the slope into the vertical channel above each outlet. Until a downward funnel reaches the entire outside wall, there is a dead region on the top surface that collects airborne superfines, gradually settling out over time. Unless a downward funnel reaches the outside wall on a regular basis, substantial amounts of superfines will accumulate on the non-flowing top surface.

The physical size of the silos and the cost to retrofit, in this instance, prohibited the conversion of each silo to mass flow. Instead, the current funnel flow patterns were retained, with the fill and discharge points in each silo modified to minimize the superfines segregation. The prediction of the optimal positioning, size, and number of inlet and discharge chutes, hoppers, etc. was examined utilizing a proprietary computer model. The result was acceptable levels of superfines, provided that some controls were placed on the fill and discharge cycles.

6.4. MIXTURE OF PELLETS AND BEADS

A multi-national chemical firm building a new plant in southeast Asia needed to blend a highly segregating mixture, consisting of relatively large plastic pellets and small beads, to a high degree of uniformity. Initially the firm went to a supplier of commercial V-blender equipment. After several weeks of testing, they found that their V-blender could not provide the required close blending uniformity. Samples taken from inside the blender indicated that the mix was not adequately blended. Furthermore, samples taken from the blender outlet indicated that significant segregation was occurring during discharge from the blender.

In order to examine this particular blending problem, a pilot scale test model of the commercial V-blender was built. This confirmed that it indeed was not capable of producing the required blend uniformity, namely that the amount of beads in the total beads/plastic pellets sample be between 13% and 17%. The results of the V-blender displayed a beads percentage ratio of between approximately 1% and 30%, with the beads discharging only over two-thirds of the total mass of material discharged.

Next, experiments were conducted with a series of inserts in the V-blender scale model, which, while significantly improving its operation, still did not meet the firm's requirements. Finally an optimized BLENSERT® tumble blender (a cone-in-cone design that has improved segregation control over a typical mass flow container), was designed and tested. This produced results that fully satisfied the firm's requirements for the material discharged from the blender. Not only did the beads discharge over the complete mass of the contents in the blender (120 lb. in this instance), but all except the first and last two samples fell very close to the mean target percent of beads of 15%. Also included in this design work was an anti-segregating surge bin to collect the blender discharge and feed an extruder, without de-mixing the blend.

After the pilot scale model study, a detailed design of a full scale unit having a live capacity of 200 cu. ft. was developed. Following the detailed design, a contract was awarded to fabricate and supply three of the units to the new plant in southeast Asia. The supplied units have been operating successfully now for over two years.

7. References

1. Carson, J.W. and Marinelli, J.: Characterize Bulk Solids to Ensure Smooth Flow, *Chemical Engineering*, April 1994, pp. 78-90.
2. Standard Shear Testing Method for Bulk Solids Using the Jenike Shear Cell, ASTM Standard D6128-97, American Society for Testing and Materials, 1998.
3. Jenike, A.W.: Storage and Flow of Solids, *Bulletin 123 of the Utah Engineering Experimental Station,* **53** (1964), No. 26, (Revised 1980).
4. Marinelli, J. and Carson, J.W. (1992) Solve Solids Flow Problems in Bins, Hoppers, and Feeders, *Chemical Engineering Progress*, May 1992, pp. 22-28.
5. Prescott, J.K., Ploof, D.A. and Carson, J.W.: Developing a Better Understanding of Wall Friction, *Powder Handling & Processing* **1** (1999), 27-36.
6. Harwood, C.F., and Ripley, T.: Errors Associated with the Thief Probe for Bulk Powder Sampling *Journal of Powder and Bulk Solids Technology*,**1** [2] (1977), 20-29.
7. Berman, J. and Planchard, J.A.: Blend Uniformity and Unit Dose Sampling *Drug Development and Industrial Pharmacy*, **2** [11] (1995), 1257-1283.
8. Berman, J., Schoeneman, A, and Shelton, J.T.: Unit Dose Sampling: A Tale of Two Thieves *Drug Development and Industrial Pharmacy*, **22** [11] (1996), 1121-1132.
9. FDA, Guide to Inspections of Oral Solid Dosage Forms Pre/Post Approval Issues for Development and Validation, January 1994.
10. Carstensen, J.T., and Rhodes, C.T.: Sampling in Blending Validation *Drug Development and Industrial Pharmacy*, **19** [20] (1993), 2699-2708.
11. Chang, R., Shukla, J. and Buehler, J.: An Evaluation of a Unit-Dose Compacting Sample Thief and a Discussion of Content Uniformity and Blending Validation Issues, *Drug Development and Industrial Pharmacy*, **22** [9,10] (1995), 1031-1035.

MICROGRAVITY SEGREGATION IN COLLISIONAL GRANULAR SHEARING FLOWS

M.Y. LOUGE, J.T. JENKINS, A. REEVES and S. KEAST
Cornell University
Upson Hall, Ithaca, NY 14853

Abstract

This article concerns granular shearing flows in microgravity where segregation of grains by size or mass is driven by spatial gradients in the fluctuation energy of the grains.

Experiments were conducted on NASA's microgravity aircraft with a shear cell shaped like a race track and containing a mixture of two types of spherical grains. A gradient of fluctuation energy was produced between the inner, energetic moving boundary driving collisions among the grains and the outer, more dissipative boundary at the periphery of the cell. The energetics were achieved by covering the two boundaries with bumps of appropriate size and spacing and by setting the distance between the flat side walls through which the flow was observed.

The grain segregation and the corresponding velocity statistics were captured by a high-speed video camera and analyzed using computer vision software. The resulting profiles of particle mean and fluctuation velocities were compared with computer simulations of the entire experimental apparatus.

1. Introduction

The size segregation of flowing or shaken grains is a commonly observed phenomenon in industrial processes and in nature. In systems that do not involve much agitation of the grains, segregation mechanisms involving gravity include the preferential downward percolation of smaller particles in relatively slow inclined shear flows [1], the upward frictional ratchetting of large particles [2], and the preferential filling of space beneath larger particles by smaller particles in a system that is occasionally shaken [3,4]. In sheared or vibrated systems of colliding grains, gravity also influences mixtures of grains of different sizes. Here, buoyant forces act to separate grains that differ in size and, consequently, in the local volume that they displace.

In reduced gravity, buoyancy is suppressed and attention can be focused on another segregation mechanism that involves highly agitated flows and is driven by spatial gradients in the energy of granular velocity fluctuations. In steady, fully-developed flows, the balance of momentum exchanged in collisional interaction among different species of grains requires that concentration gradients be balanced by gradients of particle

103

A.D. Rosato and D.L. Blackmore (eds.), IUTAM Symposium on Segregation in Granular Flows, 103–112.

fluctuation energy. Reduced gravity also eliminates the possibility that a collisional flow may condense into a slower, denser flow dominated by enduring contacts.

Because collisions inevitably dissipate energy, collisional granular shear flows have limited extent in the direction transverse to the flow and, consequently, are strongly influenced by their boundaries. Because grains, on average, slip relative to the latter, a bumpy or frictional boundary can locally convert slip energy into fluctuation energy. This production competes with the dissipation of fluctuation energy arising in collisions with the boundary. A similar competition between the working of the mean shear and collisional dissipation largely determines the level of fluctuation energy in the interior. Although boundaries with a large slip velocity tend to create high local levels of fluctuation energy, they also promote smaller mean velocities in the interior, which in turn limit the magnitude of the fluctuation kinetic energy there.

In a shearing flow bounded by two bumpy walls normal to the gradient and two flat walls normal to the mean vorticity, the size and spacing of the bumps and the distance between the flat walls can be designed to adjust the granular slip velocity and the rates of production or dissipation of fluctuation energy [5]. This permits the control of the component of the spatial gradient of the fluctuation energy that is perpendicular to the flow. Such gradients in fluctuation energy may be exploited to drive the separation by size or other properties in a binary mixture of spherical grains.

To observe this phenomenon, we have constructed a shear cell in the form of a race track in which the segregation of a sheared binary mixture of two different types of sphere is maintained by the motion of the inner boundary relative to the outer. The experiments are meant to isolate two different sub-mechanisms of collisional segregation that usually occur together: the first is associated with differences in the inertia of the grains; it is achieved with spheres of different masses, but equal diameters. The second is associated with differences in the geometry of the grains; we approximate it using a mixture of grains of different diameter and densities.

We conducted experiments with a prototype shear cell on NASA's KC-135 microgravity aircraft. Because in the absence of gravity the only available time scale is proportional to the speed of the moving boundary, we could reduce that speed and avoid particle damage without affecting the phenomenon under study. The relatively low speed also allowed us to track the grain motion with a digital video camera.

We developed a computer vision software to analyze the resulting images and compared the profiles of volume fraction and mean and fluctuating velocities with computer simulations of the apparatus. The link between the computer simulations and the physical experiments was achieved by measuring in our laboratory the parameters that characterize individual impacts [6,7].

We begin with a description of the computer simulations and the apparatus. We then report experimental results and compare those with predictions of the simulations.

2. Simulations

Computer simulations guided our design of the microgravity shear cell and informed the development of theory. In particular, they prescribed a cell geometry featuring regions of nearly fully-developed flow and showed that measurements through the side walls closely represented the state of flow in the interior.

The simulations follow the dynamics of an ensemble of two species of spheres of diameters σ_1 and σ_2 and material densities ρ_1 and ρ_2 interacting with the boundaries and among themselves through individual impacts. They are based on the algorithm described by Hopkins and Louge [8]. In that algorithm, collisions occur when a sphere overlaps slightly with another sphere or with the wall. The algorithm adjusts its time step periodically to ensure that the mean overlap is kept below a negligible tolerance. In addition, a search grid is superimposed on the flow domain to permit fast identification of near neighbors. Because this method makes it superfluous to maintain a list of future impacts, its computing time is merely proportional to the number of spheres and, consequently, it can simulate the entire shear cell on a relatively small workstation. Figure 1 shows a simulated microgravity shear cell of size modest enough to distinguish details of the flow.

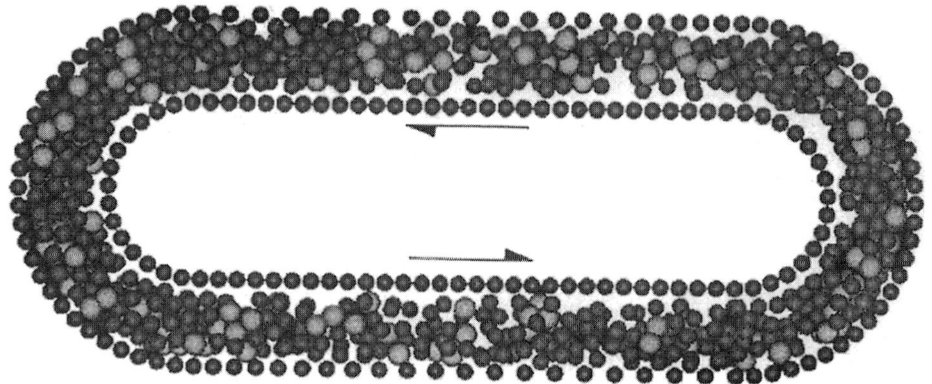

Figure 1. A shear cell of modest size with two species that differ in diameter. The outer boundary is fixed. The inner boundary moves in the direction shown.

Profiles of solid volume fraction, velocity and temperature for the two grain species were measured by dividing the flow domain into a number of averaging slices (Fig. 2). The average value $<\psi>$ of an intrinsic grain property ψ in a slice was calculated by considering a number J of instantaneous realizations of the flow, by summing all contributions from K spheres passing through the slice over all realizations, and by dividing the result by J and K. This center-averaging is consistent with the kinetic theory. Because the flow could contain two species of different masses, the second moments of velocity fluctuations were mass-weighted. For the second moment of species j along direction i:

$$T_i(j) \equiv <m(j) \, (u_i(j) - <u_i(j)>)^2 > , \qquad (1)$$

where m(j) is the mass of a species j sphere and $u_i(j)$ is its velocity in the direction i. The temperature of species j is then $T(j) = [T_1(j)+T_2(j)+T_3(j)]/3$. In the absence of gravity, all grain velocities scaled with the velocity U of the boundary. In addition, we

                    M. Y. LOUGE ET AL.

made the second moments in Eq. (1) dimensionless with U^2 and the mass of small spheres. Dimensionless quantities are denoted by a cross.

Because any rigid boundary tends to order spheres in its neighborhood, the transverse profiles of solid volume fraction exhibit spatial oscillations near the bumpy inner and outer walls with wavelength of order a sphere diameter. Because the kinetic theory ignores these fluctuations, their presence can hinder comparisons of the measured segregation with the corresponding theoretical predictions. In grain mixtures, we alleviate this difficulty by focusing on the relative number fraction ϕ rather than the volume fractions of each individual species. For example, for species 1,

$$\phi_1 = \frac{(\sigma_2/\sigma_1)^3 + \bar{v}_2/\bar{v}_1}{(\sigma_2/\sigma_1)^3 + v_2/v_1} \ . \tag{2}$$

where the overbar denotes averaging in the transverse direction. Values of ϕ_1 above unity indicate a local surplus of species 1 spheres. As long as σ_1 and σ_2 are not greatly different, the relative number fractions incorporate ratios of volume fractions that nearly oscillate in phase and, consequently, their own fluctuations are considerably reduced.

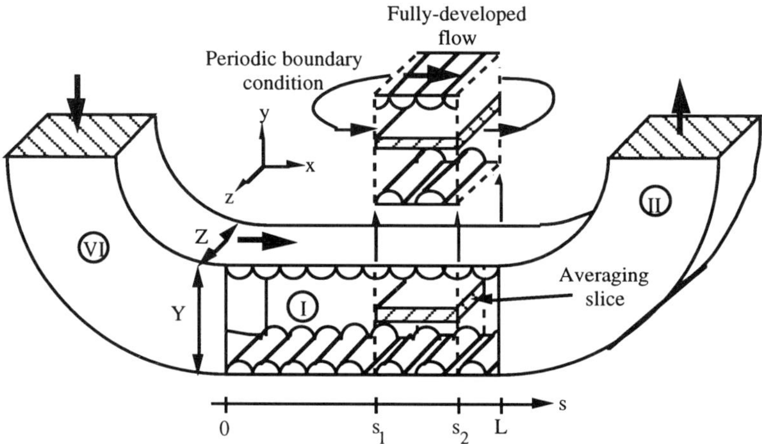

Figure 2. Sketch of the straight section (I) of the simulations with length L, width Y between centers of the cylindrical boundary bumps shown and depth Z between the two flat side walls. Bumps are affixed to both boundaries, including the curved regions (II) and (IV) and the upper straight section (III) that is omitted for clarity. Fully-developed flows are simulated using a periodic boundary condition in the "observation region" with $s_1 \leq s \leq s_2$. Dimensions are not to-scale.

In the simulations, impacts were characterized in terms of three coefficients [9]. The first is the coefficient of normal restitution e. The second is the coefficient of friction μ for sliding impacts. The last is the coefficient of tangential restitution β_0 for impacts that do not involve sliding. For collisions between two free spheres or between a sphere and one of the flat side walls, we measured these parameters with the facility described by Foerster *et al* [7]; for collisions with a cylindrical boundary bump, we adopted the method of Lorenz *et al* [6]. Table 1 summarizes the results.

<u>Table 1 - Impact properties</u>

sphere 1	σ_1 (mm)	ρ_1 (g/cm^3)	sphere 2, bump or wall	σ_2 (mm)	ρ_2 (g/cm^3)	e	μ	β_0
acrylic	3.2/3.96	1.22	acrylic	3.2/3.96	1.22	0.93	0.12	0.35
ceramic	3.18	3.86	ceramic	3.18	3.86	0.97	0.10	0.24
acrylic	3.2/3.96	1.22	ceramic	3.18	3.86	0.93	0.11	0.10
acrylic	3.2/3.96	1.22	fixed bump	3.18	-	0.97	0.22	0.28
ceramic	3.18	3.86	fixed bump	3.18	-	0.68	0.08	0.29
acrylic	3.2/3.96	1.22	aluminum	∞	-	0.94	0.14	0.51
ceramic	3.18	3.86	aluminum	∞	-	0.61	0.10	0.14
acrylic	3.2/3.96	1.22	glass	∞	-	0.83	0.12	0.34
ceramic	3.18	3.86	glass	∞	-	0.96	0.09	0.00

3. Shear Cell

With insight from the simulations, we designed the shear cell to visualize steady, fully-developed granular segregation in a rectilinear shearing flow without significant body forces. The cell has dimensions L = 419 mm, Y = 29 mm, Z = 40 mm, and its moving boundary has a radius R = 62 mm.

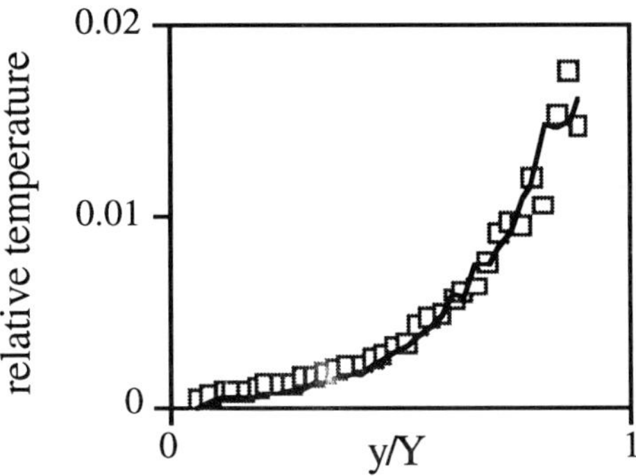

Figure 3. Comparison of simulated transverse profiles of the dimensionless granular temperature of ceramic spheres calculated from the observation region of the camera (line) and from a similar fully-developed region with periodic boundaries (symbols) with conditions of test A. The abscissa is the relative distance from the outer bumpy boundary.

Flat side walls with windows provided lateral containment and permitted the flow to be viewed. A digital Kodak EktaPro RO camera imaged a 25 mm wide region of the cell in the range $292 \le s \le 326$ mm at a frame rate F of 500 or 1000 Hz. To verify that the flow observed with the camera was nearly fully-developed, we compared transverse profiles of mean and fluctuation velocities obtained in the observation region from simulations of the entire cell against identical simulations with periodic boundaries in the direction normal to the flow (Fig. 3).

The inner boundary consists of a specialty chain to which stainless steel segments of 9.5 mm width are attached. Three closely spaced hemi-cylindrical stainless steel rods of 3.2 mm diameter perpendicular to the flow direction were vacuum-brazed onto these segments to roughen the moving boundary. The chain is driven by dynamically-balanced sprockets that produce negligible vibrations. It was run without lubrication to prevent any corruption of the frictional impact coefficients of the interior grains and the bumps. In the linear section of the cell, plastic rubbing strips support the chain while stabilizing its motion and avoiding artificially low restitution coefficients in its impacts with interior grains. The surface of the outer stationary boundary is also made bumpy with hemi-cylinders of 3.2 mm perpendicular to the flow direction and with centers separated by 4.8 mm.

The chain is driven by a dc motor capable of delivering a torque of 2 N.m at 700 RPM and connected to one of the sprockets through a timing belt. The motor is controlled through a servo amplifier receiving angular velocity feedback from a magnetic tachometer mounted on the other sprocket. On the aircraft, the z-direction perpendicular to the flat walls was aligned with the principal direction of gravitational accelerations.

4. Experiments

Table 2 summarizes experimental conditions for the flight tests in the KC-135 microgravity aircraft. The symbols $\bar{v}_1$ and $\bar{v}_2$ represent the fractions of the volume of the whole cell occupied by species 1 and 2, respectively. At these levels of overall packing, the simulations indicated that these mean volume fractions were nearly identical to those in the viewing regions.

<u>Table 2 - Experimental conditions</u>

Test	U m/s	F Hz	Spec. 1	ρ_1 g/cm^3	σ_1 mm	$\bar{v}_1$ %	Spec. 2	ρ_2 g/cm^3	σ_2 mm	$\bar{v}_2$ %	$t^\dagger$	τ s
A	1.4	1000	ceramic	3.86	3.18	10	acrylic	1.22	3.96	30	300	10
B	1.4	1000	acrylic	1.22	3.2	20	ceramic	3.86	3.18	20	130	10
C	0.7	500	acrylic	1.22	3.2	31	acrylic	1.22	3.96	10	500	15

The spheres were selected from many materials for the relatively small collisional energy dissipation that their low friction and high restitution produced. They also exhibited a finish that allowed the vision software to track their movement with accuracy. We retained Engineering Laboratories and Hoover Precision to manufacture, respectively, colored acrylic and ceramic grains of excellent sphericity ($\sigma\pm<0.3\%$) and narrow size distribution ($\sigma\pm<0.2\%$). Electrostatic charging of these dielectric solids was mitigated by maintaining high relative humidity in the apparatus. With this precaution, no evidence of such charging could be observed.

A typical KC-135 parabola included a gravitational pull of order 18 m/s^2 followed by 20 seconds of reduced gravity. A tri-axial accelerometer indicated that residual gravitational accelerations in the aircraft did not exceed ± 0.1 m^2/s. We began each test by starting the motor at the onset of reduced gravity.

The computer simulations helped us prescribe mixtures for which a constant residual gravitational acceleration $g^\dagger = g_z\, \sigma_1/U^2$ in the z-direction would produce minimum average sinking of the most massive spheres. The worst case involved ceramic spheres in Test A. There, with a residual acceleration $g^\dagger = 0.00063$, on the order the observed peak-to-peak fluctuations, the simulations predicted a sinking distance of merely 1.5 σ_1 without any apparent change in the granular temperature profile.

Using the same simulations, we also estimated the dimensionless time $t^\dagger = t\, U/Y$ required for each experiment to re-establish steady segregation after canceling that level of residual acceleration. Based on that estimate, we waited a time τ after the onset of reduced gravity to acquire images with the camera (Table 2).

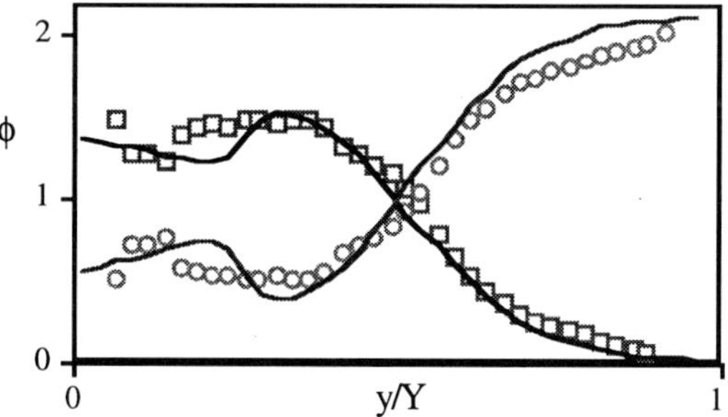

Figure 4. Comparison of simulated transverse profiles of φ estimated from area fractions observed through the side window (lines) and calculated from profiles of interior volume fraction averaged in the z-direction (symbols) for the conditions of Test B. The squares and circles represent ceramic and acrylic spheres, respectively.

The simulations also helped develop a strategy for relating the surface fractions of the two species observed through the side walls to the corresponding fractional populations in the interior. Generally, the vision algorithm detected grains that were within a sphere diameter from the window. The two species were then distinguished according to their distinct gray scale or size. Each frame was subdivided into ten horizontal strips of constant width. Strips spanned the entire length of the image to reflect the fully-developed state of the flow. By merely incrementing the observed sphere cross-sectional area intersecting each strip, the vision algorithm calculated the fraction α of the strip surface occupied by each species within the field of view and estimated the corresponding relative number fraction by substituting α for ν in Eq. (2). We tested this measurement strategy using computer simulations. As Fig. 4 illustrates, the simulated transverse profiles of φ produced by this method closely represent the state of segregation in the interior of the simulated flow.

We developed a computer image tracking analysis for this application. It consisted of five principal steps. In the first, a gradient image was computed with the Sobel edge operator from the initial digitized gray-scale image. The resulting edges were then thinned by retaining only the pixel of largest intensity in the gradient direction. Next, a binary image was generated by setting to one the intensity of pixels above a certain threshold. The result was correlated with a thin binary circle of diameter equal to that of the species of interest. Finally, a hill-climbing procedure reached the sphere centers at the peaks of the resulting correlation field. In subsequent images, the algorithm repeated those steps only in small regions adjacent to the detected centers.

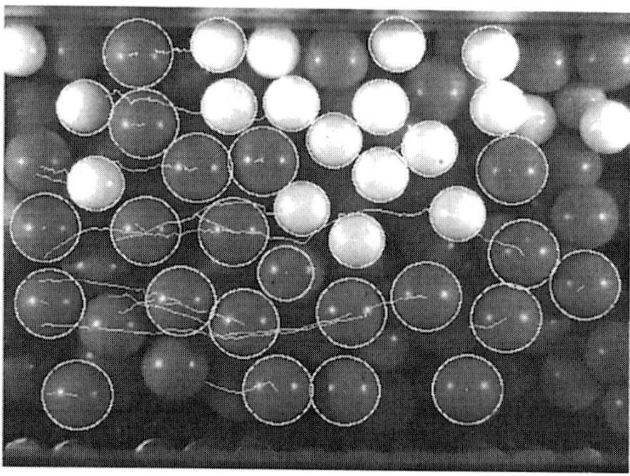

Figure 5. A typical image for the conditions of Test A. Circles and lines are superimposed to indicate the location and trajectory of detected spheres. The moving boundary can be discerned at the bottom of the picture.

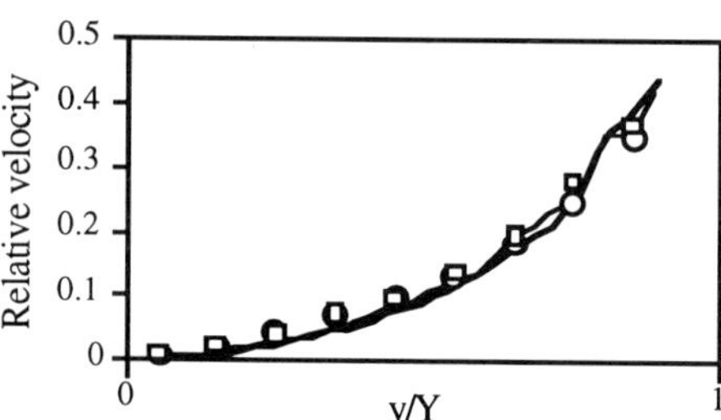

Figure 6. Transverse profiles of mean velocity relative to the chain speed for the conditions of Test A. The symbols and lines are experimental measurements and predictions of the simulations, respectively. The circles and thin line are for acrylic spheres; the squares and thick line for ceramic.

As long as the displacement between two consecutive frames remained less than a sphere radius, this method permitted unambiguous tracking of the moving spheres. Periodically, as the original spheres progressively left the field of view, the software initiated a new search in the entire image. Adequate lighting proved to be an important condition for success of the algorithm. The field of view was illuminated with two optical fiber light guides. The shutter speed of the camera, the opening of the lens and the illumination level were tuned by trial and error until grains could be reliably detected within a sphere diameter from the window.

Figure 5 is a typical image from the digital camera. Velocities were calculated from the positions of each sphere center in two consecutive images and the corresponding strip statistics were incremented as outlined earlier.

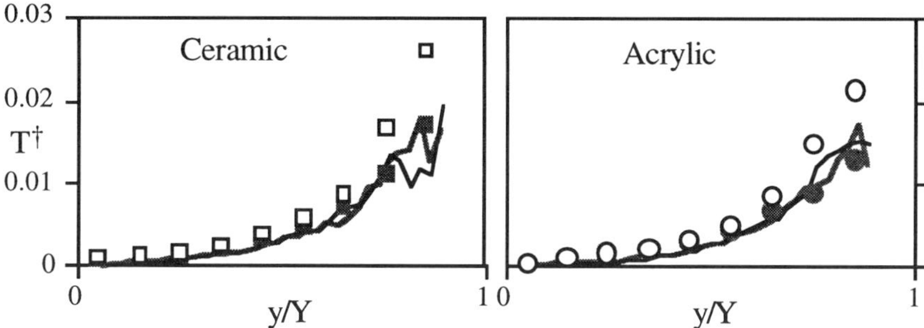

Figure 7. Transverse profiles of dimensionless second moments for the conditions of Test A. The symbols and lines are experimental measurements and predictions of the simulations, respectively. The open symbols and thin lines refer to the x-direction, while the closed symbols and thick lines refer to the y-direction.

Figures 6 to 8 compare experimental data with the predictions of the computer simulations for Test A. We observed similarly good agreement with the other two tests.

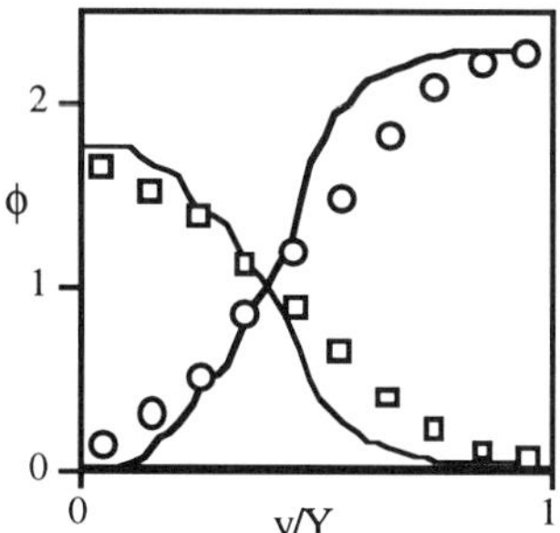

Figure 8. Transverse profiles of relative number fractions for the conditions of Test A. For symbols and lines, see Fig. 6.

5. Conclusions

The agreement between experiments and simulations demonstrated the utility of computer simulations as an equal partner to theory and physical experiment. In this event, fewer physical experiments need be done, and the simulations can be used with confidence as the basis of the theoretical modeling. Also, with advances in computing power, such simulation may eventually be employed to study the behavior of complete systems of technological importance that involve collisional granular flow.

6. References

1. Savage, S. B. and Lun, C. K. K.: Particle size segregation in inclined chute flow of dry cohesionless granular solids, *J. Fluid Mech.* **189** (1988), 311-335.
2. Haff, P. K. and Werner, B. T.: Computer simulation of the mechanical sorting of grains, *Powder Technol.* **48** (1986), 239-245.
3. Rosato, A. D., Strandburg, K. J., Prinz, F., Swendsen, R. H.: Monte Carlo simulation of particulate matter segregation, *Powder Technol.* **49** (1986), 59-69.
4. Rosato, A. D., Strandburg, K. J., Prinz, F., Swendsen, R. H.: Why the Brazil nuts are on top: size segregation of particulate matter by shaking, *Phys. Rev. Lett.* **58** (1987), 1038-1040.
5. Jenkins, J. T. and Askari, E.: Rapid granular shear flows driven by identical, bumpy, frictionless boundaries, in C. Thornton (ed.), *Powders and Grains 93*, Balkema, Rotterdam, 1993, pp. 295-300.
6. Lorenz, A., Tuozzolo, C. and Louge, M. Y.: Measurements of impact properties of small, nearly spherical particles, *Experimental Mechanics* **37** (1997), 292-298.
7. Foerster, S., Louge, M. Y., Chang, H. and Allia, K.: Measurements of the collision properties of small spheres, *Phys. Fluids* **6** (1994), 1108-1115.
8. Hopkins, M. A. and Louge, M. Y.: Inelastic microstructure in rapid granular flows of smooth disks, *Phys. Fluids A* **3** (1991), 47-57.
9. Walton, O. R.: *Granular Solids Flow Project*, Quarterly Report, UCID-20297-88-1, Lawrence Livermore National Laboratory, 1988.

7. Acknowledgements

This research was sponsored by NASA's Microgravity Science and Applications Division under contract NCC3-468. The specialty chain was donated by More Industrial, a division of Emerson Powder Transmissions Corporation.

The authors are indebted to Rowin Andruscavage, Gregory Aloe, Claudio Bazzichelli, Amelia Dudley, Patrick Florit, Joshua Freeh, Lance Hazer and Rami Sabanegh for measuring the impact properties in Table 1; to Fran McLeod for helping with the dc motor; to Edward Balaban, Taro Banno, Michael Chen, Michael Garon, Marie-Noëlle Pons and Christopher Salvestrini for developing a prototype of the vision algorithm; and to Gregory Aloe, Birgir Arnarson, Roshanak Hakimzadeh, Jeffrey Larko and Elaina McCartney for participating in the KC-135 flights.

We are equally indebted to the engineering teach at the NASA-Lewis Research Center: Joe Balombin, Christopher Gallo, Frank Gati, Roshanak Hakimzadeh, Jeffrey Larko, Pamela Mellor, Emily Nelson, Enrique Ramé, Leon Rasberry, John Yaniec and Gary Wroten; and to the KC-135 flight crews and ground support team.

SIZE-SEGREGATION AND MIXING PHENOMENA IN POWDERS FLUIDISED BY ACOUSTIC VIBRATIONS

A. P. MARLAND and L. V. WOODCOCK
Department of Chemical Engineering,
University of Bradford,
Bradford, BD7 1DP,
UK

Abstract

Acoustic vibrations generated by a loud-speaker system have been applied to fluidise binary mixtures of ballotini powders with dissimilar sizes. New phase coexistence effects have been observed akin to solution and de-mixing behaviour in binary liquid mixtures. It is postulated that the size-segregated phase above a mixed phase, which occurs when a steady-state is reached, is due to the pressure gradient. Unlike a molecular system, although the granular "temperature" may achieve uniformity, the column has a higher granular "pressure" at the bottom due to its own weight. An improved experimental method is reported together with further results for similar systems with variable starting conditions.

1. Introduction

Although powders and granular materials are essentially solids at the microscopic level, these materials can be fluidised in bulk such that they exhibit liquid-like properties [1]. Fluidisation can occur when the powder is subjected to a hydrodynamic gas stream that counteracts the gravitational force, e.g. a fluidised bed, or a shear stress that exceeds a characteristic yield stress of the static bulk powder, for example in a rotating drum. Powders can also be fluidised by mechanical vibrations. In 1975, Ratkai [2] formally investigated the effect of external vibrations on binary plastic granular beds noting a range of exotic segregation effects due to convective currents and turbulence. Soon after, in similar experiments, Harwood [3] reported size segregation effects when finer powders were mechanically vibrated.

Computer simulations [4,5] suggest that when granular beds are fluidised by vibrations of the container, at very small amplitudes comparable to particle size, the equipartition-of-energy principle that underpins the existence of classical thermodynamics (sometimes called the zeroth law) holds when the system reaches a steady-state. The combination of uniform granular "temperatures", and non-uniform pressure gradients due to the gravitational field, could give rise to quasi-thermodynamic phase behaviour. We have reported preliminary results of some simple experiments [6], which indeed show effects analogous to the thermodynamic

A.D. Rosato and D.L. Blackmore (eds.), IUTAM Symposium on Segregation in Granular Flows, 113–120.

phases in binary molecular systems. Using a primitive loud-speaker system, acoustic vibrations have been applied to search for phase behaviour of fluidised binary powder mixtures of components differing only in particle size and colour. Novel phase-coexistence effects have indeed been found [6], which could have potential applications in the processing of granular solids [7].

A series of laboratory experiments were embarked upon in search of such phase transition from segregated to mixed states or *vice versa*. The computer simulations indicated that to achieve equipartition for the fine powders in common use, say less than 300 microns in size, a much smaller amplitude and/or higher frequency, than has previously been used in mechanical vibrators, is required. Most conventional devices for shaking powders in the laboratory tend to have too large an amplitude, too low a frequency range, and the waveforms, moreover, are complex and difficult to characterise experimentally. A standard loud speaker system provides just the right frequency and amplitude range to fluidise powders in the 50- to 300-micron size range, and has well-characterised wave forms for the vibrations.

The first observations (reported previously but not in colour) with a relatively primitive loud-speaker system are shown in Figure 1 for a 50/50 mixture of 200 micron spherical ballotini (silica) and 50 micron. However vigorous one shakes the mixture it seems impossible to mix because of size segregation effects. After vibration of a bed, for about 5 minutes in this case, a steady-state is reached (Figure 1) When there is no further propensity for the particles to reorganise, an amorphous solid-like phase is seen to coexist below a granular liquid-like region with a sharp interface in-between. In the upper phase, which appears to remain fluid-like, size segregation occurs with the larger particles on top. Below the quasi- fluid/solid interface, where the gravitational pressure is higher, there rests a perfectly homogeneous mixed phase. In this static phase, packing considerations predominate, and the two components form a quasi "solution", rather like a binary amorphous metal alloy.

It is otherwise difficult to get two powders, which are so dissimilar in size, to mix homogeneously like this by any other means. In preliminary studies we found that attempts to mix binary samples with size (diameter) ratio greater than about 2:1 completely by any repetitive shaking or tumbling motions that involve large amplitude motions that generate powder flow and size segregated heterogeneities.

2. Acoustic Vibration Experiments

Since the original acoustic vibration experiments that resulted in the discovery of the phase behaviour seen in Figure 1, it became apparent that the primitive equipment and procedure needed to be upgraded and modified. More accurate data on the physical properties such as composition and diffusivities are now required, and the waveform of the vibration, particularly the amplitude, needs to be well characterised. The present modified experimental arrangement is shown in Figure 2.

A Hewlett Packard function generator is connected in series with a specially modified amplifier. This is necessary to drive the loudspeaker, without loading the generator. The signal from the function generator can be finely adjusted to the frequency and amplitude required for any particular experiment. An oscilloscope is place before the speaker so that the input signal to the speaker can be monitored (Figure 3).

The free standing structure utilised in the initial experiments [6] has been removed and replaced by a new mounting that sits directly onto the cone of the speaker. The powder sample is fixed onto the mounting system by two screws. This modification was undertaken to eliminate any lateral movement and frictional loses and to allow for greater vertical amplitude.

Whilst conducting experiments there was a need to determine the physical displacement of the speaker and hence the amplitude of the vibration. The input signal from the function generator gives the amplitude as a measurement of voltage. This is not meaningful as this measure is peculiar to the equipment being used and cannot be easily reproduced on other systems. To overcome this problem a sensor device was developed so that the frequency and physical displacement could be determined. The sensor consists of an infrared emitter diode fixed to the speaker cone and a phototransistor mounted on a multi-directional platform so that it can be adjusted and moved into any position. The signal from the sensors is passed through an amplification circuit (Figure 4). The signal from amplified circuit can then be measured either by oscilloscope or monitored by computer using a data acquisition card. The software found to be most suitable for this particular application is known as DASYLab and is necessary because of the high sample rate that is required to monitor the speaker vibration.

The sensor is calibrated so that it is possible to determine physical displacement of the speaker. This is done by placing the two sensors a measured distance apart and measuring the corresponding output voltage. This is repeated over a number of different distances until a satisfactory data series has been measured. Once this first set of data has been collected, the process is repeated to establish consistency. A sensor calibration graph (Figure 5) enables a polynomial data fit for subsequent sensor calibration. Once the data series has been established it is then be fitted to a suitable expression. The best fit for the data was a fourth order polynomial expression, from which the physical displacement is easily determined. Then, using DASYLAB software, a maximum and minimum D.C. output signal is recorded. Inputting this information into the dependent variable in the polynomial equation (Figure 5) physical distance or height is obtained. The difference between the maximum and minimum is the physical displacement of the speaker cone and hence, the amplitude of the vibration.

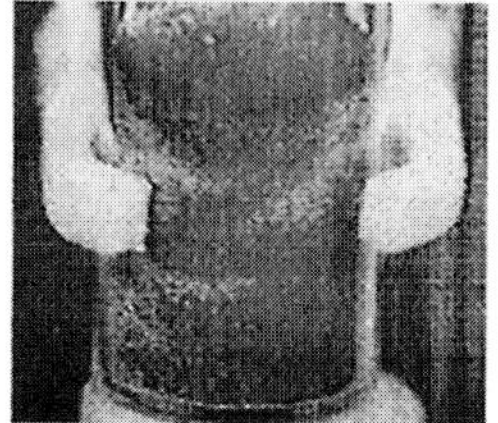 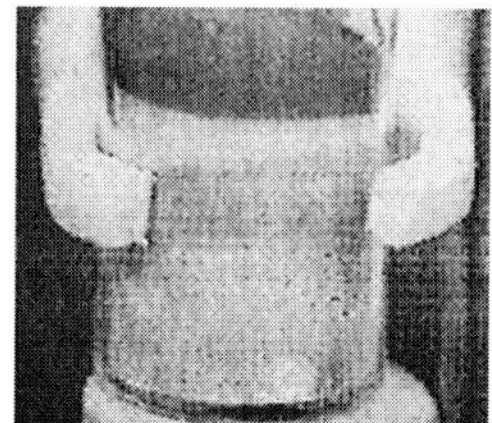

Figure 1. The original system [Reference 6] for the binary mixture of ballotini; the larger particles are 200 micron (blue) and the smaller particles 50 micron (white) (Left) before the commencement of vibrational fluidisation and (Right) after 5 minutes when steady state is established.

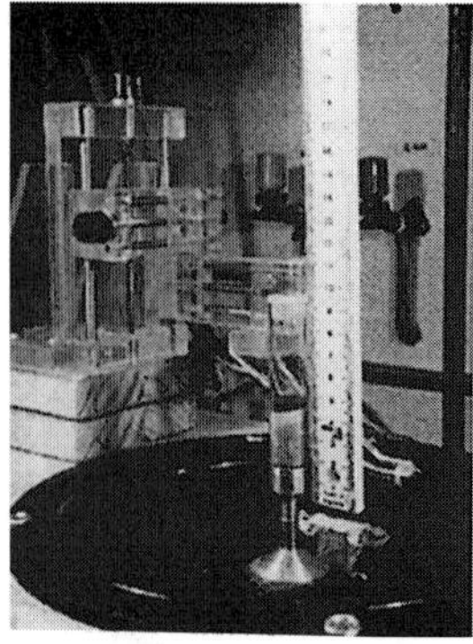

Figure 2. A close up of the speaker system, showing a sample after vibrational fluidisation and exhibiting 3-phase coexistence.

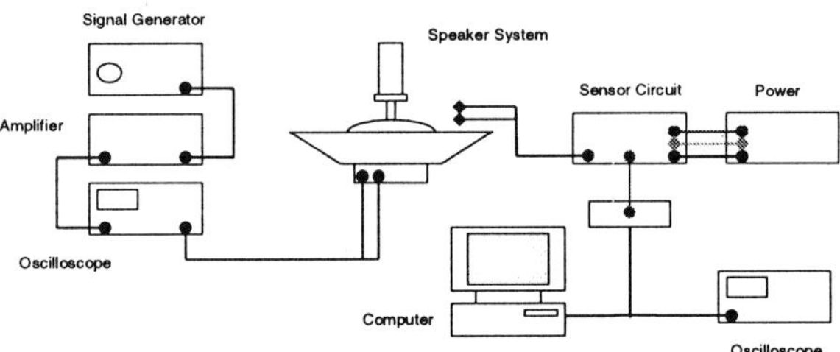

Figure 3. Schematic diagram of the apparatus used to excite binary powders by acoustic vibrations, comprising of a signal generator, oscilloscope to monitor wave form, loud speaker, with a sample mounted on the cone, and sensor device.

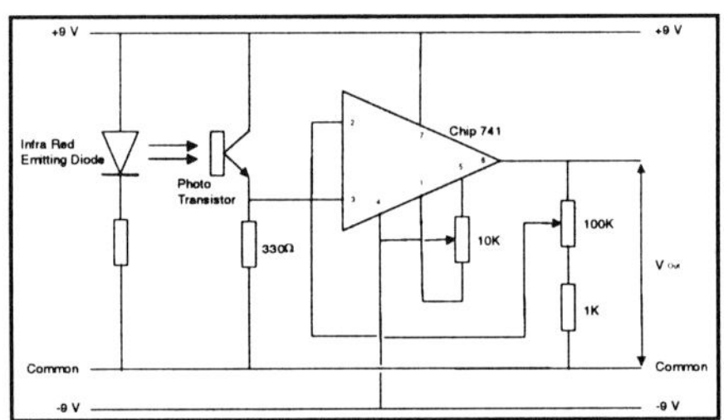

Figure 4. Circuit diagram for the speaker vibration sensor

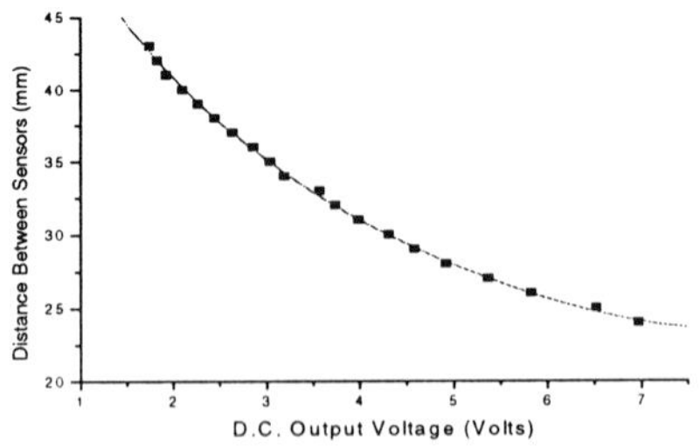

Figure 5. Characteristic calibration curve for the speaker

3. Experimental Results

Binary mixtures using different colours and sizes of ballotini are produced within small glass test tubes, which are fitted with stoppers. The glass tubes have the dimensions of 2cm diameter and 7.5cm height. Approximately 7 grams (determined by experimenting with the amount of material in the glass cylinder), of two different sizes of ballotini differing in colour are combined to produce a typical system. A range of starting conditions have been used. Once the particle system has been produced it can be placed onto the mount which is located directly on the speaker cone. The signal output from the signal generator can be set to a constant amplitude and a constant frequency, which are chosen for the particular experiment.

All particulate systems, when excited, tend to show two distinct regions, the upper region is almost fluidised and is definitely mobile whereas the lower region tends to be a more solid-like amorphous mass. When the vibration begins, the structure reorders by setting up convection currents. The role of convection currents as a mechanism for size segregation has recently been predicted theoretically [8]. Once this initial reordering has taken place, two distinct phases within the system begin to emerge. This convective behaviour generally gives way to a steady-state situation as seen before in Figure 1. In many cases this is seen as, not two regions or phases but three, as before the upper region remains fluid like but contains within itself two perfectly segregated phases, large particles on top and small particles at the bottom, however the lower region remains solid-like but with a significant difference and this is that it has become mixed.

Since it was generally found that the rate of size segregation in the upper phase, and the degree of size segregation at steady state are both enhanced by increasing the size ratio, we have embarked upon a series of experiments on a binary system of 100 micron ballotini particles (white) mixed with 600 micron particles (red) (as seen in Figures 6 to 8), i.e. with a size ratio 6:1.

The first experiment we report started with a sample which was clearly as far away as possible from "equilibrium" on fluidisation. In a 50/50 mixture by weight, the larger particles (red) were placed on the bottom of the sample and the smaller particles (white) were placed on top (Figure 6). The system was then vibrated at the frequency 60 Hz and an amplitude of 0.25mm. The resulting movement showed a remarkable dissolution of the small particles by the large particle, as evidenced by movement upwards of the original interface, and the transient appearance of an intermediate second interface, which moved downwards until it reached the bottom. At this stage, after about 5 minutes, the solid-like "solution" appeared to be "saturated" and there remained an excess of smaller particles in an apparent stable pure phase on top. On continuing the fluidised state for a further hour, there is no change in phase behaviour. The system appears to be in a steady state.

This final state may be the stable phase for this system, or it could be a metastable state in which a fraction of the larger spheres would be on top, as in Figure 1, but there is no mechanism whereby they can get there from the 2-phase state shown in Figure 4b. Further experiments are presently being undertaken resolve these questions about the stability, or metastability, of what appear to be quasi-thermodynamic states.

Subsequent experiments with the same sample have been carried out in which the starting sample was to some degree mixed (Figures 7 and 8). In the first experiment the particles were mixed to start with by manual shaking (Figure 7) and in a second experiment the particles were mixed by stratifying the initial filling of the sample tube. It is found that all these experiments result in a phase behaviour sometimes 2-phase and sometimes 3-phase and the amount of the phases can be very sensitive to the frequency and amplitude of vibrations. Increasing the amplitude appears to increase the ratio of fluidised layer(s) to solid-like layers, whereas increasing the frequency appears to favour the amount of mixed to segregated phases, analogous to increasing the temperature of a partially demixed liquid-liquid solution.

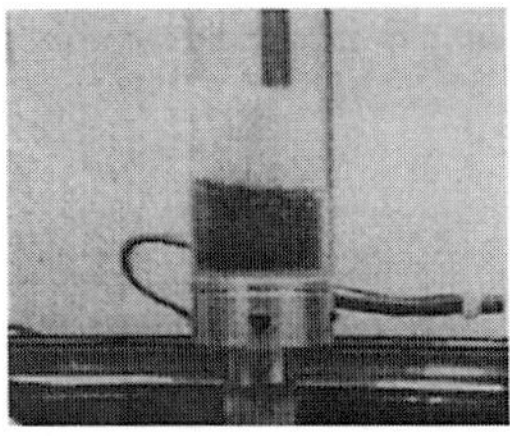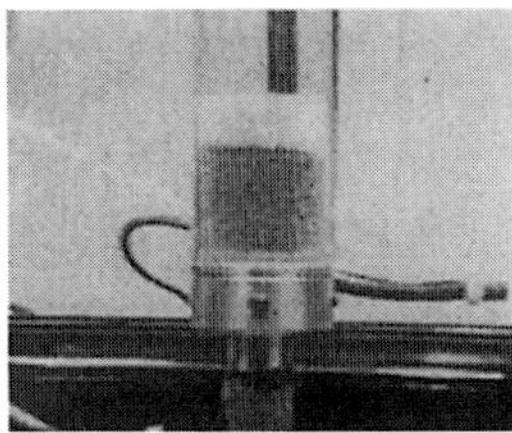

Figure 6. The experimental system for a binary mixture of approximately 6:1 (600 micron- red, 106 micron - white) (Left) before the commencement of vibrational fluidisation starting with inverted 2-phases size segregation, and (Right) after 5 minutes vibrational fluidisation with an amplitude of 3V (358 Micron) and frequency of 80Hz.

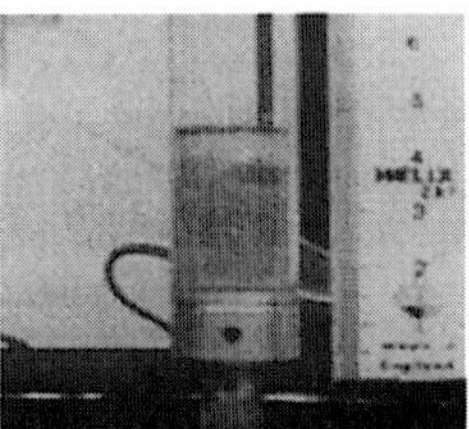

Figure 7. The experimental system for a binary mixture of approximately 6:1 (600 micron- red, 106 micron - white) (Left) before the commencement of vibrational fluidisation starting with the 2 components partially mixed by shaking, and (Right) after 5 minutes vibrational fluidisation with an amplitude of 3V (358 Micron) and frequency of 80Hz.

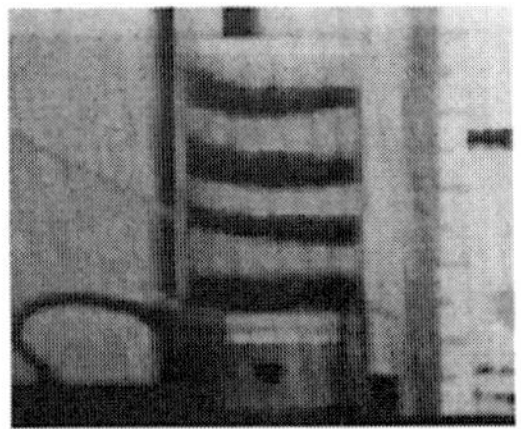

Figure 8. The experimental system for a binary mixture of approximately 6:1 (600 micron- red, 106 micron - white) (Left) before the commencement of vibrational fluidisation starting with the 2 components in a stratified size segregated state, and (Right) after 5 minutes vibrational fluidisation with an amplitude of 3V (358 Micron) and frequency of 80Hz.

3. Interpretation

The following explanation for the two distinct phases seen in these experiments is proposed. At steady state there is a tendency for the system of particles to distribute its kinetic energy in accordance with the equipartition principle. Thus, the powder system has a uniform "granular temperature" (in reality a kinetic energy) like a molecular fluid. Unlike a molecular fluid, however, the effect of the gravitational force of the quasi-thermodynamic properties is not negligible, and its effect is to impose a steep gradient upon the granular pressure. Once fluid, it ceases to behave like a static powder where the pressure depends on the dimension of the container in the fluidised state pressure is given by the integral of density with depth, like a liquid. The highest pressure is at the bottom, i.e. the pressure of its own weight. Here, the powder is the analogue of a molecular glass, whereas at lower pressures, above, the powder is the analogue of a molecular liquid. The sharp interface could be tentatively identified with the granular temperature and pressure at which a molecular fluid, in corresponding states, might exhibit a pressure-induced glass transition, i.e. an "isothermal" transition from supercooled fluid to an amorphous solid induced by increasing the pressure.

The equipartition-of-energy principle of kinetic theory requires the system to continuously redistribute the kinetic energies of all the particles evenly between all the available degrees-of-freedom irrespective of shape or size. The particles continually receive their energy from external boundaries, dissipate the energy by inelastic collisions, and at the same time redistribute the residual kinetic energies according to the principle of equipartition. When granular materials are vibrated, at a characteristic frequency and sufficiently small amplitude, the properties of these materials can behave like quasi-thermodynamic state functions. Consequently, one might expect to see powders mix at high granular energies, and perhaps demix at low granular energies, and that there should be a property of the powders analogous to the entropy of mixing.

In the upper phase, size segregation occurs if convective shear flow is present at steady state. In fact, this is observed in all the binary systems investigated. For a static situation there is no more reason for the large spheres to be on top than for the smaller size fraction, as both mass density and packing densities are identical; i.e. there is no apparent driving force in these binary mixtures for segregation when the powders are at rest. Dilatancy is a measure of the extent to which the density of spheres of a given diameter decreases with the increase in granular temperature when a powder is fluidised either by shear or by vibrations. At constant packing density, dilatancy implies an increase in the normal pressures with shear.

Under ambient pressure conditions, such as the convective flow seen here in the uppermost region of the acoustic- fluidised powders, the volume is not fixed and the increase in normal granular pressures is manifested as dilatancy or volume expansion. In the presence of gravity, where the gravitational potential energies become influential, these steady fluid motions can eventually give rise to total segregation with the least dilatant size fractions, smaller particles, lower down, and the more dilatant larger particles flowing above, to minimise the gravitational potential.

The homogeneous mixed phase underneath, seen in all the experiments to date, occurs because the system can achieve its most efficient packing density in compliance with the higher granular pressure; rather like an atomic solid solution, or an amorphous binary metal alloy. In atomic systems, thermodynamics requires the two particles to mix on the microscopic distance scale of particle sizes because the local geometric packing, albeit amorphous, of the solution, is more efficient than that of the two components separately. An alternative way of saying this is that at the dilatancy effect prevails at lower pressures but at higher pressures packing efficiency prevails. What's interesting and intriguing is that the transition between these two states in the same experimental system is sharp.

4. Conclusions

In conclusion, we can speculate that it will eventually be possible to excite these powder systems under zero gravity conditions, perhaps in space micro gravity. The use of acoustic vibrations to excite powders will reveal that every binary system has a quasi-thermodynamic phase behaviour and a phase diagram like molecular liquids, except the state variables are granular "temperature and pressures." That phase diagram would depend on the two system-state variables granular temperature (determined by vibration frequency/amplitude) and the volume of the container. Given that such a phase diagram exists for every powder and binary powder, it then should be possible to predict the effect of the same fluidised system condensed at ambient pressure, but with a gradient of granular pressure proportional to the weight above, caused by the gravitational force.

5. References

1.　Jaeger, H.M., Nagel S.R., and Behringer R.P.: The physics of granular materials, *Physics Today*, (1996), 32-38.
2.　Ratkai, G.: Particle flow and mixing in vertically vibrated beds, *Powder Technol.* **15** (1976), 187-192.
3.　Harwood, C. F.: Powder segregation due to vibration, *Powder Technol.* **16** (1977), 51-57.
4.　Knight, T.A. and Woodcock L.V.: Test of the equipartition principle for granular spheres in a saw-tooth shaker, *J. Phys. (A) Math. Gen.* **29** (1996), 4365-4386.
5.　Knight, T.A. and Woodcock L.V.: Test of the equipartition principle for granular spheres in a saw-tooth shaker; the effect of gravity, *Proceedings of IV International Conference on the Micro-mechanics of Granular Media*, Durham NC, 1997.
6.　Castle, J., Marland A.M., and Woodcock, L.V.: New Phase Coexistence effect is powders fluidised by acoustic vibrations, *Proceedings of IV International Conference on the Micro-mechanics of Granular Media*, Durham NC, 1997.
7.　Kestenbaum D.: Sand castles and cocktail nuts, *New Scientist* **154** (1997), 25-28.
8.　Knight, J.B., Jaeger, H.M. and Nagel, R.: Vibration-induced size segregation in granular media; the convective connection, *Phys. Rev. Letters* **70** (1993), 3728.

SEGREGATION OF BINARY MIXTURES IN A VERTICAL SHAKER

S. S. HSIAU and H. Y. YU
Department of Mechanical Engineering
National Central University
Chung-Li, Taiwan 32054
R.O.C.

Abstract

Segregation phenomena for different binary mixtures were experimentally studied in a vertical shaker. The larger particles tend to move upwards in the shaker for a certain range of vibrational accelerations. The segregation effect was more significant for the mixture with larger size difference. The segregation coefficient was measured and found to increase with the vibrational acceleration until reaching a maximum and then to decrease with the acceleration. The bed expansion height was also measured. It was found that the greatest segregation effect occurred when the bed transformed from a dense state to a loose state. The binary mixtures with different material densities were also tested. The factor of size difference is much more important in the segregation process than the factor of density difference. However, the segregation effect could be enhanced if the larger particles have the greater material density than the smaller particles.

1. Introduction

Vibrated granular beds are used widely in industries to mix and dry granular materials. Shakers are also used to sort particulate materials according to particle size in pharmaceutical, powder metallurgy, and food industries. Some fundamental aspect of the research has been reviewed by Jaeger and Nagel [4]. In this paper, segregation phenomena of binary mixtures in a vibratory particulate systems are experimentally studied.

Segregation may result from the differences in the particle size, in particle density, in the properties and the angle-of-repose of the materials [1, 5, 12], and in the granular temperature gradient [2]. Segregation is important and is broadly discussed in industrial fields involving powders and granular materials [1, 5]. Some studies about this complicated phenomenon have been reviewed by Williams [12] and Savage [9]. The segregation caused by the granular thermal diffusion was also theoretically studied by Hsiau and Hunt [2]. Recently there are many studies concerning the segregation phenomenon in a shaker. [6, 8].

Most experiments and computer simulations about segregation in vibrated granular bed use only one larger particle with background of smaller particles. The purpose is to

A.D. Rosato and D.L. Blackmore (eds.), IUTAM Symposium on Segregation in Granular Flows, 121–127.

study the segregation process microscopically. However, in most cases of practical applications, there is not only one larger particle in the bed. Recently Hsiau and Yu [3] used the equal-mass and well-mixed binary mixture to study the size effect on segregation in a vertical shaker. The binary mixtures were of the same material (glass beads) with the same total mass but different sizes. They measured the segregation coefficient and the solid fraction in the bed at different amplitudes of vibrational accelerations. They found that the binary mixtures had the maximum segregation ratio when the vibrational acceleration was 1.75 times gravitational acceleration. For a fixed vibrational acceleration, the maximum segregation coefficient occurs when the solid fraction has a sudden decrease. The greater size ratio also resulted in the greater segregation effect. This paper studies both effects of the density difference and the size ratio of the binary mixtures on segregation in a vertical shaker.

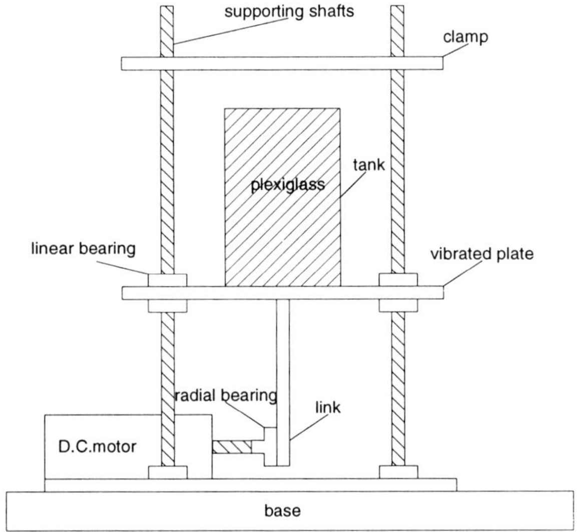

Figure 1. Schematic drawing of the shaker.

2. Experimental Setup

The segregation experiments are performed in a rectangular container, which is vertically driven by a radial bearing that is off-centered from the shaft extension of the variable-speed d.c. motor. The schematic drawing of the shaker is shown in Figure 1. The container size is 20 cm in width, 1.9 cm in depth and 29 cm in height. The container is made of plexiglass for visualization purposes. The vibrational amplitude z_0 can be changed by using different radial bearings. The amplitude of 5 mm is used in the present study. The vibrational frequency f is adjusted by changing the rotational speed of motor. A HT-4000 OND SKKI tachometer is used for measuring the rotational speed of the motor. The amplitude of the vibrational acceleration a_0 is found from $a_0 = z_0\omega^2$, where ω is the angular frequency, $\omega = 2\pi f$. The highest accelerational amplitude in this experiment is up to 3.5 g, where g is the gravitational acceleration. The box was cleaned after every two experimental runs to provide smooth internal wall conditions.

The experiments used soda lime glass beads (G) with density of 2490 kg/m^3 and steel balls (S) with density of 8290 kg/m^3. The diameters of the beads include 1 mm, 2 mm, 3 mm and 4 mm. For size differences, there are six combinations: 1 mm & 2 mm, 1 mm & 3 mm, 1 mm & 4 mm, 2 mm & 3 mm, 2 mm & 4 mm, 3 mm & 4 mm. Let α and β denote the larger and the smaller sizes of particles respectively. For each combination of size difference, four kinds of binary mixtures were prepared before the experiments: αG-βG, αS-βS, αS-βG, and αG-βS. The mass of glass beads is 0.08 kg and that of steel balls is 0.264 kg so that both materials occupy the same volume. The binary mixture was well-mixed in the box in each experiment. The granular bed was started to vibrate at a preset frequency. A set of image processing system was used to record the motions of the granular bed. The image processing system included a Delsa CCD image sensor (up to 220 films per second) with a 55-mm Micro-Nikkor f/2.8 lens, an image grabber board (Dipix P360F Power Grabber), and a 150 Watt tungsten halogen light source.

In Hsiau and Yu's [3] experiment, different sizes of glass beads with equal total mass (volumes) were used. The larger particles tended to move upwards and the segregation phenomena occurred. If there were N_u of particle α in the upper half part of the bed and N_l of particle α in the lower half part, the segregation coefficient C_s was defined by

$$C_S = (N_u - N_l)/(N_u + N_l) \tag{1}$$

and could be measured by counting particles from the image. The segregation coefficient of 1 denoted complete segregation of the binary mixture and $C_s = 0$ represented that the material was fully mixed. When the bed vibrated, the granular bed compressed and expanded continually. The particles in the upper levels were more dilute than the lower part during expansion. Therefore, it was difficult to decide the interface between the upper half and the lower half parts. In this experiment, the recording speed of the image processing system was adjusted to synchronize with the vibration so as that the images were always recorded when the bed was in a compressed state. When counting particles in the image, the error was within a particle diameter; and the error for the segregation coefficient was under 5% in this experiment. The maximum segregation coefficient for a binary mixture was called the maximum segregation coefficient $C_{s,max}$. Some binary mixtures could segregate completely ($C_{s,max} = 1$). For the binary mixtures which could only segregate partially, the maximum segregation coefficient $C_{s,max}$ was recorded.

3. Results and Discussions

Figure 2 shows the maximum segregation coefficient ($C_{s,max}$) as a function of the vibrational acceleration (a_0) for the binary mixtures of 1 mm G & 2 mm G, 1 mm S & 2 mm S, 1 mm S & 2 mm G and 1 mm G & 2 mm S. The segregation effects are the same for the binary mixtures of 1 mm G & 2 mm G, 1 mm S & 2 mm S. Both binary mixtures compose of the same materials. There is no segregation for the amplitude of vibrational acceleration less than 1.0g or greater than 2.7g. The segregation coefficient is increased with the acceleration, and reaches the highest value at around 1.75g.

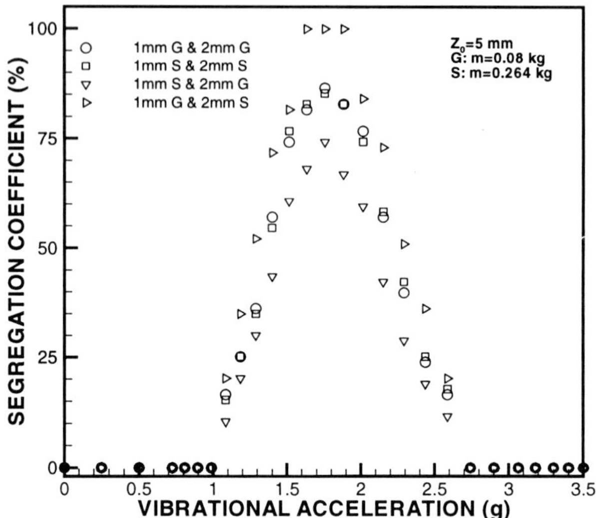

Figure 2. Segregation coefficient as a function of vibrational acceleration for different density mixtures of 1 mm and 2 mm.

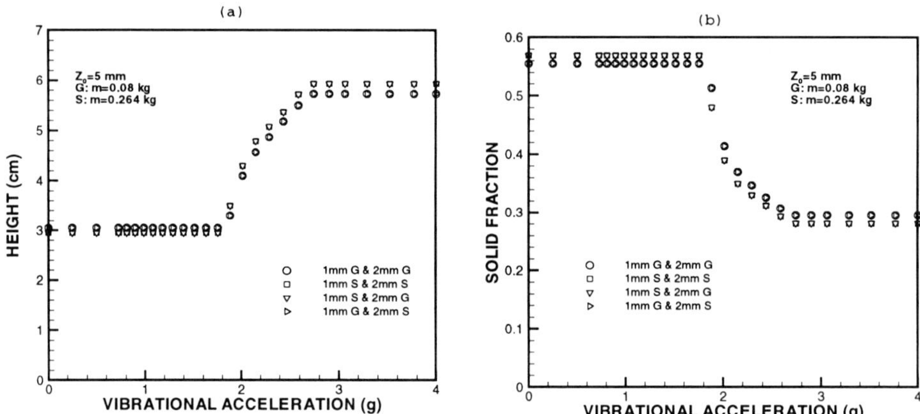

Figure 3. (a) Bed height, (b) solid fraction, as a function of vibrational acceleration for different density mixtures of 1 mm and 2 mm.

The granular bed expands when the bed is vibrated. The bed height was measured by the image processing system. The solid fraction v could be found from the total particle mass divided by the average particle density, the cross-sectional area of the container and the bed height. Figure 3 shows the bed expansion height and the solid fraction of the bed for the four binary mixtures and the results are similar for all four cases. When the acceleration is beyond $1g$, particles have no relative motion, so the bed is not in an expanded condition and no segregation occurs. When the acceleration is between $1g$ and $1.75g$, there are some small relative motions among particles. The motions cause the voids among particles to rearrange but the energy is not high enough to expand the bed. During this range of vibrational accelerations, some larger voids are formed; therefore, the smaller particles fall more easily down through the voids resulting in segregation. The higher the acceleration, more and larger voids are formed and then the segregation coefficient is increased. When the acceleration is higher than $1.75g$, the particle motion becomes significant and the bed starts to expand causing the solid fraction to decrease. The motion of the particles becomes more random, so larger voids are formed in the bed. At this stage, the probability of larger particles falling through the voids in the bed increases. Therefore the material become more difficult to segregate and the segregation coefficient is decreased. When the acceleration increases up to $2.7g$, the voids are large enough that the probabilities for larger and smaller particles to fall through the voids are the same. The segregation phenomenon does not exist and the material is in well-mixed condition. The bed also stops expanding indicating that the material has already transformed from a "dense" state to a "loose" state. From Figure 2, if the materials of the binary mixture are different, the segregation effect is also different. From the figure, if the density of the smaller particles is smaller (glass) and the density of the larger particles is greater (steel), the segregation effect is enhanced, and the material could reach a completely segregated condition when the vibrational acceleration ranges from $1.6g$ to $1.8g$. On the other hand, if the smaller particles have the greater density comparing with the larger particles, the segregation effect is decreased. Similar conclusion was shown in Hsiau and Hunt's study [2]. They employed dense-gas kinectic theory to analyze the "granular thermal diffusion" effect of binary mixtures in a shear flow. Dense-gas kinetic theory was introduced to analyze the granular flow since the similarity between the motions of molecules in a gas and the motions of particles in a granular flow. Similar to the thermal temperature in a gas, the term "granular temperature" was used to quantify the mean-square value of the fluctuating velocities [7]. Although the granular temperature plays a similar role as the thermal temperature in the gas kinetic theory, it does not have the dimension of temperature (K or °C) but has the dimension of specific kinetic energy (m^2/sec^2).

In Hsiau and Hunt's paper [2], they analyzed a basic phenomena called "granular thermal diffusion." The basic idea is that the partial pressure gradient in granular material should be balanced by the average momentum transfer resulting from the collisions between two groups of particles. In a usual case, the different-species collisions and the greatest relative velocity determine the direction of the momentum transfer. This case happens when the lighter (smaller) particles coming from the region with higher granular temperature collide with the heavier (larger) particles coming from the lower granular temperature region. This kind of collision results in a net momentum transfer from the lighter (smaller) particles to the heavier (larger) particles, in the opposite direction of the granular temperature gradient. From the constitutive equations derived from dense-gas kinetic theory [2], the lighter-species partial pressure increases in the opposite direction of that of the momentum transfer which is in the direction of granular temperature gradient. Since the partial pressure is increased with partial number density, this effect causes the lighter (smaller) particles concentrating in the place with higher granular temperature. Hence, if the smaller particles have smaller material density comparing with the larger particles, the granular thermal diffusion phenomena and the segregation effect are more apparent.

The vibrated granular bed receives kinetic energy from the base plate, and the energy decays from the bottom to the top because the collisions among particles dissipate the energy. Therefore the granular temperature in the bottom is higher than the top [10, 11]. According to the granular thermal diffusion theory, the larger and heavier particles tend to move upwards because the upper bed has smaller granular temperature. The above segregation phenomena was demonstrated in the current experiments.

The results of the maximum segregation coefficients as functions of vibrational accelerational amplitudes for different size-ratios of binary mixtures (1 mm & 3 mm; 1 mm & 4 mm; 2 mm & 3 mm; 2 mm & 4 mm; 3 mm & 4 mm) are similar to Figure 2. Every binary mixture tested could not segregate when the amplitude of vibrational acceleration is less than $1g$ or greater than $2.7g$; and the mixture reaches the maximum segregated condition when the accelerational amplitude is $1.75g$. All cases show that the smaller particles with lower density result in the greater segregation effect. If the binary mixture has the same species, then the segregation effect is only dependent on the size ratio; that is, the binary mixtures of αG-βG and αS-βS have the same segregation results. Therefore, the influence of size difference on segregation [1, 12] is more important than the density difference. The results also show that the larger size ratio between two materials cause **a** higher degree of segregation. For the greater size difference, the complete segregation accelerational range is also larger ($1g$ to $2.7g$ for 1 mm & 4 mm and 1 mm & 3 mm binary mixtures (not for αG-βS cases); $1.6g$ to $1.8g$ for 1 mm G & 2 mm S and 2 mm G & 4 mm S binary mixture). For 2 mm & 3 mm and 3 mm & 4 mm combinations, since the size differences are not significant, the mixtures could not reach the complete segregated condition. For 1 mm & 2 mm and 2 mm & 4 mm combinations which are of the same size ratio, the segregation curves are similar. It indicates that the size ratio is the most important factor to influence segregation.

4. Conclusions

This study found that the size difference is the most important factor causing segregation in a vertical vibrated bed. The greatest segregation coefficient occurs during the phase transition from a dense state to a loose state. The larger particles having greater density than the smaller particles could enhance the segregation effect. The larger size difference causes the easier segregation of the material. The material needs an appropriate amount of vibrational energy to be segregated. If the energy is too small, it is not enough to segregate the material. However, too much energy also results in the re-mixing of the material and segregation does not occur. The formation of the voids is the most important mechanisms causing the segregation. The granular temperature generated from the vibrational energy should be studied to theoretically understand the segregation phenomena.

5. Acknowledgements

The authors would like to acknowledge the support from the National Science Council of the R.O.C. for this work through Grants NSC 83-0410-008-018 and NSC 84-2211-E-008-036.

6. References

1.   Bridgewater, J.: Fundamental powder mixing mechanisms, *Powder Technol.* **15** (1976), 215-236.
2.   Hsiau, S. S. and Hunt, M. L.: Granular thermal diffusion in flows of binary-sized mixtures, *Acta Mechanica* **114** (1996), 121-137.
3.   Hsiau, S. S. and Yu, H. Y.: Segregation phenomena in a shaker, *Powder Technol.* **93** (1997), 83-88.
4.   Jaeger, H. M. and Nagel, S. R.: Physics of the granular state, *Science* **255** (1992), 1523-1531.
5.   Johanson, J. R.: Particle segregation ...and what to do about it, *Chemical Engineering* **85** (1978), 183-188.
6.   Jullien, R. and Meakin, P.: Three-dimensional model for particle-size segregation by shaking, *Physical Review Letters* **69** (1992), 640-643.
7.   Ogawa, S.: Multi-temperature theory of granular materials, in S. C. Cowin, M. Satake (eds.), *Proceedings of the U.S.-Japan Seminar on Continuum-Mechanical and Statistical Approaches in the Mechanics of Granular Materials*, Gakujutsu Bunken Fukyu-Kai Publishers, Tokyo, 1978, pp. 208-217.
8.   Rosato, A. D., Prinz, F., Strandburg, K. and Swendsen, R.: Monte Carlo simulation of particulate matter segregation, *Powder Technol.* **49** (1986), 59-69.
9.   Savage, S. B.: Interparticle percolation and segregation in granular materials: A review, in A. P. S. Selvaduri (ed.), *Developments in Engineering Mechanics*, Elsevier, Amsterdam, 1987, pp. 347-363.
10. Savage, S. B.: Streaming motions in a bed of vibrationally fluidized dry granular material, *Journal Fluid Mechanics* **194** (1988), 457-478.
11. Warr, S., Huntley, J. M. and Jacques, G. T. H.: Fluidization of a two-dimensional granular system: experimental study and scaling behavior, *Physical Review E* **52** (1995), 5583--5595.
12. Williams, J. C.: The segregation of particulate materials. A review, *Powder Technology* **15** (1976), 245-251.

MIXING AND SEGREGATION IN VERTICALLY VIBRATED GRANULAR LAYERS

A. ALEXEEV, V. ROYZEN, V. DUDKO, A. GOLDSHTEIN,
and M. SHAPIRO[⊥]
Laboratory of Transport Processes in Porous Materials
Faculty of Mechanical Engineering
Technion - Israel Institute of Technology, Haifa 32000, Israel

Abstract

Vibrational motion of 2D layers composed of identical inelastic solid disks is investigated experimentally and characterized in terms of the dimensionless acceleration. Several vibrational regimes with different degrees of vibrofluidization are studied by means of the layers' videorecording and tracking the motion of one larger disk immersed into each bed of smaller particles. It is observed that depending on the vessel's vibrational acceleration the larger disk either ultimately rises on top of the layer, or vigorously moves throughout it. These regimes respectively correspond to the particles' segregation and mixing. In a certain narrow range of the vibrational acceleration the layer is observed to re-pack and move as a single block without any relative particles' motion. This acceleration range is well described by the model of an absolutely plastic body moving above a vibrated plate.

1.   Introduction

Segregation and mixing are two competing processes, which prevail in different industrial operations handling granular materials, such as ceramics, plastics, detergents, fertilizers, glass, pharmaceuticals, processed food and animal feeds. When the ultimate goal of a process is segregation, e.g. separation according to size, density, or else, any conditions leading to mixing should be avoided. On the contrary, a well-mixed granular mixture can deteriorate by segregation occurring as a result loading, transportation or other external influence. Thus, for efficient design of equipment dealing with granular mixtures, it is important to know what are conditions for which either segregation or mixing prevails.

Mixtures of granular materials may be classified into two major groups, those consisting of free-flowing particles and those containing cohesive or interactive

---

⊥ Author for correspondence: E-mail → mersm01@tx.technion.ac.il

129

A.D. Rosato and D.L. Blackmore (eds.), IUTAM Symposium on Segregation in Granular Flows, 129–139.
© 2000 *Kluwer Academic Publishers. Printed in the Netherlands.*

constituents. Free-flowing mixtures allow the individual particles to move independently, whereas in cohesive mixtures interparticle bonding forces exist, allowing particles to move in clusters. The present work deals with mixing and segregation in free-flowing materials.

It is commonly believed that any external vibrations lead to segregation of mixtures containing particles of various sizes, shapes, specific gravitates etc. [1-5]. One well known results of vibrations is the so-called 'Brazil Nuts' effect [5], which is segregation of granules according to their sizes. This phenomenon is widely used in various technological processes an example is vibro-classification. Segregation in vibrated granular layers is a process occurring during vibrofluidization. By vibrofluidization we will mean any state of a vibrated granular bed when granules prevail in relative motion. The intensity of this relative motion depends on the intensity of the vibrational regime, depth of the granular layer, physico-mechanical properties of the granules and the surrounding gas. There exists an upper limit on the depth of granular bed of large particles beyond which it cannot be fluidized by vertical vibrations [6].

Other examples of vibrofluidization-induced processes are convection [7, 8, 9], surface waves and arching [10, 11, 12], shock waves [13]. Vibrofluidization of granular materials used in several industrial processes including vibro-milling, vibro-classification etc. Nevertheless, no industrial mixers exist providing high quality mixing only by vibrations. The main reason is that vibrofluidization leads to segregation [1-5]. Experiments show that increasing intensity of vibrations leads to layer fluidization and stimulates segregation [14].

It is known from experiments [7, 8, 14] and computer simulations [9] that the basic mechanism responsible for vibration-induced size separation is convective (circulating) motion of granular beds. On the other hand, two studies [15, 16] report that circulation motion may cause mixing of particles with different sizes. A possible explanation of the controversy in the segregation-mixing effects between these [15, 16] and other studies is, probably, the effect of the air drag force. This force plays an important role in beds consisting of relatively small particles with diameter less than 1mm [17, 18], and can significantly modify regimes caused by 'pure' vibrations. A large particle placed on the bottom of a vessel beneath a bed of small particles is usually driven up by convective streams induced by vertical vibrations causing segregation, when the air forces are negligibly small. The non-convective segregation was registered for small amplitude and high acceleration vibrational regimes only [14]. However, motion of a large particle in opposite direction may be observed as well. It was shown in [7, 8] that downward motion of a large particle occurs in cells with slanted walls. Here we discuss the experimental evidence of downward motion of a large particle within a rectangular cell [19] and demonstrate that this phenomenon is caused by a circulating flow of small granules.

In the present paper we experimentally investigate vibrofluidization regimes leading to either segregation or mixing of different granular materials. In particular we study the physical mechanisms underlying vibrational mixing. We experiment with 2D granular layers, vertically vibrated with high amplitudes and low frequencies. We use layers consisting of large enough particles uninfluenced by the air. Since no standard equipment, providing controllable vibrations with amplitudes and frequencies in the required ranges is available, a special vibrational stand [20] was designed and constructed, which allows videotracking of the particles' motion within the granular layer.

2. Experimental setup

The experimental setup consists of a vibrational stand, which provides vertical vibrations of a vessel with a weight of up to 25 kg with controllable frequencies and changeable amplitudes in the respective ranges $0 < f < 20$ Hz, $0.5 < A < 5$cm. According to the kinematic scheme, the vibrostand oscillations were non-sinusoidal. For all results reported below the vessel maximal acceleration in the downward direction was about fifty percent larger than the upward maximal acceleration. Further on we will characterize the vessel acceleration with the single parameter $\Gamma = A(2\pi f)^2 / g$.

Vessels of several kinds were used, all made of steel frames with transparent glass walls. The width and height of the vessels are 35 and 50 cm, respectively. For all experimental regimes the particle do not reach the upper bound of the vessels. Therefore we consider the vessels as open from above. For the investigation of the disk-like particles, the gap between the vertical transparent walls was about 1 mm larger than the disk thickness, to allow their free two-dimensional motion with minimal friction. Here we describe the experiments with one large disk (with diameter 18mm, 27mm, 36mm) immersed into layers composed of 6-12 monolayers of small identical disks (with diameters 11mm) made of the same material, aluminum or brass. Such relative large particles' diameters were used to reduce the influence of the air on their motion. The coefficients of restitution e of these materials were measured and found to be about 0.5 for the impact velocities around 1m/s, in agreement with the literature data. This value is typical for the granular materials processed in the majority of vibrational technologies; it significantly differs from almost elastic materials used in the laboratory experiments [11, 12, 13]. The disks motion within the vessel was recorded by videocamera. A more detailed description on the experimental measurement system can be found in [19, 21].

3. Experimental results and discussions

The two-dimensional vessels were filled with 6-12 monolayers of smaller identical disks and one large disk made of the same material. The small disks were painted in different colors, which allowed their easy tracking with the video camera. Each series of experiments was performed at constant amplitude with the large disk initially placed on the bottom of the vessel beneath the small disks.

Figures 1, 2, 3, and 4 present photographs of the characteristic vibrational regimes observed in the study. At low frequencies the layers remain attached to the bottom (non-detaching regime, Fig. 1a). At higher frequencies they detach from the bottom and prevail in a free flight during some parts of the vibrational period (detaching regime). Further increase of the vibrational acceleration leads to diminution of the time of contact with the vessel to the extent which allowed viewing the layer-vessel interaction as bouncing (bouncing regimes, Figs. 2, 3). Such bouncing is accompanied by the layers' periodic expansion and contraction, and also distortion of their free

   A. ALEXEEV ET AL.

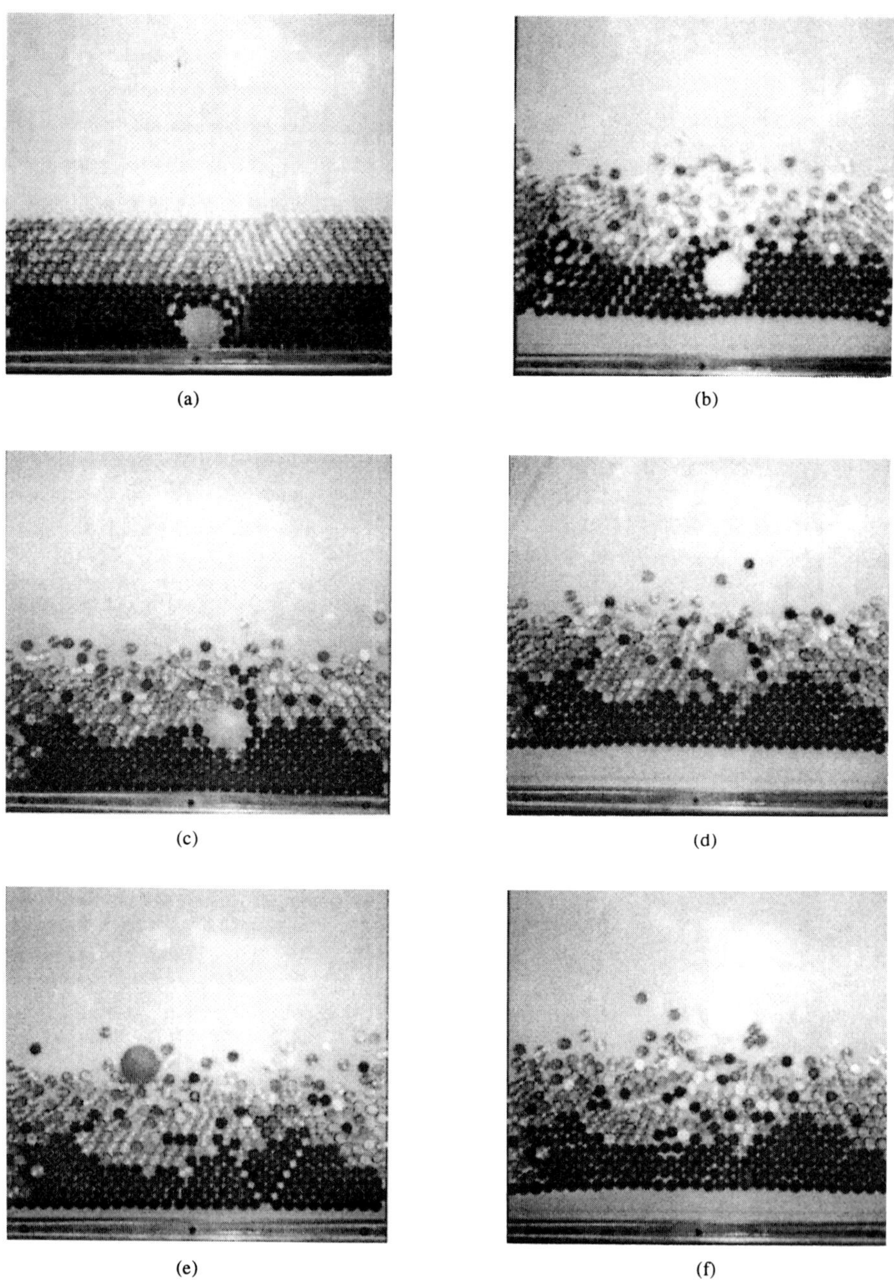

Figure 1. Photographs of segregation regime (A=25 mm, Γ=2.7) at different moments t of the experiment: a) t=0 (initial state of the granular layer); b) t=5 s; c) t=10 s; d) t=15 s; e) t=20 s; f) t=25 s

(a)   (b)

Figure 2. Photographs of surface waves (A=25 mm, Γ=6.8)

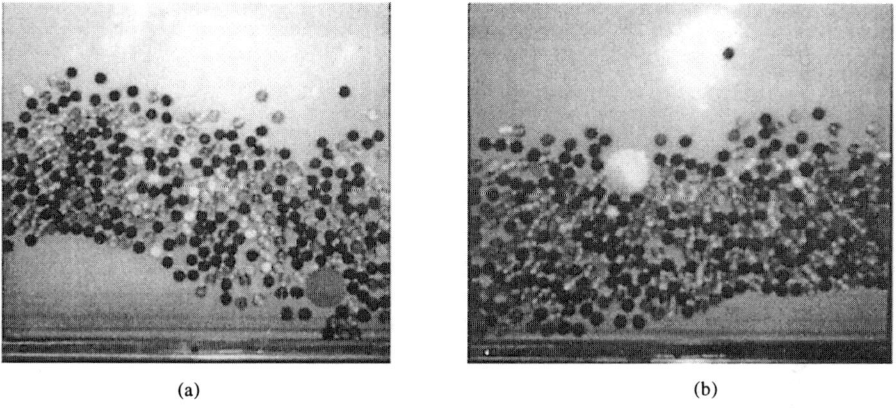

(a)   (b)

Figure 3. Photographs of bending waves (A=25 mm, Γ=7.3)

surfaces (so-called surface waves) or their transverse bending (so-called arching), first registered by Doudy et al [12]. In [13] it was suggested that the dispersion relation is the same as for the gravitational waves in fluids. This relation was examined in [10]. A tendency was reported for clustering the data around the theoretical curve, though the data scatter was significant. This scatter, probably, means that the standard expression, which is valid for an inviscid fluid, should be modified to apply to a description of vibrated granular layers. In [11] it was suggested to characterize the surface waves in terms of the Froude number, *Fr*. For *Fr* ~ 1 a good correlation between a phenomenological model and experimental data was found [11]. Other phenomenological models are discussed in [10].

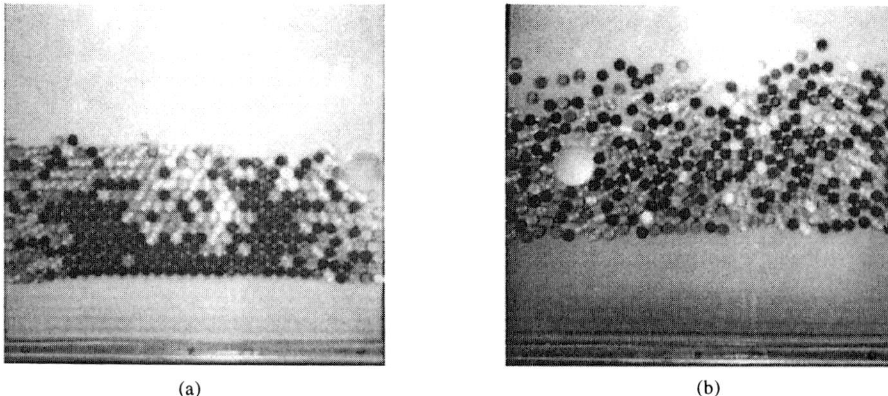

(a)　　　　　　　　　　　　　　(b)

Figure 4. Photographs of repacking phenomenon (A=25 mm): a) Γ=5.07 - first repacking, b) Γ=8.4 - second repacking.

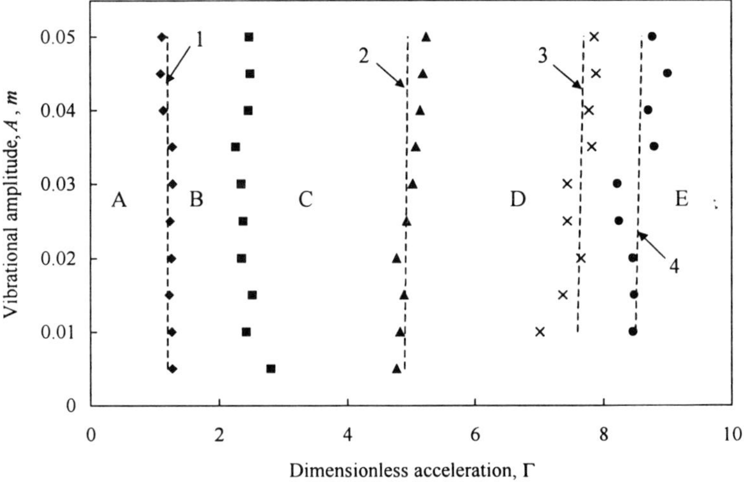

Figure 5. Kinematic map of vibrofluidization regimes for vibrated 2-D layers composed of six to twelve monolayers). A - non-detaching regime; B - detaching regime (no relative motion); C - segregation regime; D, E - mixing regimes. Triangles, circles and crosses correspond to first, second repacking regimes and regime of most vigorous vibromixing, respectively.

One interesting peculiarity of the layer motion in the detaching regimes is the so-called repacking phenomenon, first observed in [19]. It was shown that for each amplitude there exists a certain narrow acceleration range, lying within the bouncing domain, where the particles' intensive relative motion terminates and they collapse into a solid block (see Fig. 4a), which periodically jumps off the vessel bottom. With increasing acceleration this repacking disappears. This phenomenon is also observed at a higher acceleration, although in a less pronounced manner, since the layer motion in this regime is significantly affected by the transverse waves (see Fig. 4b). The frequencies where the particles' repacking had been registered are further related to the predictions of the plastic body model [19]. According to the plastic body model, a granular layer upon touching the vessel instantaneously obtains its velocity. From this moment on and till the detachment, the layer and the vessel have equal velocities without rebound. The layer detachment from the vessel occurs when its acceleration, a is less than the gravitational acceleration, g, i.e., $a < -g$.

The above regimes were also characterized by the relative motion of the particles within the layer and by the path of the large disk immersed in the layer. In the non-detaching and detaching regimes (A and B, see Fig. 5) this disk remains essentially on the vessel's bottom, where it was initially placed. Moreover, in these regimes no relative particle motion occurs. The layer moves as a single solid block in accordance with the plastic body model.

Beginning from vibrational acceleration about 2.4, such block-like motion is destroyed. This acceleration is almost independent of the vibrational amplitude and layers height in the ranges used in our experiments (see Fig. 5). The acceleration 2.4 bounds from above regime B; beyond this value regime C prevails. The latter can be characterized as a vibrofluidization regime when a large particle moves upward. When the acceleration reaches 2.4, a relative motion of the small particles begins within the upper part of the layer, although the lower particles continue to move together (i.e. with zero relative velocity). With increasing vibrational acceleration this relative motion occurs deeper in the granular layer. When this motion reaches the particles surrounding the large disk, it is gradually brought to the upper layer surface (Figs. 1b,c, d), where it continues to move without diving back into the layer (Figs. 1e,f). This is a manifestation of the "Brazil nuts" [5] effect, which is segregation of larger particles on top of vibrated layers composed of smaller ones. Thus, in order to achieve segregation, a granular layer should be fluidized, i.e. brought to the state when the particles prevail in relative motion. In this state effective contact time of a granular layer significantly differs from the time predicted by the single plastic body model [19], the forces acting on the vessel's bottom are significantly decreased. In spite of this, the layer's porosity changes insignificantly and can hardly be measured with a good accuracy by optical methods. The segregation regime C prevails with increasing acceleration until the repacking regime is reached.

The path of the large disk within the layer in the segregation regime C is presented in Fig 6a. The positions of the disk were registered once per vibrational period at the moment when the layer touched to the vessel's bottom. The disk rises up during about 9 s. After this it remains on the top surface of the layer. An increase of acceleration increases the layer dilatation, which is most significant when Γ is close to the value Γ_r prevailing in the repacking regime. In this acceleration range, the disk sometimes dives into the layer although not deeply.

Increasing acceleration to $\Gamma_r=5$ terminates all particles' motion both in the granular layer and on its upper surface. The particles collapse into a single block similar to what was observed in the regime B. In contrast to the latter, in the repacking regime the layer

136   A. ALEXEEV ET AL.

motion is unstable. So, small deviation from the repacking acceleration Γ_r or even an external disturbance breaks repacking and resumes the layer fluidization. We observed that the repacking stability decreases with decreasing of the vibrational amplitude. For the vibrational amplitude $A=5$ mm, the repacking regime prevails during several vibrational periods. Then it is destroyed and resumes again. When the stable motion of the repacked layer is broken insignificant fluidization occurs. It means that some amount of kinetic energy is stored within the granular layer. This energy is dissipated by interparticle collisions and, as a result, the granular layer is repacked once again. Such layer motion repeats itself. In passing we note that increasing of particles' restitution coefficient breaks down the repacking as well. Our additional experiments with glass spheres ($e=0.95$) showed that these particles constantly wave on the upper surface of the layer even at the repacking acceleration.

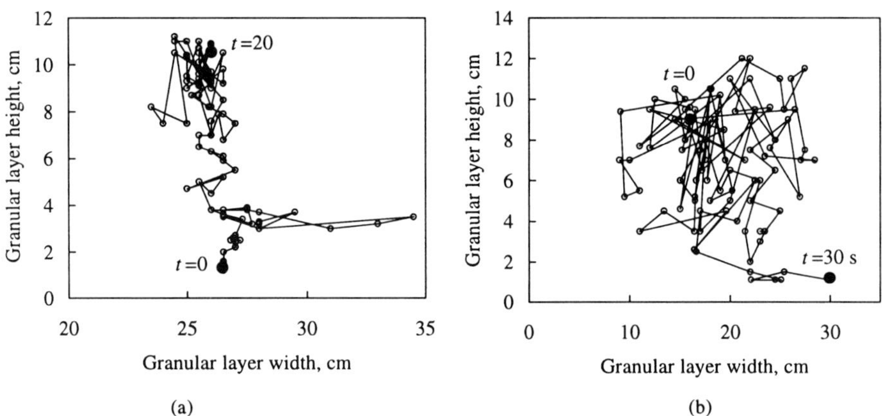

(a)   (b)

Figure 6. Trajectory of large disk (diameter 27 mm) motion within the granular layer composed of 9 monolayers of aluminum particles with diameter 11 mm: a) segregation regime, amplitude $A=25$ mm, $\Gamma=3.6$, b) mixing regime, amplitude $A=45$ mm, $\Gamma=6.5$.

When the vibrational acceleration is further increased beyond Γ_r and exceeds 5.1, the layer expands again and the transverse waves affect dramatically the large particle's motion. These waves cause intensive circulation motion of particles within the granular layer. Such circulation motion was registered experimentally [11] as well as numerically [9] for homogeneous granular beds. In the present paper we demonstrate that these circulations make the large particle dive deeply into the layer and move in a chaotic manner (Fig. 6b), sampling all possible positions within the layer with about equal probability. That is, the effect of these waves is to eliminate the segregation, thereby providing conditions potentially beneficial for particle mixing. This mixing regime D prevails in the acceleration range till the second repacking is reached.

Figure 5 depicts the layer kinematic map plotted in terms of the vibrational amplitude, A and the dimensionless acceleration, Γ. One can see that the boundaries of the regimes are almost independent of the amplitude. One can thus characterize the basic layer regimes with respect to the large particle motion in the following universal way:

A: $0 < \Gamma < 1.2$ non-detaching regime,

B: $1.2 < \Gamma < 2.4$ detaching regime without relative motion between the large disk and the layer,

C: $2.4 < \Gamma < 5$ segregation regime (upward motion of the large disk),

D: $5 < \Gamma < 8.5$ vibromixing regime (chaotic motion of the large disk),

E: $\Gamma > 8.5$ vibromixing regime (chaotic motion of the large disk).

The representation of the vibrational regimes given in Fig. 5 is valid for sufficiently large amplitudes, as predicted by the condition that the parameter $V = A\omega/(gh)^{1/2}$ is of order one [13] (more recently this parameter was recalled as Froude number [11]). In the present case this condition is already fulfilled for $A \geq 5$ mm for all layer heights h tested. However, these regimes are to be tested for wider ranges of the layer heights and granular properties. For small amplitudes the second repacking is very difficult to achieve, since it occurs for very large frequencies. According to the plastic body model predictions, for 1 mm amplitude the second repacking occurs at about $f \sim 46$ Hz, which is beyond the capability of our vibrational stand.

The first and the second repacking regimes were registered within narrow ranges around the respective dimensionless accelerations $\Gamma=5\pm0.05$ and $\Gamma=8.5\pm0.05$. These results, apparently, depends on the specific (non-sinusoidal) law of vibration of our stand. Speaking more generally, both characteristic "repacking" accelerations are predictable by the plastic body model (see Fig.5). Namely, at these accelerations the relative velocity between the body and the vessel at the moment of their contacts is minimal [13]. These acceleration will characterize the threshold of period doubling bifurcation [11] predictable by the plastic body model (see Fig. 5, dashed lines 2 and 4). For these regimes the maximal amplitude of surface waves was shown to be zero [10]. This minimum of fluidization (no relative motion) is also shown to correlate with the minima of the work produced by the vibrated piston [13]. Another important phenomenon, which may be described by means of the plastic body model, is vigorous mixing. We found that the lower acceleration threshold of period doubling bifurcation (see Fig. 5, dashed line 3) correlates with the acceleration of vigorous mixing regime (see Fig. 5, crosses).

The repacking phenomenon was registered in 2-D [10] and 1-D [22] for monosize granular systems. We could not distinguish any significant change in this critical acceleration also for polydisperse granules.

The side videorecording allows observation of the transverse waves propagation within the granular layers [10-13]. One can distinguish two types of the waves prevailing therein: One is the *surfaces* waves shown in Figs. 2a,b. The waves of the second type cause bending of the *whole* layer (Figs. 3a,b). Thus, we can distinguish these two types according to the state of the lower surface of the layers. Experiments showed [10-13] that such layer bending leads to a vibrational regime in which the layer contacts the vessel *once* per vibrational period. At the same time in the surface wave regime, the layer contacts the vessel once per two or even more vibrational periods.

We observed that the waves of both types lead to circulative motion of the particles within the layers – a process, which controls the trajectory of the large disk motion. However, the rate of the particles' circulations is also governed by the prevailing wave type. For the vibrational regimes characterized by surface waves, the particles' circulations concentrate in the upper part of the layer wherein such waves prevail. In contrast, the bending waves lead to particles' circulations throughout the whole layer (in agreement with [11]). Indeed, our experiments revealed that there exists a correlation between the type of waves prevailing in the layer and the intensity of the motion of the large disk within the bed, characteristic of the mixing process. The most vigorous "vibromixing" has been registered in the vibrational regimes where the bending waves prevail. These regimes are marked in the map shown in Fig. 5 by cross symbols.

Our preliminary experiments with wider granular layers showed that both upper and lower surfaces of granular layers, vibrated in the bending wave regime can have sinus-like shapes. We observed that in each part of the bent layer one circulation loop appears between its upper- and lowermost points, causing intensive motion of the large disk through the layer. We found also that the distance between the minima or maxima (wave length) strongly depends on the vibrational acceleration. We observed the layers having 1/2 periods (like those in Figs. 3), one, and 1.5 periods. It rules out any possibility that the bending waves are caused by the influence of side walls of the vessel.

4.   Conclusions

We observed new and interesting phenomena in the behavior of vibrated 2D layers of granular materials with collisional properties close to those of real granular materials. These are repacking regimes (no relative motion of granules under action of intensive vibration) and de-segregation, i.e. "vibromixing", regimes (chaotic motion of a single large particle). These and other vibrational regimes: non-detaching regime, segregation regime (upward motion of the large disk) are indicated in the map plotted in terms of vibrational amplitude and acceleration. We found that the physical mechanism underlying vibromixing of particles with different sizes is circulation of particles within the granular layer caused by surface and bending waves. Both lead to particle mixing, however, with different intensities. We found that the bending waves (arching) are most preferable for efficient vibromixing. These regimes need further experimental and theoretical investigation.

5.   Acknowledgment

This research was supported by the Israel Science Foundation administered by the Israel Academy of Sciences and Arts, by the Center of Absorption in Science and the Gilleady Program for Immigrant Scientists Absorption.

6.   References

1.   Ahmad, K. and Smalley, I. J.: Observation of particle segregation in vibrated granular systems, *Powder*

Technology **8** (1973), 69-75.

2. Williams, J. C.: The segregation of particulate materials, *Powder Technology* **15** (1976), 245-251.
3. Parson, D. S.: Particle segregation in fine powders by tapping as simulation of jostling during transportation, *Powder Technology* **13** (1976), 269-277.
4. Harwood, C. F.: Powder segregation due to vibration, *Powder Technology* **16** (1977), 51-57.
5. Rosato, A., Prinz, F., Standburg, J. K. and Svendsen, R.: Monte Carlo simulation of a particulate matter segregation, *Powder Technol.* **49**[59] (1986); Rosato, A., Prinz, F., Standburg, J. K. and Svendsen, R.: Why Brasil nuts are on top: Size segregation of particulate matter by shaking, *Phys. Rev. Lett.* **58**[7] (1987), 1038-1041.
6. Clement, E. and Rajchenbach, J.: Fluidization of a bidimentional powder, *Europhys. Lett.* **16**[2] (1991), 133-138.
7. Knight, J. B., Jaeger, H. M. and Nagel, S. R.: Vibration-induced size segregation in granular media: The convection connection, *Phys. Rev. Lett.* **70**[24] (1993), 3728-3731.
8. Ehrichs, E. E.., Jaeger, H. M. , Karczmsan, G. S., Knight, J. B., Kuperman, V. Yu. and Nagel, S. R.: Granular convection connection observed by magnetic resonance imaging, *Science* **267** (1995), 1632-1634.
9. Lan, Y. and Rosato, A. D.: Convection related phenomena in granular dynamics simulations of vibrated beds, *Phys. Fluids* **9**[12] (1997), 3615-3624.
10. Clement, E., Vanel, L., Rajchenbach, J. and Duran, J.: Pattern formation in a vibrated granular layer, *Phys. Rev. E.* **53**[3] (1996), 2972-2975.
11. Wassgren, C. R., Brennen C. E. and Hunt, M. L.: Vertical vibration of a deep bed of granular material in a container, *J. Appl. Mech.* **63**[3] (1996), 712-719.
12. Doudy, S. Fauve, S. and Laroshe, C.: Subharmonic instabilities and defects in a granular layer under vertical vibrations, *Europhys. Lett.*, **8**[7] (1989), 621.
13. Goldshtein, A., Shapiro, M., Moldavsky, L. and Fichman, M.: Mechanics of collisional motion of granular materials. 2. Wave propagation through vibrofluidized granular layers. *J. Fluid Mech.*, **287** (1995), 349.
14. Vanel, L., Rosato, A. and Dave R.: Rise-time regime of a large sphere in vibrated bulk solids, *Phys. Rev. Lett.* **78**[7] (1997), 1255-1258; Vanel, L., Rosato, A. and Dave R. Size dependent segregation in a 3D vibrated bed, in R.P. Behringer and J.T. Jenkins (eds.), *Powders & Grains 97*, A.A. Balkema, Rotterdam, 1997, pp. 385 – 387.
15. Chlenov, V. A. and Mikhailov, N. D.: *Vibrofluidized Beds.* Nauka, Moscow, 1972.
16. Brone, D. and Muzzio, F. J.: Size segregation in vibrated granular systems: A reversible process, *Phys. Rev. E.* **56**[1] (1997), 1059-1063.
17. Thomas, B.: Shallow vibrated particulate beds - bed's dynamics and heat transfer, Ph. D. Dissertation, Virginia Polytechnic Institute and State University, Blacksburg, Virginia, 1988.
18. Eccles, E. R. A. and Mujumdar, A. S.: Particle flow patterns and mixing in aerated vibrated beds of small particles, *Powder Handling & Processing* **4**[1] (1992), 39-45.
19. Alexeev A., Royzen V., Dudko V., Goldshtein A. and Shapiro M.: Dynamics of vertically vibrated two-dimensional granular layers, *Phys. Rev. E,* **59**[3] (1999), 3231–3241.
20. Royzen, V., Dudko, V., Goldshtein, A. and Shapiro, M.: Apparatus for vibromixing and measurements of kinematic and dynamic parameters of granular materials, Proc. 26th *Israel Conference on Mechanical Engineering*, Haifa, 1996, pp. 225-227.
21. Royzen, V., Dudko, A. Alexeev, A., Goldshtein, A. and Shapiro, M.: Vibromixing of granular materials: Kinematics and dynamics of vibrofluidization, *Proc. Second Israel Conference for Conveying and Handling of Particulate Solids,* Jerusalem, 1997, pp.11-23.
22. Alexeev, A., Goldshtein, A. and Shapiro, M.: Vibrofluidization of highly dissipative granular systems: one-dimensional computer simulations, *Powder Tech.*, Submitted (1999).

MIXING AND SEGREGATION OF GRAINS STUDIED BY N.M.R. IMAGING INVESTIGATION – APPLICATION TO A TURBULENCE MIXER: THE TURBULA BLENDER

P. PORION
*Centre de Recherche sur la Matière Divisée, CNRS
and Université d'Orléans, 45071 Orléans cedex 2, France*

N. SOMMIER
*Lab. Physique Pharmaceutique, Université Paris-Sud,
92296 Châtenay-Malabry cedex, France*

P. EVESQUE
*Lab. MSSM, Ecole Centrale Paris,
92295 Châtenay-Malabry cedex, France*

Abstract.

Magnetic Resonance Imaging (MRI) techniques have been used to study the mixing and segregation processes of granular materials in Turbula mixer using binary mixtures of sugar beads of different diameters d. When the ratio R of bead diameters is approximatively $R = 1$ mixing process is observed, but segregation occurs as soon as $R > 1.1$ or $R < 0.9$. Furthermore, segregation develops faster than mixing and is induced by surface effect. We report in this paper, a qualitative analysis of these phenomena. Effects of some parameters (beads diameter ratio, speed rotation, mixing time, concentration) have been checked on segregation and mixing processes.

1. Introduction

Segregation of granular materials such as sand and powders is a significant problem when processing granular media in pharmaceutical and food-processing industries. Indeed, heterogeneous mixtures of granular materials tend to segregate by size, even in situations where one might expect mixing such as shaking or rotation. Many experiments have been dealing with

141

A.D. Rosato and D.L. Blackmore (eds.), IUTAM Symposium on Segregation in Granular Flows, 141–152.
© 2000 *Kluwer Academic Publishers. Printed in the Netherlands.*

the segregation which occurs in rotating cylinders or in drum mixers. For instance, the well known axial segregation was reported already in 1939 by Oyama [1]; such axial bands of segregation appear after a few tens of rotation when a binary mixture of different size is rotated in a drum blender and it is now a standard problem in the field of granular-media studies [2, 3, 4, 5, 6]. More recently, Magnetic Resonance Imaging (MRI) studies have been used to analyse the axial and radial segregations inside the bulk of granular media [7, 8] or to study granular flow [9, 10] and convection [11, 12, 13]. These experiments have demonstrated the large possibilities of the MRI to access to the three-dimensional geometries in such materials. In this paper, we use a more sophisticated mixer called Turbula which generates an intricate flow pattern (this is why its flow is called "turbulent"), and we report observations on the generation of mixing and segregation in this blender using MRI. Sugar beads used in pharmaceutical industry have been used here; and they have been doped or not with liquid, to make them visible or invisible to MRI. This paper reports the first step of our study; it gives a qualitative analysis and description of the findings. Experimental technique and procedures are described in Section 2; whereas NMR images and experimental data are reported in Section 3.

2. Magnetic Resonance Imaging (MRI) Experiments

2.1. MRI BACKGROUND

MRI method is a non-invasive technique based on nuclear magnetic resonance (NMR) which measures relaxation-weighted proton density. In a conventionnal NMR experiment, the sample is placed in a static homogeneous magnetic field B_0. The proton's sample with identical nuclei in an identical environment (e.g. the protons in water) have the same resonance frequency ω_0 related to B_0 by the Larmor's relation (Eq. 1) where γ, the proportionality coefficient is the gyromagnetic ratio of the nucleus.

$$\omega_0 = \gamma B_0 \tag{1}$$

In the MRI experiments, in order to obtain spatial localisation of the spins/molecules, the value of the magnetic field is a function of the space vector $\mathbf{r} = (x, y, z)$. In other words, the resonance frequency depends on the localisation of the molecules in the sample. So in MRI [14, 15], spatially resolved magnetic resonance is made possible by superimposing a time dependent linear gradient field $\mathbf{G}(\mathbf{r}, t)$ onto the static magnetic field B_0 (Eq. 2), allowing to define a local Larmor frequency $\omega(\mathbf{r}, t)$:

$$\omega(\mathbf{r}, t) = \gamma B(\mathbf{r}, t) = \gamma \left[B_0 + \mathbf{G}(\mathbf{r}, t) \cdot \mathbf{r} \right] \tag{2}$$

In these conditions, if one considers a local spin density $\rho(\mathbf{r})$, the NMR signal $dS(\mathbf{G},t)$ from a dV element in a gradient field $\mathbf{G}(\mathbf{r},t)$ may be written as (Eq. 3):

$$dS(\mathbf{G},t) = \rho(\mathbf{r})dV \exp\left[i(\gamma B_0 + \gamma \mathbf{G} \cdot \mathbf{r})t\right] \tag{3}$$

and the integrated signal amplitude $S(t)$

$$S(\mathbf{G},t) = \iiint \rho(\mathbf{r}) \exp\left[i\gamma(\mathbf{G} \cdot \mathbf{r})t\right] d\mathbf{r} \tag{4}$$

Using the concept of the reciprocal space vector, $\mathbf{k}$, introduced by Mansfield [16], the integrated signal amplitude $S(t)$ (Eq. 4) can be written as (Eq. 5):

$$S(\mathbf{k}) = \iiint \rho(\mathbf{r}) \exp\left[i2\pi(\mathbf{k} \cdot \mathbf{r})\right] d\mathbf{r} \quad \text{with} \quad \mathbf{k} = \frac{\gamma \mathbf{G} t}{2\pi} \tag{5}$$

In this $\mathbf{k}$-space formalism, using the Fourier transform and its inverse, the signal, $S(\mathbf{k})$, and the spin density, $\rho(\mathbf{r})$, are mutually conjugated. Then, the spin density $\rho(\mathbf{r})$ is the Fourier transform of $S(\mathbf{k})$:

$$\rho(\mathbf{r}) = FT\left[S(\mathbf{k})\right] = \iiint S(\mathbf{k}) \exp\left[-i2\pi(\mathbf{k} \cdot \mathbf{r})\right] d\mathbf{k} \tag{6}$$

This last equation (Eq. 6) is the fundamental relationship of MRI. Of course, in practice, the experimental signal $S(\mathbf{k})$ is not a perfect representation of the Fourier transform of $\rho(\mathbf{r})$. Indeed, in equation (Eq. 4), it is assumed that the NMR signal is simply proportional to the spin density $\rho(\mathbf{r})$, whereas in reality, the signal is affected by several parameters as the spin-spin relaxation time, T_2, dipolar and scalar coupling interactions, or the translation of the spin in the presence of magnetic field gradients.

2.2. EXPERIMENTAL PRECEDURE

2.2.1. *Material Preparation*
Sugar beads from pharmaceutical industry have been used with the following different diameters d (350, 500, 600, 700, 800, 980, and 1100 μm). Their mixing and segregation with a reference diameter (i.e. $d_{ref} = 980\,\mu$m) has been studied. The reference grains have been doped with liquid in order to be sensitive to MRI. The other beads are not NMR-visible. The reference beads appear in grey or black in the picture, the others in white. So, the beads diameter ratio R defined as d_{ref}/d has been varied in the range $R = 0.89$ to $R = 2.80$ corresponding respectively to the 1100 μm and 350 μm beads.

The cylindrical container (axis Oz, diameter 52 mm, and height 65 mm) is filled successively with the reference beads and with the other set of beads in half-half proportion. The filling rate of the container is 2/3 of the height. So, the initial packing is made of two horizontal layers of beads, the bottom one being the reference bead layer (980 μm). A first image corresponding to the initial state ($t = 0$) is then acquired using MRI.

The container is then placed in the Turbula blender (Fig. 1) which is then run during a given time t with a given rotation speed V among the set $\{22, 32, 48, 68, 100 \text{ rpm}\}$. The Turbula blender is currently used in chemical and pharmaceutical industries; it imposes an intricate three-dimensional motion that causes the material to be swept by an intense turbulence [17]. After this mixing time t_{mix}, the container is removed and placed in the MRI where a second image is taken.

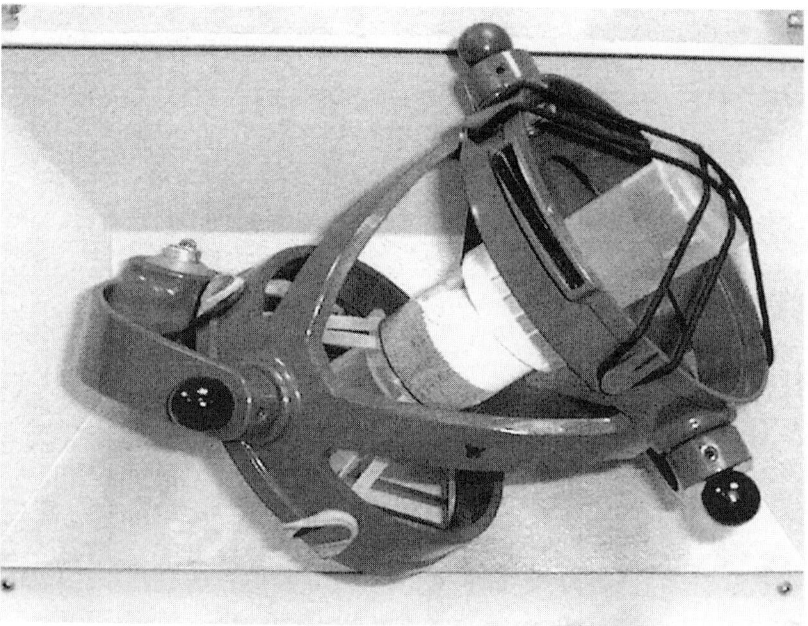

Figure 1. Turbula blender. One can observe the modifications of the experimental set-up in order to insert the small cylindrical container (diameter 52 mm and height 65 mm) at the centre of the device.

2.2.2. *Apparatus and MRI Experimental Conditions*

Proton NMR measurements were performed on a Bruker DSX100 spectrometer operating at 100 MHz, using a 2.35 T superconducting magnet. All the experiments were done at room temperature (296°K). We have used Bruker imaging probehead (Mini 036).

Labelling Oz the cylinder axis, MRI has been performed in two different kinds of experiment. First a 1d profile has been done along the Oz cylinder axis by integrating on the xy plane, with a z-resolution of 137 μm the total field of 70 mm corresponds to 512 points. Secondly, two different series of 16 images have been taken, one corresponding to 16 cuts parallel to the xy plane, the other to 16 cuts parallel to the xz plane. So, each series corresponds to a set of sixteen images at different depths; the thickness of a cut is 3.5 mm so that the whole 16 cuts correspond to 56 mm; each 2d image contains (64 × 64 pixels) and correspond to a field of 60 mm × 60 mm and to a spacial resolution of 937.5 μm (Fig. 2). The 2d MRI experiments have been performed using selective excitations and an usual two-dimensional Fourier imaging (spin-warp method). The 16 slices allow a complete analysis of the spacial repartition of the different grains. And a good enough signal-to-noise ratio requires a 20 minutes acquisition time about for the set of sixteen images.

Interestingly, 1d profiles have been taken, but they do not allow to study mixing and segregation since xy integration washes out completely the segregation pattern, excepted at the top and bottom surfaces (Fig. 3). But the depletion at the middle of the pile is not visible.

3. Results and Discussion

3.1. MIXING PROCESS - IDENTICAL BEADS

Mixing has been studied first, using the 980 μm beads and a Turbula speed of 22 rpm. The MRI results as a function of rotation number N_T defined as $N_T = V.t_{mix}$ are reported in Fig. 4 and Fig. 5. One can observe the complexity of the flow.

These images show clearly that the MRI technique is a nice tool for observing how an initial segregated binary mixture reaches an homogeneous state. With our experimental conditions, the steady state is achieved after about $t_{mix} = 10$ min. or $N_T = 220$.

3.2. SEGREGATION PROCESS - DIFFERENT BEADS

Experimental conditions were always ($V = 22$ rpm and $t_{mix} = 10$ min.). For each ratio of size R (ranging from 0.89 to 2.80), we have studied the

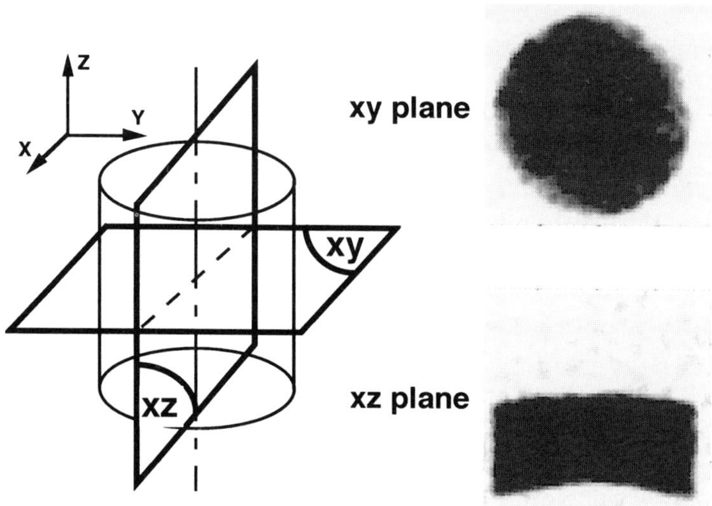

Figure 2.   Sketch indicating the orientation of the MRI images used and NMR images examples of 3.5 mm slice through the centre of the container in the xy and xz planes corresponding respectively to axial and sagittal cuts.

transient of segregation and the stationary state obtained after few rounds. We report successively these two kinds of results.

3.2.1. *Diameter Ratio Effects*

Stationary patterns obtained after 10 minutes of rotation are reported in the Fig. 6 as a function of the size ratio R ranging from 0.89 to 2.80. The pictures corresponds to the 6 central cuts of each sample. No doubt, samples with R smaller and larger than 1 exhibit segregation with the larger beads located near the edge of the container and the smaller one at the centre; sample with $R = 1$ does not exhibit segregation. Furthermore, the larger the R the larger the segregation. It is known that the beads diameter ratio is an important parameter in the segregation process. For the case $R = 0.89$ where the reference beads are not the biggest beads, one can observe that the 980 μm beads are concentrated at the centre of the container and the MRI-invisible 1100 μm beads are at the periphery. We can also remark that the shape of the border line between the two granular species is intricate but it is about the same for the different R values. It seems that this shape is induced by the intricate motion of the Turbula blender.

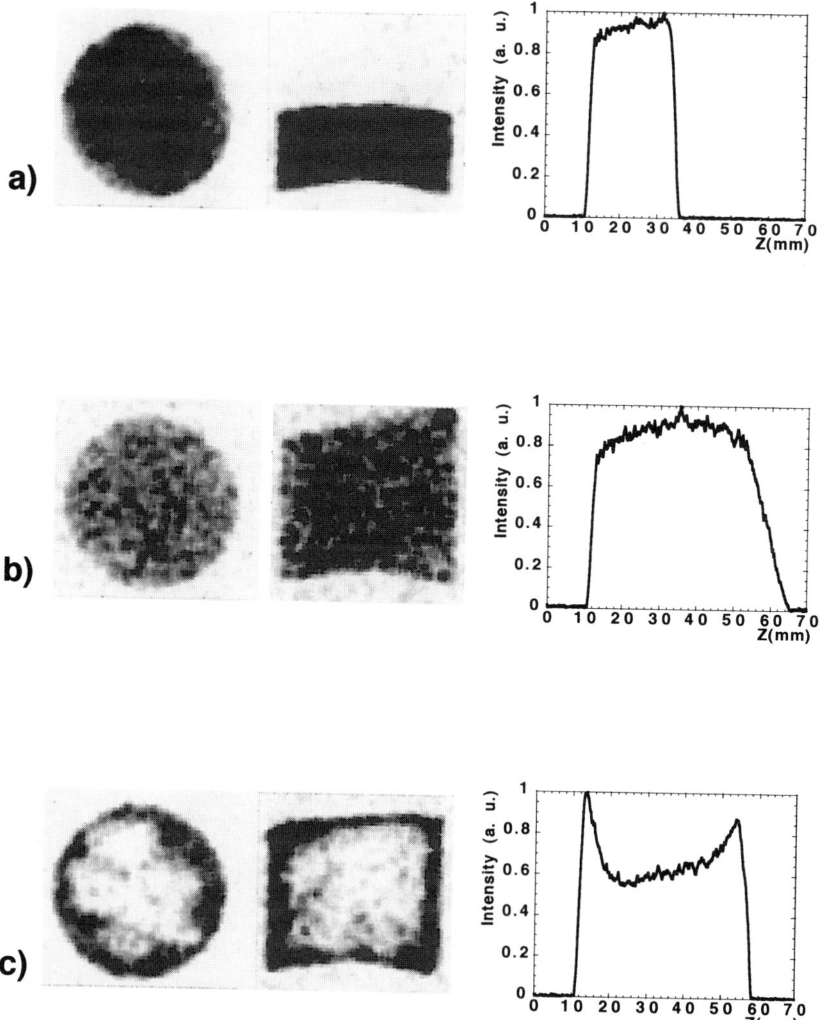

Figure 3. Comparaison between the 2d images and the 1d profiles corresponding to: a) $N_T = 0$ for 980 μm - 980 μm sample, b) $N_T = 220$ for 980 μm - 980 μm sample, and c) $N_T = 220$ for 350 μm - 980 μm sample.

3.2.2. *Transient Mixing and Segregation*

Transient of mixing and segregation has been studied for the binary mixture 350 μm - 980 μm (corresponding to a diameter ratio R equal to 2.80). The Fig. 7 displays a series of six xy cuts corresponding to the central part of the cell at different times corresponding to different numbers of rounds N_T. It

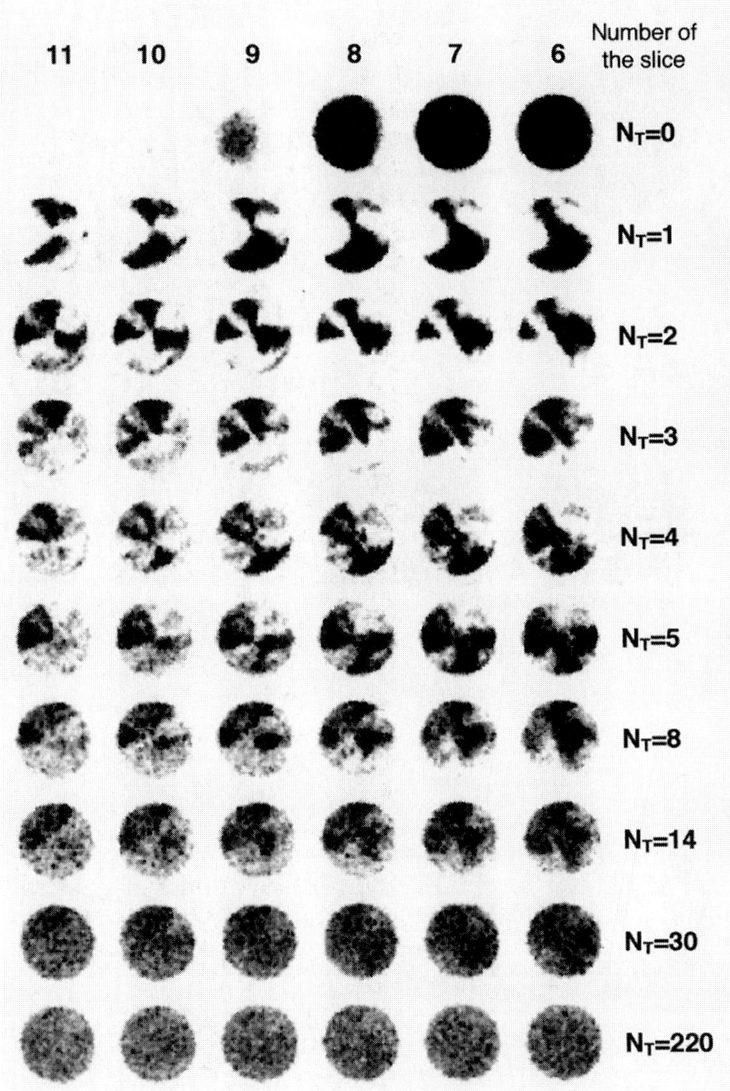

Figure 4. MRI study of mixing of $980\,\mu$m doped beads with $980\,\mu$m undoped beads in half-half proportion as a function of the number of rounds N_T (rotation speed $V = 22$ rpm). Only the 6 central images of xy cuts are displayed. Comparing the first two lines demonstrate the complexity of the flow pattern. Mixing requires 220 rounds about.

is worth comparing the first few distributions with the ones of Fig. 4. They both show up approximately the same flow pattern demonstrating that the flow pattern is not strongly affected by the presence of two different bead sizes. Nevertheless, segregation appears already after 5 or 8 rounds which

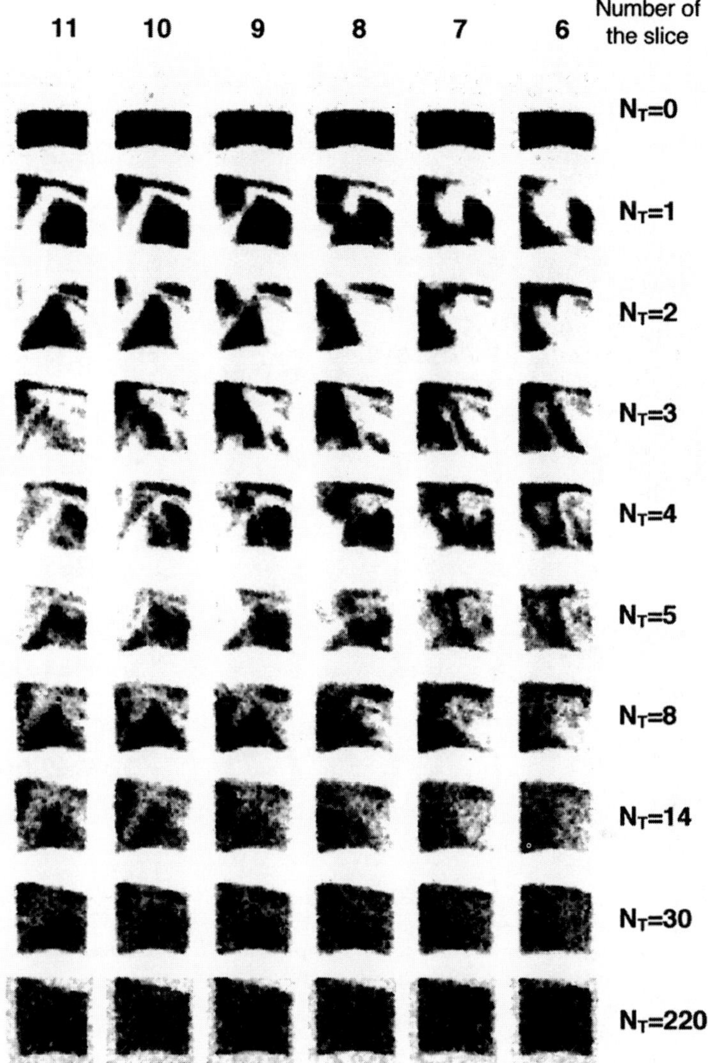

Figure 5.   MRI study of mixing of 980 μm doped beads with 980 μm undoped beads in half-half proportion as a function of the number of rounds N_T (rotation speed $V = 22$ rpm). Only the 6 central images of xz cuts are displayed.

is quite shorter than the mixing time observed in Fig. 4.

So, analysis of the Fig. 7 shows clearly that the segregation process in the Turbula blender is very efficient, i.e. more efficient than mixing. Indeed, after approximately ten rounds in the Turbula, a binary mixture with larger R is segregated already. This might be surprising at first sight; however

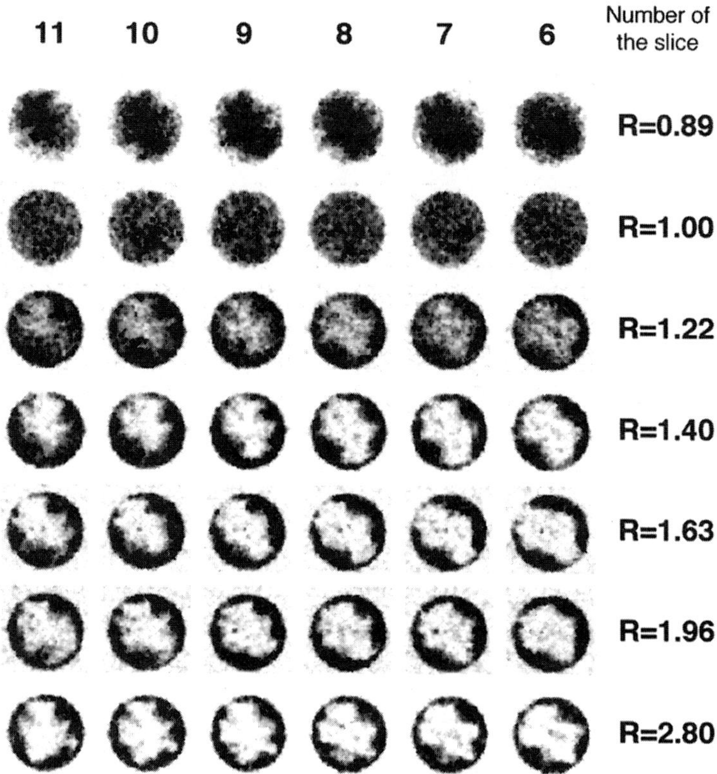

Figure 6. Stationary segregation pattern obtained after 10 minutes rotating with different size ratio R and a concentration $c_{ref} = 0.5$. (rotation speed $V = 22\,\mathrm{rpm}$). Larger beads are located near the container edges. Sample with $R = 1$ does not exhibit segregation. $R = 0.89$ exhibits inverse segregation pattern.

the reason for this comes from the selective efficiency of the interaction between the beads and the container walls and the free surfaces which both "attract" the large beads. This introduces a new process which enhances the dynamics.

4. Conclusion

MRI was used to non-invasively measure concentration of sugar beads in Turbula blender. We show that its technique can be used to analyse the 3d geometries in such granular media and it allows to know the concentration in all points of the sample. So, MRI appears to be a powerful technique

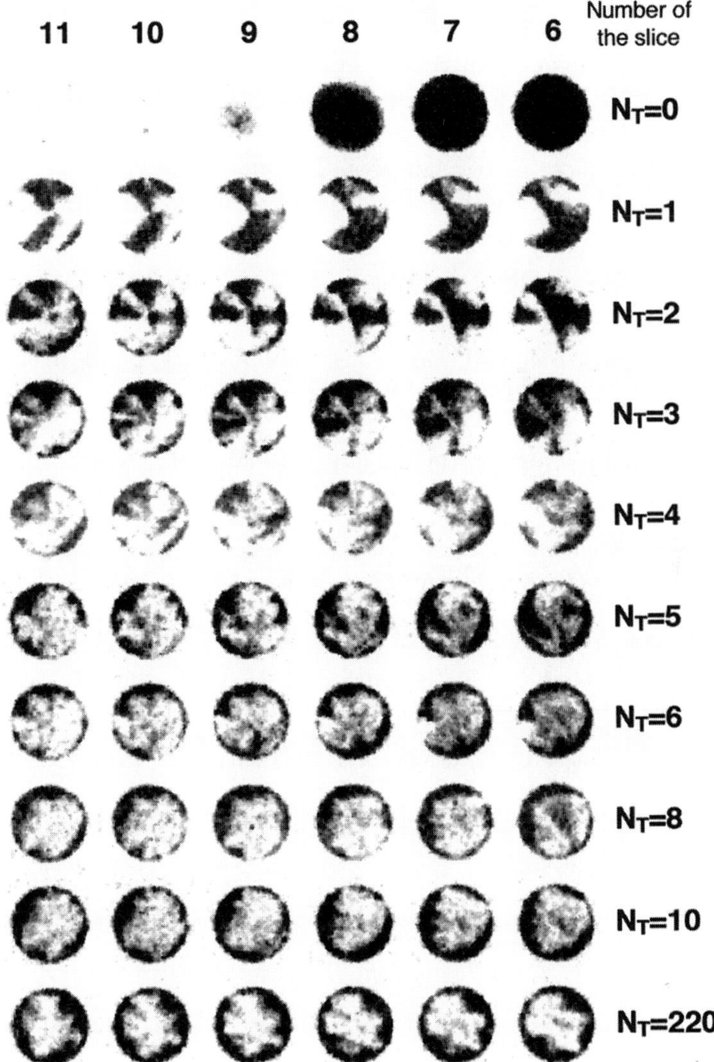

Figure 7. Transients of segregation for a binary mixture $350\,\mu\mathrm{m}$ - $980\,\mu\mathrm{m}$ obtained after different number of rounds N_T at $22\,\mathrm{rpm}$ ($c_{ref} = 0.5$). The six images of each line correspond to xy distributions of the central part of the cell. The three first sets exhibit same pattern as in the $980\,\mu\mathrm{m}$ - $980\,\mu\mathrm{m}$; they demonstrate that complexity of the flow has not changed. The other sets demonstrate that segregation is a process faster than mixing.

to obtain three-dimensional spatial informations compared to other non-invasive techniques such as X-rays [18] or radioactive tracer beads [19].

Acknowledgements

It is a pleasure to acknowledge Drs G. Couarraze, P. Tchoreloff, B. Leclerc, F. Agnely and S. Geiger (Université Paris-Sud) for helpful discussions, and L. Pothuaud (C.R.M.D.) for the numerous discussions concerning the image processing development. We also acknowledge P. Sanial from the NP Pharm company for the sugar beads samples. The Bruker spectrometer and mini-computers used in this study were purchased thanks to grants from CNRS and Région Centre (France).

References

1. Oyama Y. (1939), *Bull. Inst. Phys. Chem. Res. (Tokyo)*, Rep. **5**, p. 600; Oyama Y. and Ayaki K. (1956), *Kagaku Kikai* **20**, p. 6.
2. Donald M.B. and Roseman B. (1962), *Br. Chem. Eng.* **7**, p. 749.
3. Williams J.C. (1976), *Powder Technol.* **15**, p. 237.
4. Bridgwater J. (1976), *Powder Technol.* **15**, p. 215.
5. Das Gupta S., Khakhar D.V., and Bhatia S.K. (1991), *Chem. Eng. Sci.* **46**, p. 1513.
6. Savage S.B. (1993), in D. Bideau and A. Hansen (eds.), *Disorder and Granular Media*, North-Holland, Amsterdam, p. 255.
7. Hill K.M., Caprihan A., and Kakalios J. (1997), *Phys. Rev. Lett.* **78**, p. 50.
8. Hill K.M., Caprihan A., and Kakalios J. (1997), *Phys. Rev. E* **56**, p. 4386.
9. Nakagawa M., Altobelli S.A., Caprihan A., Fukushima E., and Jeong E.-K. (1993), *Exp. Fluids* **16**, p. 54.
10. Nakagawa M., Altobelli S.A., Caprihan A., and Fukushima E. (1997), in R.P. Behringer and J.T. Jenkins (eds.), *Powders and Grains 97*, Balkema, Rotterdam, p. 447.
11. Ehrichs E.E., Jaeger H.M., Knight J.B., Karczmar G.S., Kuperman V. Yu., and Nagel S.R. (1995), *Science* **267**, p. 1632.
12. Kuperman V. Yu., Ehrichs E.E., Jaeger H.M., and Karczmar G.S. (1995) *Rev. Sci. Instrum.* **66**, p. 4350.
13. Caprihan A., Fukushima E., Rosato A.D., and Kos M. (1997) *Rev. Sci. Instrum.* **68**, p. 4217.
14. Callaghan P.T. (1991), *Principles of Nuclear Magnetic Resonance Microscopy*, Clarendon Press, Oxford.
15. Kimmich R. (1997), *NMR Tomography Diffusometry Relaxometry*, Springer-Verlag, Berlin.
16. Mansfield P. and Grannell P.K. (1973), *J. Phys. C*, **6**, p. L422; Mansfield P. and Grannell P.K. (1975), *Phys. Rev.*, **12**, p. 3618; Mansfield P. (1976), *Contemp. Phys.*, **17**, p. 553.
17. Spring M.S. (1988), *Encyclopedia of Pharmaceutical Technology*, Vol. 2: Biodegradable Polyester Polymers as Drug Carriers to Clinical Pharmacodynamics, p. 172-188, Marcel Dekker (eds.), New-York.
18. Baxter G.W., Behringer R.P., Fagert T., and Johnson G.A. (1989) *Phys. Rev. Lett.*, **62**, p. 2825.
19. Harwood C. (1977) *Powder Technol.*, **16**, p. 51.

VELOCITY FLUCTUATION MEASUREMENTS OF GRANULAR MEDIA IN A ROTATING CYLINDER BY NUCLEAR MAGNETIC RESONANCE IMAGING

A. CAPRIHAN, J.D. SEYMOUR, S.A. ALTOBELLI, and
E. FUKUSHIMA
New Mexico Resonance,
2425 Ridgecrest Dr., SE, Albuquerque, NM 87108 USA

Abstract

The concept of velocity fluctuations (granular temperature) is important in describing the flow and mixing of granular particles. Recent theories model the axial segregation of granular mixtures in a horizontal rotating cylinder, as an interaction between the surface flow and the random diffusive motion of granular particles. We discuss the Magnetic Resonance Imaging (MRI) technique for characterizing the stochastic properties of this random diffusive motion of granular media in a rotating horizontal cylinder.

1. Introduction

The nuclear magnetic resonance (NMR) technique for non-invasive diffusion measurements has been known for a long time [1], and its adaptation to spatially image diffusion and flow is discussed by Callaghan [2]. We apply this method to study random motion of granular particles undergoing flow and collisions. Our system consists of oil filled 2 mm spherical particles rotated in a 70 mm i.d. horizontal cylinder. Our previous magnetic resonance imaging (MRI) studies in a similar system include velocity measurements [3] and a study of segregation, when a mixture of different size particles is rotated [4]. In segregation/mixing studies of mixtures, either in the rotating cylinder or in a vibrated layer, an important phenomena is that of time correlated motion of particles causing segregation. A diffusion coefficient describing this particle migration has been used to develop segregation theories [5]. The concept of velocity fluctuations and temperature has also been an important parameter in theories of granular flow developed on the basis of kinetic theory analysis [6]. The theory of Brownian motion has been used in conjunction with kinetic theories to relate asymptotic diffusive behavior to granular temperature [6]. Thus, a non-invasive MRI technique to spatially image the stochastic properties of the particle motion appears to be a valuable measurement. We have previously made MRI images of the apparent diffusion coefficient [7]. In these measurements there was a problem of relating the MRI measured diffusion parameter to the conventional definition of diffusion. In this paper we present some results from an improved MRI experiment and explain the significance of the MRI measurement. MRI can make 3D velocity fluctuation measurements within

A.D. Rosato and D.L. Blackmore (eds.), IUTAM Symposium on Segregation in Granular Flows, 153–161.
© 2000 *Kluwer Academic Publishers. Printed in the Netherlands.*

the bulk of opaque materials, unlike a recently proposed technique of diffusing-wave spectroscopy [8], which makes these measurements closer to the surface.

2. Theory

We consider that a particle velocity is a sum of a mean velocity V and a random fluctuating component u. We will restrict u to one direction. This stochastic component will be considered to be a stationary Gaussian Markov process with zero mean and autocorrelation function $R_u(\tau) = \langle u(t+\tau)u(t)\rangle$. The position of the particle at any time Δ, will be defined by $x(\Delta)$. Then a time dependent diffusion coefficient is defined by

$$D(\Delta) = \frac{\langle (x(\Delta) - x(0))^2 \rangle}{2\Delta}, \tag{1}$$

and the diffusion coefficient D, which is the long time limit of $D(\Delta)$, is defined by

$$D = \lim_{\Delta \to \infty} D(\Delta) \tag{2}$$

The two diffusion coefficients can be calculated from the velocity autocorrelation functions from

$$D(\Delta) = \int_0^\Delta (1 - \frac{\tau}{T}) R_u(\tau) d\tau \tag{3}$$

and

$$D = \int_0^\infty R_u(\tau) d\tau. \tag{4}$$

The average correlation time for a stochastic process is defined by

$$\tau_c = \int_0^\infty R_u(\tau) d\tau \Big/ R_u(0), \tag{5}$$

and the granular temperature by $\langle u^2 \rangle = R_u(0)$. It follows that

$$D = \langle u^2 \rangle \tau_c \tag{6}$$

The three quantities, diffusion coefficient, correlation time, and granular temperature are inter related, with only two of them being independent.

In the case of Wiener process, $R_u(\tau) = 2D_w \delta(\tau)$ we have,

$$D(\Delta) = D = D_W. \tag{7}$$

The velocity of a Brownian particle with Ornstein-Uhlenbeck process has the autocorrelation function $R_u(\tau) = \langle u^2 \rangle e^{-|\tau|/\tau_a}$, then $\tau_c = \tau_a$, D is given by Eq. (6) and

$$\frac{D(\Delta)}{D} = (1 - \frac{\tau_c}{\Delta}(1 - e^{-\Delta/\tau_c})) \tag{8}$$

For short correlation times, that is as $\tau_c \rightarrow 0$, the Ornstein-Uhlenbeck process reduces to Weiner process with uncorrelated motion.

The NMR method consists of measuring an apparent time dependent diffusion coefficient, which approximates $D(\Delta)$ but includes specifics of the NMR experiment, and then calculating the parameters of $R_u(\tau)$ by measuring $D(\Delta)$ for several values of Δ. The method can be applied for any parametric form of $R_u(\tau)$ but will be illustrated for $R_u(\tau) = \langle u^2 \rangle e^{-|\tau|/\tau_c}$.

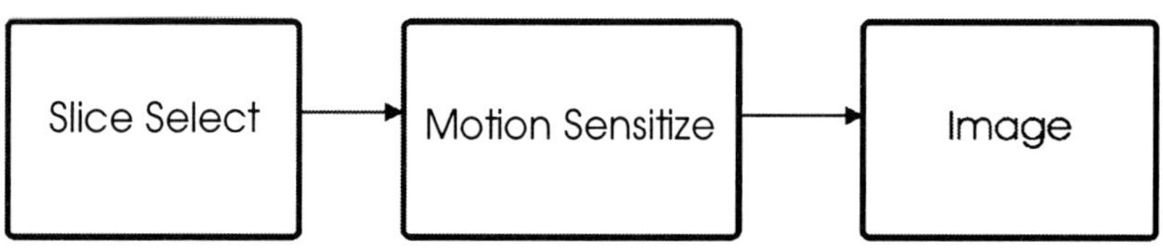

Figure 1.

The MRI experiment to measure diffusion consists of slice selection, followed by sensitizing the magnetization phase to motion, and concludes by imaging the slice (Fig. 1). Magnetic field gradients for motion sensitivity should have zero area and otherwise can have various patterns to probe motion properties [2]. The most common pattern is that of a bipolar gradient (Fig. 2a), also known as a pulsed-gradient spin-echo (PGSE) method [1], we call it a single pulsed gradient sequence (SPG). In Fig. 2b we have a flow compensated sequence, a sequence with its first moment zero ($\int tg(t)dt = 0$).

We will call it a compensated pulsed gradient sequence (CPG). In Fig. 2c we have a SPG sequence, repeated twice, and call it a repeated single pulsed gradient sequence (RSPG). If necessary, these sequences can be extended to n pulses to probe more complex velocity autocorrelation functions.

A. CAPRIHAN ET AL.

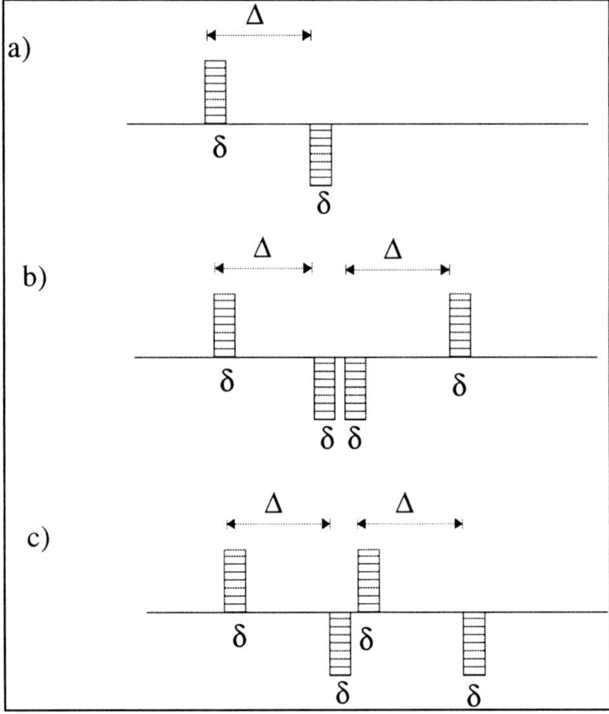

Figure 2. Motion encoding gradient patterns for a) the SPG NMR experiment, b) the CPG experiment and c) the RSPG experiment.

For SPG the image intensity at any voxel is given by

$$E(g, \Delta, \delta) = Ke^{-\gamma^2 g^2 \delta^2 (\Delta - \delta/3) D_{s1}(\Delta)} \tag{9}$$

where δ is the gradient width, Δ the gradient separation and g the gradient amplitude. $D_{s1}(\Delta)$ is called the apparent time dependent diffusion coefficient. Its spatial distribution can be calculated by repeated imaging for different values of g, for fixed values of Δ and δ, followed by a voxel by voxel least-squares fit in equation (9) to calculate $D_{s1}(\Delta)$.

For RSPG the apparent time dependent diffusion coefficient $D_{s2}(\Delta)$ is defined by

$$E(g, \Delta, \delta) = Ke^{-2\gamma^2 g^2 \delta^2 (\Delta - \delta/3) D_{s2}(\Delta)}, \tag{10}$$

where the factor of two in the exponent normalizes for two gradient pulses. Similarly for CPG we define $D_{c2}(\Delta)$ by the equation

$$E(g,\Delta,\delta) = Ke^{-2\gamma^2 g^2 \delta^2 (\Delta-\delta/3) D_{c2}(\Delta)} \tag{11}$$

The functional form of these three apparent diffusion coefficients can be calculated in terms of velocity autocorrelation parameters (2). If $R_u(\tau) = \langle u^2 \rangle e^{-\frac{|\tau|}{\tau_c}}$ then,

$$D_{s1}(\Delta) = D(1 + \frac{a_{ou}}{\delta^2(\Delta-\delta/3)}) \tag{12}$$

$$D_{s2}(\Delta) = D(1 + \frac{a_{ou}+b_{ou}/2}{\delta^2(\Delta-\delta/3)}) \tag{13}$$

$$D_{c2}(\Delta) = D(1 + \frac{a_{ou}-b_{ou}/2}{\delta^2(\Delta-\delta/3)}) \tag{14}$$

where $D = \langle u^2 \rangle \tau_c$,

$$a_{ou} = -2\tau_c^2\delta + \tau_c^3(1-e^{-\frac{\delta}{\tau_c}})(2 - e^{-\frac{\Delta}{\tau_c}} + e^{-\frac{\Delta-\delta}{\tau_c}}) \tag{15}$$

and

$$b_{ou} = \tau_c^3(1-e^{-\frac{\delta}{\tau_c}})^2(1-e^{-\frac{\Delta}{\tau_c}})^2 . \tag{16}$$

As $\delta \to 0$, the apparent time dependent diffusion coefficient $D_{s1}(\Delta)$ tends to the time dependent diffusion coefficient $D(\Delta)$. Thus for narrow gradient pulses NMR measures $D(\Delta)$. This observation has made this technique very popular for probing diffusive motion.

For long Δ, as $\Delta \to \infty$, all three apparent diffusion coefficients tend to the diffusion coefficient D. Also for short τ_c, $\tau_c \ll \Delta$, all three apparent diffusion coefficients tend to D. In other words if the motion is uncorrelated then all three apparent diffusion coefficients measure the same quantity D. For $\tau_c \gg \Delta$, it follows that $D_{c2}(\Delta) = 0$, in other words there is no attenuation in image intensity because of motion. This is the distinguishing feature of the compensated PGC2 sequence. It compensates for correlations in motion and increases sensitivity for correlation time measurements. In addition for $\tau_c \gg \Delta$, we have $D_{s1}(\Delta) = D_{s2}(\Delta)/2 = \langle u^2 \rangle \Delta / 2$. We also have

$$D_{s1} = (D_{s2} + D_{c2})/2 \qquad (17)$$

Thus the three experiments of Fig. 2 are not independent. We can extract the information of the third one if any two experiments are done. Equation (17) is true only if the process is stationary. Note that for SPG the motion is probed over a time Δ (Fig. 2) and for RSPG and CPG it is probed over a time 2Δ. If all three experiments are done and Eq.(17) holds then we are assured that particle motion statistics that are present over Δ are also present over 2Δ. If this condition was not true then we would know that motion is not stationary over 2Δ.

Our method consists of measuring $D_{s1}(\Delta)$, $D_{s2}(\Delta)$, $D_{c2}(\Delta)$ for several values of Δ and from equations (12) to (14) calculate D and τ_c by non-linear least squares fitting. It is not necessary to measure all three apparent diffusion coefficients, D and τ_c could also have been calculated from $D_{s1}(\Delta)$ measured for several values of Δ [9]. In a number of particle dynamics studies there is usually an upper limit to Δ below which we can consider the process to be stationary. The measurement of $D_{c2}(\Delta)$ gives additional independent measurements for the same maximum Δ.

3. Experiment

The MRI experiment consisted of a 70 mm i.d. cylinder half-filled with 2 mm MRI sensitive particles and rotating at 22 RPM. The MRI experiment was done in a 30 cm, 1.9 T Oxford magnet driven by a TECMAG imager/spectrometer. We took a 20 mm slice and made an axial image with 1mm by 1mm in-plane resolution. The statistical properties of the axial component of velocity fluctuations were measured. This was done by choosing gradients in Fig. 2 to be in the axial direction. The diffusion coefficient and correlation in other directions could have been measured by choosing other gradient directions. The mean velocity in the axial direction was measured to be zero by MRI. The in-plane particle velocities in the horizontal and the vertical direction were also measured by MRI. The three apparent diffusion coefficient images $D_{s1}(\Delta)$, $D_{s2}(\Delta)$, $D_{c2}(\Delta)$ were measured for a fixed $\delta = 2.048$ ms and for Δ equal to 2.75 ms, 6.31 ms, and 8.06 ms respectively. This gives us nine apparent diffusion coefficient images from which to calculate D and τ_c by non-linear least squares fitting. We verified experimentally that equation (17) was satisfied within measurement errors thus giving us some confidence that particle fluctuations are stationary within the measurement time.

In Fig. 3 the main cause of intensity loss in the flowing layer in the diffusion weighted image is the diffusive motion of the particles in the axial direction. It is this sensitivity to diffusive motion which is used to calculate the apparent diffusion coefficients and subsequently the diffusion coefficient and the correlation times for in the axial direction.

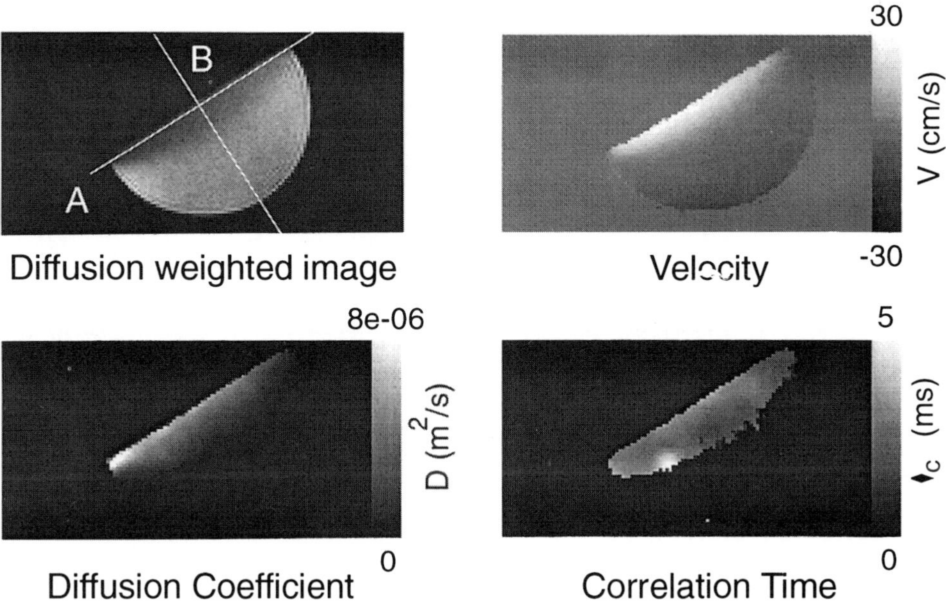

Fig.3. Diffusion weighted, velocity, diffusion coefficient, and correlation time images are shown. The velocity component is parallel to the flowing layer (direction (A)) in the velocity image, while the diffusion coefficient and correlation times are for the axial direction perpendicular to mean velocity. The rotation rate is 22 RPM.

Profiles of diffusion coefficient, velocity, and the correlation time parallel to the flowing layer (direction A, Fig. 3) are shown in Fig. 4 and transverse to the flowing layer (direction B, Fig. 3) are shown in Fig. 5. The velocity profile along the flowing layer (Fig. 4) is a skewed parabola with its maximum shifted downwards. The diffusion coefficient increase from a minimum at the top of the flowing layer to maximum at the bottom. The correlation time shows more complex behavior with local minima and maxima but increases globally along the flowing layer. In Fig. 5 profiles are transverse to

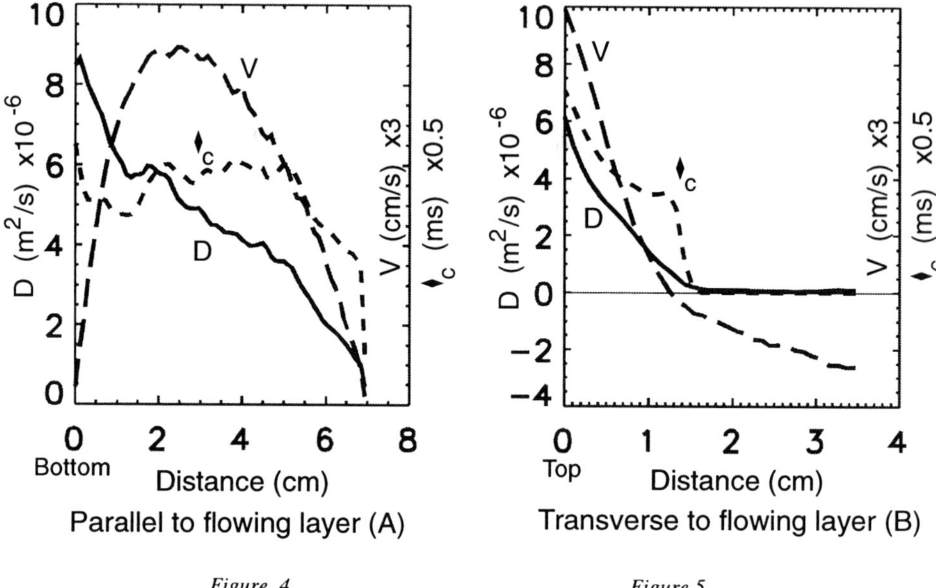

Figure 4   *Figure 5*

the flowing layer (direction B, Fig. 3). The velocity profile is linear in the rigid body motion part and increases to a maximum value almost linearly in the flowing layer. The diffusion coefficient is a very small number in the rigid body motion part and it increases linearly at first and then quadratically as we approach the top of the flowing layer. In the rigid body motion region the correlation time was put to zero as the apparent diffusion coefficient data was small and correlation time calculations not reliable. In the flowing layer it seems to increase quadratically to a maximum at the top of the layer.

At the top and center of the flowing layer (where the two directions intersect, Fig. 3) the velocity is 29 cm/s along the flow direction, the diffusion coefficient is 5.5e-06 m^2/s and a correlation time of 3.7 ms. This gives a granular temperature, $\langle u^2 \rangle = 14.8$ cm^2/s^2. This gives a mean free path of 0.14 mm which is 7% of particle diameter. The self-diffusion coefficient of water and oil is of the order of 10^{-9} m^2/s, which is about three orders of magnitude smaller than the diffusion coefficient for the particles.

4. Conclusions

We have shown that with properly designed MRI experiments it is possible to measure correlation time, diffusion coefficient, and granular temperature in flowing granular systems. The MRI technique can non-invasively probe within the material and gives us the spatial distribution of these quantities in an image like format. This technique does require that particles be NMR sensitive and because several images have to be made before the desired parameters can be calculated, the flow properties should be steady during this time. The total imaging time is a function of signal-to-noise ratio and for these experiments it was of the order of an hour.

5. References

1. Stejskal, E. O. and Tanner, J. E.: Spin Diffusion measurement: Spin echoes in the presence of a time-dependent field gradient, J. Chem. Phys. **42** (1965), 288-292.
2. Callaghan, P. T.: *Principles of Nuclear Magnetic Resonance Microscopy*, Chaps. 6-8, Oxford Univ. Press, 1991, New York.
3. Nakagawa, M., Altobelli, S. A., Caprihan, A., and Fukushima, E.: Non-invasive measurements of granular flows by magnetic resonance imaging, *Exp. Fluids* **16** (1993), 54-60.
4. Hill, K. M., Caprihan, A., and Kakalios, J.: Axial segregation of granular media rotated in a drum mixer: Pattern evolution, *Physical Review E* **56** (1997), 4386-4393.
5. Savage, S. B.: Disorder, diffusion and structure formation in granular flows, in D. Bideau and A. Hansen (eds.), *Disorder and Granular Media*, Elsevier Science Publishers B.V., 1993, pp. 255-284.
6. Savage, S. B. and Dai, R.: Studies of granular shear flows, wall slip velocities, 'layering' and 'self-diffusion', *Mech. Materials* **16** (1993), 225-238.
7. Yamane, K., Nakagawa, M., Altobelli, S. A., Tanaka, T., and Tsuji, Y.: Steady particulate flows in a horizontal rotating cylinder, *Phys. Fluids* **10** (1998), 1419-1427.
8. Menon, N. and Durian, D. J.: Particle motions in a gas-fluidized of bed, *Phys. Rev. Lett.* **79** (1997), 3407-3410.
9. Stepisnik J.: NMR measurement and Brownian movement in the short-time limit, *Physica B* **198** (1994), 299-306.

AXIAL MIGRATION OF TWO SPECIES OF PARTICLES IN A HORIZONTAL CYLINDER

MASAMI NAKAGAWA*, AKINORI AWAZU**
and HIRAKU NISHIMORI**
*Particulate Science and Technology Group
Division of Engineering, Colorado School of Mines
Golden, Colorado 80401
**Department of Mathematical Sciences, Osaka Prefecture University
Sakai, Japan 559-8531

Abstract

Axial segregation in a long horizontal rotating cylinder has generated much interest. Recent developments of the MRI (Magnetic Resonance Imaging) technology have given much needed insights into the dynamics as well as the internal structures of complex behavior. We have seen a number of axial segregation/mixing experiments and simulations but issues associated with the mechanism(s) of the axial segregation are still under investigation. In this paper we simply summarize axial migration phenomena that occur in an almost completely filled rotating cylinder.

1. Introduction

Through a series of non-invasive MRI experiments we observed that the radial segregation occurred along the entire axis of a long cylinder after a few rotations, and it transformed into axial segregation after an extended period of rotation [8]. A great deal of work has been done on radial segregation partially due to its occurrence in a very short time and, consequently, due to its industrial concerns in different particle processes. In 1939, Oyama conducted a landmark experiment on axial segregation and extensively reported the external appearances of the different axial segregation patterns [10].

Previously, most of the experiments were conducted using a half-filled cylinder [2-10]. In gaining further insights in axial segregation, we tried to isolate the influence of the dynamic angle of repose to see if there might be other causes besides the difference in the dynamic angle of repose for the axial segregation. To accomplish this, we almost completely filled the cylinder so that we could virtually eliminate the influence of the dynamic angle of repose. We conducted a series of segregation experiments using almost completely filled cylinder, and to our surprise, we observed

A.D. Rosato and D.L. Blackmore (eds.), IUTAM Symposium on Segregation in Granular Flows, 163–169.

axial migration and segregation as we did before for partially filled cylinder experiments (see Fig.1).

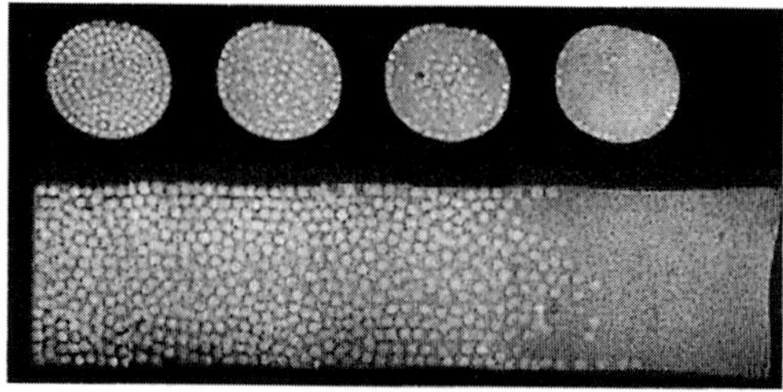

Figure 1. MRI of axial segregation of 1 and 4 mm particles in a rotating cylinder (6.9 cm diameter and 27cm length). Initially these particles were well mixed; however, after an extended rotation period, they axially segregated. Smaller particles were too small to be identified individually and they are seen as gray shaded regions.

Now in the following figures, we will show a similar migration experiment but with a much higher filling ratio. Based on the previous results, we have compelling evidence that the axial migration is not solely due to surface flow that is commonly associated with the granular flow in a partially filled rotating cylinder. We have monitored by MRI that the initially segregated axial bands of small and large particles migrate into each other's region and form annular core structures as shown below in Fig. 2.

Figure 2. Experimental cylinder showing two, three and four initial bands.

Figure 3a shows six cross-sectional views of the initial configuration. One can identify a rather sharp interface between the small and large particles that are initially axially segregated. As can be seen in Fig. 3b, within 1.5 hours of rotation, it is evident that the large particles formed one layer of a surface ring that traveled to reach the other end of the cylinder before the small particles did. The small particles, on the other hand, formed an annular ring and traveled in the opposite direction as the large particles did. For the next few hours, both large and small particles continued to travel in this way. After 24 hours of rotation, the concentration of the large particles in the middle increased but the annular core of the small particles was still intact. During the axial migration process, the large particles advanced into the region occupied by the small particles keeping a flat migrating front. In general, when two viscous fluids penetrating into each other's region, their advancing front becomes round due to shearing action. However, in this slowly evolving axial migration, the advancing front keeps its flat interface until it reaches the other end of the cylinder. This peculiar behavior might be similar when a tooth past is slowly extruded into a pool of less viscous fluid. In that case, the paste advances maintaining the flat traveling front.

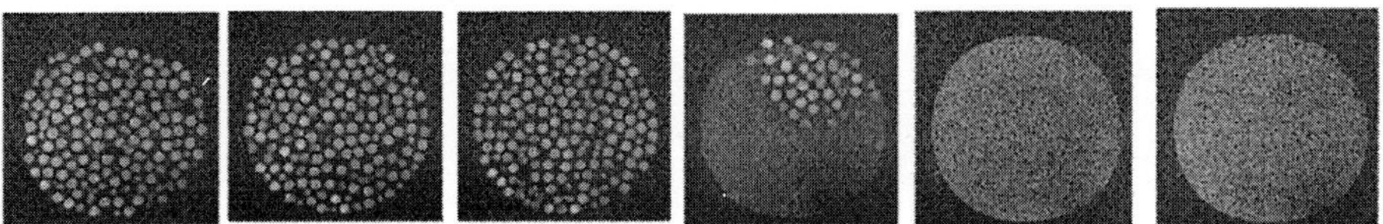

Figure 3a. Cross-sectional views of the initial configuration.

The implication of this finding is that even though these particles are made of the same material and subjected to the same external disturbances, they exhibit vastly different fluid-like behavior due to their size differences.

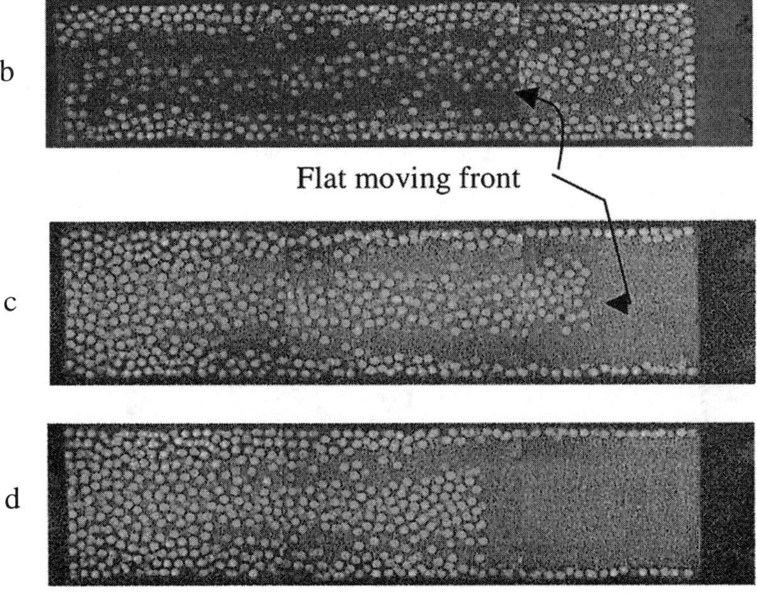

Figure 3. b) 1.5 hours, c) 2 hours, d) 24 hours of rotation

3. Numerical Simulations

In order to capture the details of global behavior we also conducted discrete particle dynamics simulations. Simulations of particle behavior in a rotating two-dimensional disk have been extensively conducted in the past. Axial migration behavior discussed above is inherently three-dimensional and a similar discrete dynamic simulation in three-dimensional is very time consuming. Therefore, we select a thin computational cell that rotates about the cylinder axis (see Fig. 4). This cell is almost completely filled with particles to reproduce the earlier experiment. The dynamics of a typical particle i can be described in the following equations:

$$\ddot{r}_i = -\sum_{j=1}^{N} \theta\left(d_i + d_j - \left|r_i - r_j\right|\right)\left\{\nabla V(d_i + d_j - \left|r_i - r_j\right|) + \eta(v_i - v_j)\right\} + F_i \tag{1}$$

$$V\left(d_i + d_j - \left|r_i - r_j\right|\right) = k\left(d_i + d_j - \left|r_i - r_j\right|\right)^2 \tag{2}$$

where di and r_i are the diameter and the position vector of the ith particles, respectively. This collision model used by Taguchi in his simulation [12] was adopted and it contains a viscous drag term that is proportional to the relative particle velocity v_i- v_j. F_i is the

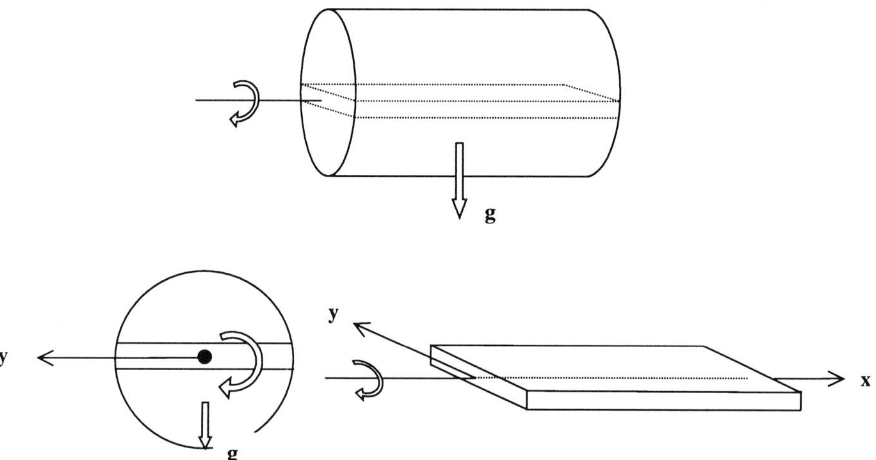

Figure 4. Computational cell used in the discrete element simulation

external forces acting on the ith particle. The first term in (1) is a step function to represent a particle contact. The particle size ratio used in this simulation is 1:2 and the number ratio is 4:1.

Figure 5 shows two different radial segregation structures depending on the non-dimensionalized acceleration defined as $A\omega^2/g = \Gamma$ where A is the cylinder radius and ω is the rotation rate. When $\Gamma < 1$, then the larger particles go outside leaving the smaller particles behind in the core region. However, this trend reverses when $\Gamma > 1$.

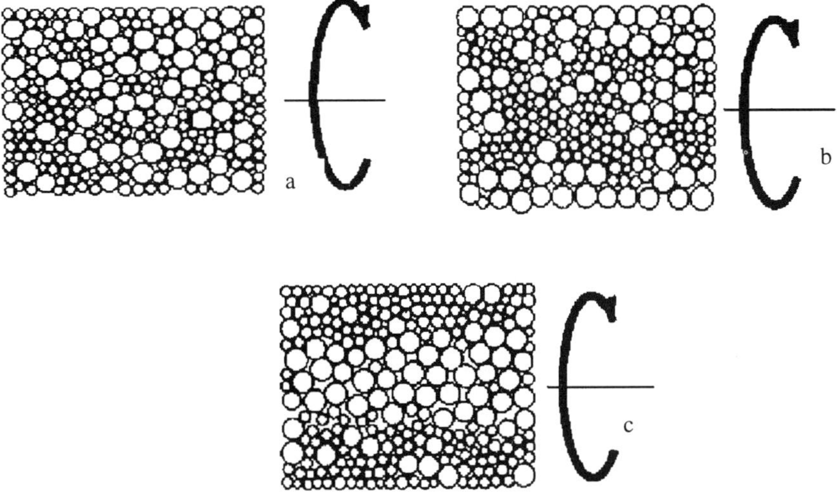

Figure 5. (a) initial packing; (b) $\Gamma < 1$; (c) $\Gamma > 1$.

We continued to rotate the cell after it showed radial segregation and found that there was always a fluctuation in the particle concentration in the axial direction. In order to investigate this, the number of particles was doubled to 500. In Fig. 6 fluctuations in the axial concentration are shown for the different simulation sizes.

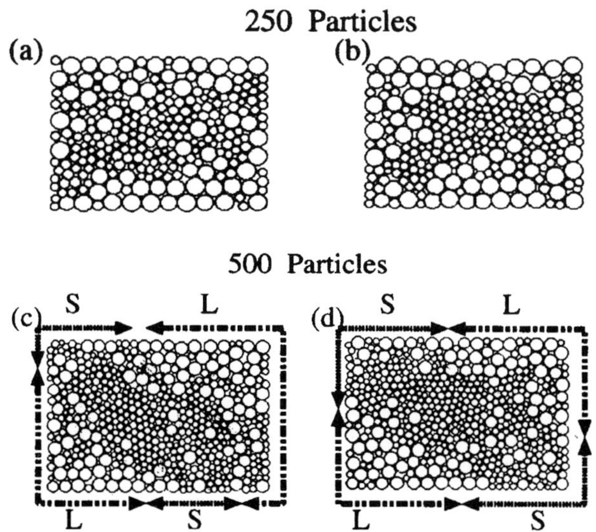

Firgure 6. Species fluctuations in the axial direction.

Figure 7 shows the particle trajectories at the initial stage before radial segregation was formed. A typical particle exhibits a very confined movement with local fluctuation. On the contrary, after forming non-uniform concentration in the axial

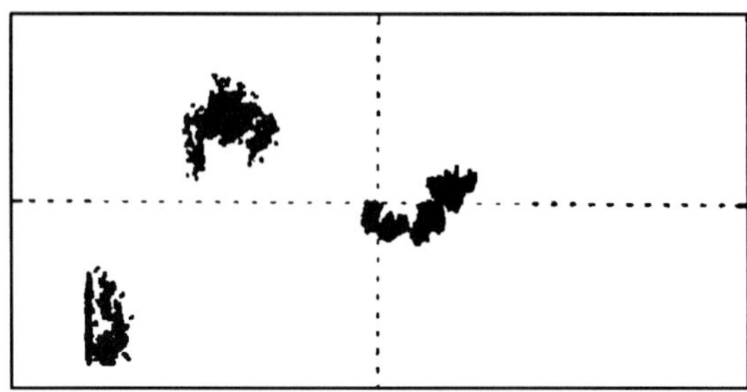

Figure 7. Trajectories of typical particles before radial segregation is formed.

direction, the particles show much more global convective motion as can be seen in Fig. 8. Figures 8a, b and c show the motion of small particles in the inner core, that of small particles circulating from inner to outer core and that of large particles circulating clockwise, respectively.

For the first time, a global circulatory movement of the particles was observed. At this stage, we need to confirm if a steady convective motion like the one shown here is possible in the long horizontal cylinder or not.

4. Concluding Remarks

We also conducted MRI measurements of the interface axial propagation of two different size particles in a half-filled cylinder [11]. In that experiment, we rotated 1mm and 4 mm particles in a cylinder at 11.4 rpm. Detailed images will be shown at the time of presentation, however, judging from the MRI data, the propagating front is round and cone-shaped. The cascading layer is almost completely occupied by the large particles whereas the small particles do not seem to surface at all. The distance of the tip of the migrating front of small particles and the free surface remains more or less constant. At the moving interface, due to the percolation mechanism, the smaller particles travel under the free surface and move in the axial direction driven by the concentration gradient. The smaller particles advance more easily into the region occupied by larger particles since there are more voids in the cascading region for the small particles to move into. We also observed that the large particles traveling on the surface always reached the other end of the cylinder before the small particles did. Eventually, this also lead to an extended radially segregated core.

Upon comparing the shape of the propagating front between the partially and almost completely filled cases, one can suspect that different mechanism(s) are dominant in different cases. We also introduced a direct numerical simulation currently being conducted by Nishimori [1] and their preliminary results suggest that there is a secondary convective flow along the axial direction.

5. Acknowledgements

The author was supported in part by NASA through Contract No. NAG3-1970. The author would like to acknowledge his students' dedication in carrying out these experiments with him. They are Jamie L. Turner and Jennifer M. Reiter. He also acknowledges the member of Particulate Science and Technology Group at Colorado School of Mines for their fruitful discussions. Finally, he would like to thank the NMR (New Mexico Resonance) group for their valuable supports in conducting MRI experiments.

References

1. Awazu, A., and Nishimori, H.: Private communications (1999).
2. Bridgewater, J.: Fundamental powder mixing mechanisms, *Powder Technol.* **15** (1992), 215-236.
3. Chicharro, R., Peralta-Fabi, R., and Velasco, R.M.: Segregation in dry granular systems, in Behringer & Jenkins (eds), *Powders & Grains 97*, Balkema, 1997, Rotterdam, pp. 479-482.
4. Choo, K., Molteno, T.C.A. and Morris, S.W.: Traveling granular segregation patterns in a long drum mixer, *Phys. Rev. Lett.* **79** (1997), 2975-2978.
5. Clement, E., Rajchenbach, P., and Duran, J.: Mixing of a granular material in a bidimensional rotating drum, *Europhys. Lett.* **30** (1995), 7-12.
6. Donald, M.B. and Roseman, B.: Mixing and demixing of solids particles-I. Mechanics in a horizontal drum mixer, *Br. Chem. Eng.* **7** (1962), 749-753.
7. Hill. K.M., Caprihan, A., and Kakalios, J.: Bulk segregation in rotated granular materials measured by magnetic resonance imaging, *Phys. Rev. Lett.* **78** (1998), 50-53.
8. Nakagawa, M., Jamie. L.M., and Altobelli, S.A.: Segregation of granular particles in a nearly packed rotating cylinder: a new insight for axial segregation, in H.J. Herrmann et al.(eds.), *Physics of Dry Granular Media*, Springer, 1998, pp. 703-710.
9. Nakagawa, M.: Axial segregation of granular flows in a horizontal rotating cylinder, *Chem. Eng. Sci.* **49** (1994), 2544.
10. Oyama, Y.: *Bull. Inst. Phys. Chem. Res. Jpn. Rep.* **18** (1939), 600.
11. Ristow, G.H. and Nakagawa, M.: "Shape dynamics of interfacial front in rotating cylinders." *Phys. Rev. E* **59** (1999), 2044-2048.
12. Taguchi, Y-h. (1992) "New origin of a convective motion: elastically induced convection in granular materials." *Phys. Rev. Lett.* **69** (1992), 1367-1370.

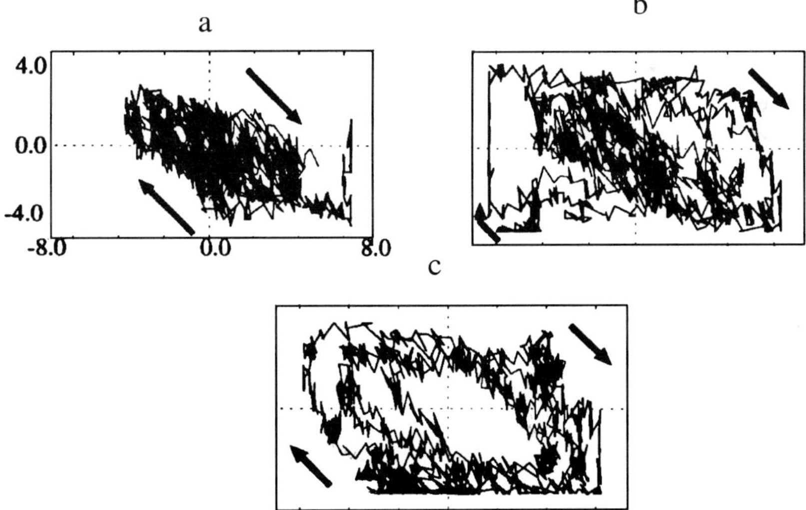

Figure 8. Trajectories after having formed nonuniform axial concentration.

SIGNATURES OF CHAOS IN 2D TUMBLING MIXERS

Interplay between Chaos, Segregation, and Mixing

K.M. HILL[1], J.F. GILCHRIST[1], D.V. KHAKHAR[2], J.M. OTTINO[1]
[1] *Department of Chemical Engineering*
Northwestern University, Evanston, IL 60208
[2] *Department of Chemical Engineering*
Indian Institute of Technology, Bombay 400076, India

Abstract

A prototypical problem in segregation of granular materials is the radial segregation of different particles in a rotating drum mixer. When particles differing in size (or density) are rotated in a partially filled circular drum, the smaller (denser) material quickly segregates toward the axis of rotation. In long rotating drum mixers, this is often followed by segregation into bands along the axis of rotation, the so-called axial segregation. Radial segregation has been viewed as a necessary predecessor to axial segregation and this has motivated numerous studies of radial segregation in two-dimensional mixers. Studies to-date have been restricted to circular cylinders. We have performed segregation experiments in circular *and non-circular* two-dimensional mixers; the new element is that non-circular mixers induce chaotic advection from a continuum viewpoint and chaotic advection interacts in non-trivial ways with density and size segregation. In non-circular mixers the final state of segregation consists of patterns that resemble computationally derived Poincaré sections for the mixers, with islands acting as attractors for the smaller or denser particles depending on the mix. The patterns are sensitive both to the shape of the mixer and to the fill level, particularly near 50%, where several bifurcations occur. Furthermore, there is a sensitivity to the type of mixture (different size or different density), particularly when the mixer is filled just over 50%, where certain flow instabilities can be reinforced.

1. Introduction

One canonical system for studying segregation of granular materials due to shear flow is a circular cylinder. Segregation occurs in the top flowing layer and the cumulative effect leads to segregation perpendicular to the axis of rotation, *radial segregation*. One of the most striking examples of segregation is the separation of different components of a mixture into bands of relatively pure monodisperse materials along the axis of rotation, known as *axial banding* or *axial segregation*. Axial segregation is complex and imperfectly understood. In some cases, band formation is dependent on the speed of

A.D. Rosato and D.L. Blackmore (eds.), IUTAM Symposium on Segregation in Granular Flows, 171–180.
© 2000 *Kluwer Academic Publishers. Printed in the Netherlands.*

rotation, and when the speed is reduced the bands remix, i.e., the systems exhibit a reversible axial segregation [1]. In some cases, the initial pattern of bands is not stable, but continues to evolve in a complex fashion [2,3]. Yet, in other cases the particles do not segregate into bands at all. However, in almost all cases radial segregation precedes axial banding. Thus, it is commonly believed radial segregation is a necessary predecessor to axial segregation [4,5].

This has motivated a number of studies of segregation and mixing in two dimensional cylinders [6-8]. To our knowledge all studies to date have been conducted in circular containers. When non-cohesive materials are used the flow is steady and independent of time. As it turns out, this is a very special case [9].

In contrast, in non-circular mixers the flowing layer is not constant but grows and shrinks in time with a frequency no less than twice that of the mixer rotation. Thus, the flow is time-periodic and a non-linear mapping may be used to represent the particle motion. Repeated iterations of the mapping can induce chaotic trajectories.

2.   Mixing and segregation in two dimensional containers – a basic model

Mixing and segregation in 2-d drum mixers depends on many factors including the type of components involved, the speed of rotation, the filling amount, and the mixer shape. Segregation and mixing experiments have shown that all can be important components for the quality, degree, and speed of segregation. Important flow differences have been found when properties of the materials are varied, such as their cohesiveness [10] and shape [11]. When the speed of rotation is changed, different flow patterns are observed, from intermittent flow (periodic avalanching) or slumping at very slow speeds, to a steady surface flow at intermediate speeds (the continuous or rolling regime) to centrifugal flow at very high speeds. The fill level alters the speed of mixing. This happens in both avalanching flow [12] and continuous flow [13]. In this study we are interested primarily in studying the effect of *mixer shape* on segregation. We restrict our studies to non-cohesive spherical materials in the rolling regime.

The flow model is shown for a circular mixer in Fig. 1. The free surface has length *2L* and the flowing layer has thickness δ(*x*) that varies along the length. A theoretical

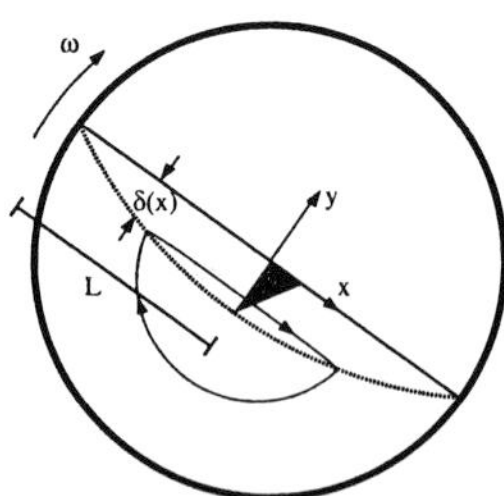

Fig. 1 Schematic view of continuous flow in a rotating cylinder. The dotted curve denotes the interface between the continuously flowing layer and the region of solid body rotation. The depth of the flowing layer is exaggerated. The mixer is rotated with angular velocity, ω and the velocity profile within the layer is nearly simple shear. A typical particle trajectory is shown, illustrating that without collisional diffusion the particles simply circulate around the mixer.

analysis indicates that $\delta(x)$ can be approximated by a parabola [13]

$$\delta = \delta_0(1-(x/L)^2) \, , \tag{1}$$

where δ_0 is the thickness at $x = 0$. Material is fed into the flowing layer for x<0 and leaves the layer for x>0. The velocity field in the flowing layer is given by:

$$v_x = 2u(1 + y/\delta) \tag{2}$$

$$v_y = -\omega x(y/\delta)^2 \tag{3}$$

where u is the average velocity at the center of the flowing layer and is given by

$$u = \omega L^2/(2\delta_0) \, . \tag{4}$$

The v_y-component is chosen to satisfy mass conservation. This model captures the essence of flow in the layer. Additional computations indicate that different velocity fields, for example $v_x \sim y^{3/2}$, $v_y \sim xy^{5/2}$, give essentially the same mixing patterns, demonstrating that geometrical effects (i.e. the shape of the container) control the most of the physics.

The case of the cylinder is relatively uninteresting. Figure 2.a shows a Poincaré section or stroboscopic plot for the circular mixer. As the flow is steady (i.e. it is not a function of time), the "streamlines" or circulation paths are time-invariant, the circulation paths are concentric paths. Without diffusion (which will be considered shortly) mixing takes place along circulation paths or streamlines.

On the other hand, in a non-circular mixer the flowing layer is not constant. For example, the length of the flowing layer $L(t)$ for a square mixer is expressed as

$$L(t) = \begin{cases} \dfrac{1}{|\cos\theta|} & \text{if } \theta < \dfrac{\pi}{4} \text{ or } |\pi - \theta| < \dfrac{\pi}{4} \text{ or } \theta < \dfrac{7\pi}{4} \\[2ex] \dfrac{1}{|\sin\theta|} & \text{otherwise,} \end{cases} \tag{5}$$

where $\theta = \omega(t)$. Experiments show that the layer maintains geometric similarity, so that $\delta_0(t)/L(t)$ may be approximated to be a constant. This implies that $u \sim L$; thus the speed u changes with mixer orientation; the longer the layer, the faster and deeper the flow. From this, equations (1)-(4) can be modified to include the mixer shape.

Figure 2b shows a Poincaré section for the square mixer, significantly different from that of the circle. Regular islands, from which the particles cannot escape without diffusion, are surrounded by a particle trajectory marked with black dots. A *chaotic sea*, where the most effective mixing can occur [14] dominates the circulation pattern. The Poincaré section is sensitive to fill level (2c, 2d), as will be discussed.

To include collisional diffusion we use the result obtained by Savage [15]:

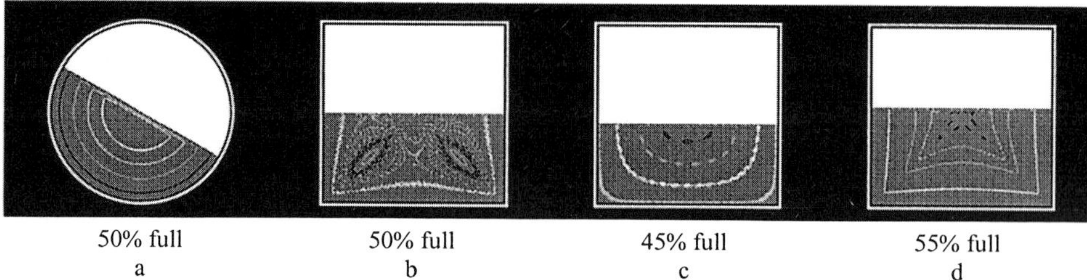

50% full	50% full	45% full	55% full
a	b	c	d

Fig. 2: Poincaré sections or stroboscopic plots for particles in mixers, fill level as noted. Each shade of grey represents one initial starting condition for a particle. After the starting position is plotted for each particle, one dot is plotted every ½ revolution for the circle and every ¼ revolution for the squares. As is standard in fluid mechanics, the flow is interpreted in a continuum sense and *collisional diffusion* – the random-like motion of individual particles as they collide with other particles – is ignored in these particle trajectory calculations.

$$D_{coll} = f(\eta)d^2 \frac{dv_x}{dy}, \qquad (6)$$

where D_{coll} is the granular self-diffusivity in shear flow. Here, dv_x/dy is the velocity gradient across the layer, and d is the particle diameter. The prefactor $f(\eta)$ is a function of the solids volume fraction, η; in our simulation we take $f = 0.025$ as calculated by Savage [15] from particle dynamics simulations for spherical particles. In mixers with circular cross sections, collisional diffusion is the only form of radial mixing. In square mixers, the combination of collisional diffusion and chaotic advection provides a more efficient mixing mechanism than does diffusion alone in a circular mixer, measured for systems of identical particles by Khakhar et al. [16]. The interaction between chaos and the tendency for different particles to segregate, however, is more complicated.

3. Experiment

To determine the effects of the changing mixer shapes on segregation in granular materials, experiments are conducted using a variety of non-cohesive spherical beads (Quackenbush Co.®) with sizes ranging from 0.8 to 2 mm and densities of 2.5 and 7.8 g/cm³ (glass and steel, respectively). We use elliptical, square and circular mixers, though only the results from the circular and square mixers are reported here. Both have a depth of 6 mm so that a quasi-two-dimensional flow is established, and both have surface area of ~600 cm². The faceplate of the mixer is Plexiglas while the rear plate is fashioned of aluminum and can be grounded to minimize electrostatic effects.

To study the equilibrium structures that arise in these segregating systems, particles differing in size or density are initially well mixed. A computer-controlled stepper motor (Compumotor®) rotates the mixers at approximately 1 revolution per minute so that a steady flow with a flat free surface is developed. The Froude number $F = \omega^2 L/g \approx 0.0002$ for the circle. Images were obtained with a Kodak® CCD camera for quantitative image analysis.

4. Observations

4.1. EQUILIBRIUM STRUCTURE – HALF FULL

As seen in Fig. 3, the shape of the container plays a significant role in determining the equilibrium segregation structures for the different systems. Equilibrium segregation structures for half-filled mixers are shown in for systems of spheres differing in density (3a, 3b) and size (3c). As previously observed, for the classic radially segregated structure in the circular mixers (3a) the dynamics in the flowing layer cause the more dense (or smaller) particles to move to lower levels in the layer leading to a segregated *time-invariant* core region that coincides with the streamlines.   However, the equilibrium segregation structures obtained in non-circular mixers do not follow this simple rule. In the square mixer (3b, 3c), the region of highly concentrated dense (or small) particles is also located away from the center of rotation and is nearly separated into two separate regions. The precise shape of the segregated region is periodic in time and depends on the instantaneous orientation of the mixer. The segregated regions for half-filled systems are similar to the regular islands in the Poincaré sections (2b, 2c). They are not identical though; for example, there is some asymmetry of the segregated structures. This will be discussed shortly.

4.2. TIME DEPENDENCE

The development of the mixed systems into the complicated segregated structures is as follows.   Rather quickly, apparent within a half rotation, a well-mixed system segregates into a state resembling radial segregation (see Fig. 4). This is true for all cases studied, whether or not the patterns move beyond the radially segregated structure. For the more complicated structures shown, the radially segregated pattern evolves more slowly – generally within three to five rotations – until the patterns resembles the Poincaré sections at the corresponding orientation. This pattern remains stable for tens of rotations and becomes the final state of segregation.

These results indicate that segregation patterns are influenced both by segregation in the flowing layer and also by the underlying Poincaré sections. Radial segregation

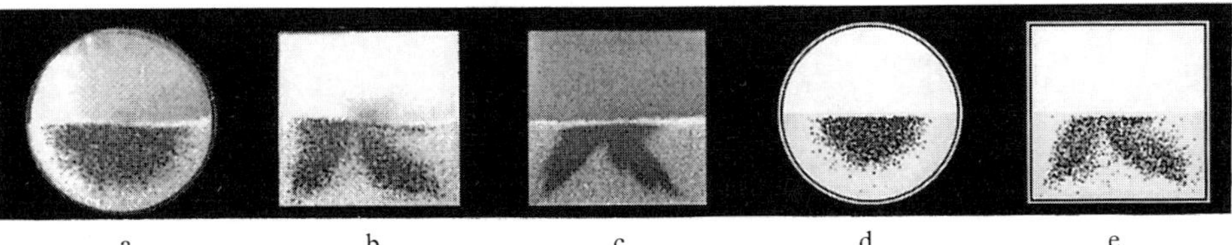

a   b   c   d   e

Fig. 3: Equilibrium segregation structures for half-full quasi 2d tumbling mixers. The mixers in Figs. 3a, 3b contain a 25/75 mixture of 2mm steel (dark) and glass (light) beads. Fig. 3c contains a 25/75 mixture of 0.8 mm (dark) and 2.0 mm (light) glass beads. Results for different sized beads in a circular mixer are similar to 3a. Results from a simulation for different density particles are shown in 3d, 3e. (Similar results are seen for an elliptical mixer (Hill et al. 1999)).

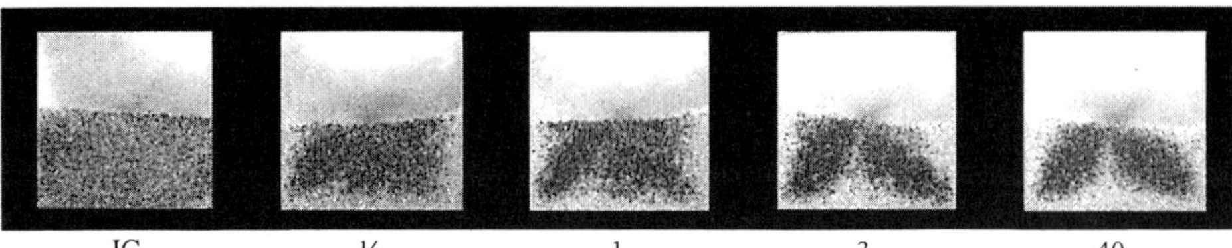

IC   ½   1   3   40

Fig. 4 Images taken as a binary mixture of 2mm glass and steel beads evolves from its initially well-mixed state (IC) to its final segregation pattern in a half-filled square mixer. The number of rotations completed at a speed of 1 rpm is noted below each image.

occurs almost immediately upon initial rotation for all systems studied, and, therefore we argue that this is due to the same segregation mechanism in the flowing layer. In circular mixers, the heavier particles simply sink to lower levels in the layer leading to a segregated time-invariant core region that coincides with the streamlines. Since the shape of the mixer is rotationally invariant, the important dynamics lie solely in the differential flow leading to this radial segregation. In the non-circular mixers, however, the dynamics depend on the orientation of the mixer, and, therefore, the initial radial structure is not invariant. Regular regions in the Poincaré section become attractors for the denser particles. Thus, the heavier particles that sink into islands are likely to be mapped back to the same region; hence as the system evolves, the segregated islands around the elliptical points grow and become more distinct.

This hypothesis is supported by computational results. It has been shown previously that radial segregation of different density particles in a circular drum can be modeled by using an effective buoyancy force causing a differential velocity of the particles in the direction normal to the flowing surface [8] The segregation velocity for the more dense particles, for example, is

$$v_{1y} = -2\beta(1-\overline{\rho})D_{coll}(1-f)/d \ . \tag{6}$$

Here, D_{coll} is the collisional diffusion (for details see [8]); $\beta \approx 2$ is the dimensionless segregation velocity determined by fitting Monte Carlo and "soft-particle" computations; f is the number fraction of the more dense particles; $\overline{\rho} < 1$ is the density ratio, and d is the particle diameter. (Note: segregation is considered important only in the y-direction, normal to the layer; convection dominates in the x-direction.) Incorporating the segregation dynamics into the details of the flow described in Section 2 yields equilibrium segregation patterns shown in Figs. 3d, 3e. The strong correlations between the results from the segregation model and the experiments indicate there is a good basis for the argument combining chaos and segregation to describe the final mixing pattern. The results indicate that segregation has an unanticipated benefit: heavier particles tag the regular regions in the Poincaré sections.

4.3 FILLING LEVEL DEPENDENCE

Both the Poincaré section and the equilibrium segregation structures in the square mixer depend strongly on the degree of filling, particularly about the half-full level. (The pattern in the circular mixer remains essentially one of circulation paths concentric with that of the mixer.) Figures 2c and 2d illustrate that several bifurcations occur in the

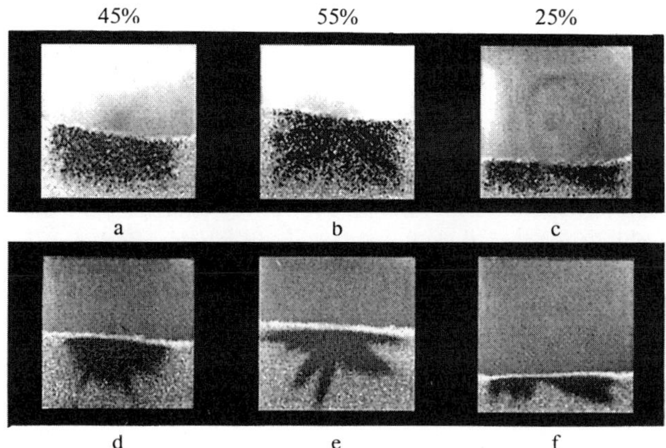

Fig.5 Segregation structures for different filling fractions for 2mm steel and glass beads (a-c) and 0.3 mm and 2.0 mm glass beads (d-f). Filling fractions are noted at the top of the columns. The tendency for the mixtures of different-sized materials to form irregular streaking patterns is apparent at all filling levels, but is particularly noticeable when the mixer is 55% full. An equilibrium is never reached for these structures, but continues to change with continued rotation.

Poincaré section for the square mixer near the half full level. The locations of the regular regions (around the periodic points) are particularly sensitive to fill level about 50%. Figure 5 illustrates some of the sensitivity of the segregation structure to the filling level. For mixtures of spheres differing only in density the complicated segregation patterns disappear at just below (45%) and just above (55%) the half-full level. In these cases, the patterns revert to the simple radial structure one might expect from earlier experiments with the circular mixers. As the filling fraction moves farther away from the half full level, the radial segregation pattern is maintained until the filling level is ~ 25% full or ~75% full. For these filling fractions, the segregated islands return to the equilibrium segregation patterns (shown in 5c for 25% full). For filling fractions of 25% and 75%, the circulation patterns, that is the Poincaré sections (not shown), indicate a return of the regular islands partway between the corners of the square and the center of the flowing layer.

For mixtures of spheres differing in only in size, an additional instability is apparent. At 45%, the segregation structure is similarly radially segregated but there is some evidence of additional streaks in the segregation structure (5d). These are particularly striking when the mixer is filled ~55% (5e), where a striping pattern dominates. We believe this represents additional phenomena, unrelated to the Poincaré sections, as will be discussed below. Further from the half full level, the segregation patterns return to those more closely resembling the equilibrium segregation structures for different density (same size) mixtures, as shown in 5f, though some streaks are always apparent.

Analysis

Two types of analyses have been performed for these systems. One of the most common measures of mixing is the standard deviation of the concentration fluctuations about the mean concentration – the so-called intensity of segregation. For all the systems reported here, the value of the standard deviation of the mean quickly rises over the first half revolution and then stabilizes (Fig. 6). In other words, this measure of segregation captures the initial radial segregation but, unfortunately, indicates that all systems considered segregate nearly equally. Thus, this measure fails to capture any of the aspects arising from spatial structure. In Fig. 6, for example, the standard deviation of the mean is graphed for the pattern development in a half-filled circular mixer and for a square mixer at different filling amounts. While the initial value for the circular mixer is higher than the values for the square, they all appear to stabilize to approximately the same equilibrium value.

Figure 7 demonstrates that a different method of analysis detects some differences in pattern development. In this graph we have plotted p^2/a for the same systems plotted in Fig. 6. Here, p corresponds to the length of an iso-concentration line $c(x,y)$=K, where K is a suitable threshold that outlines the segregated structures; a is the area enclosed by $c(x,y)$. (We have done this for K=0.75, i.e., for regions of segregation containing 75% or greater concentration of dense beads.) Thus, p^2/a is a non-dimensional measure of this segregated structure. For a perfect semi-circle this value is $p^2/a=(2/\pi)(2+\pi)^2 \approx 17$. Any deviation from straight radial segregation will increase the value of p^2/a. As shown in Fig. 7, this measure yields different results for the distinct patterns formed by different filling levels of the square. The value for the square filled 3/8 full and 7/16 full and a circle remains constant, between 20 and 25. The equilibrium values for a quarter- and a half- filled square rise to nearly 40. The value of this measure oscillates depending on the orientation of the mixer; the measurements in this graph were done with the orientation of the mixer shown in Fig. 4.

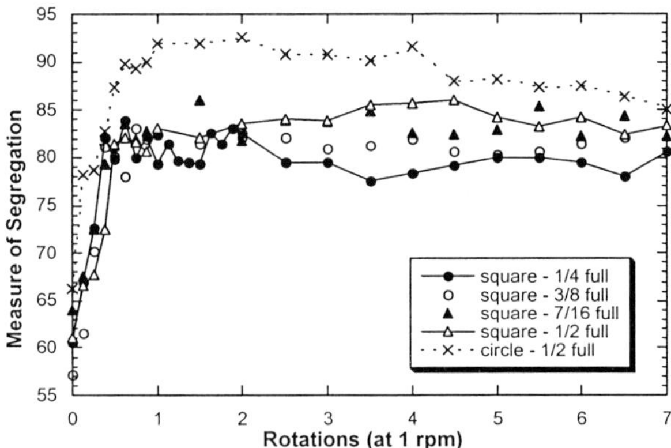

Fig. 6: Standard deviation of the mean concentration for square mixers at different fill levels and for a half –full circular mixer. The system measured consisted of 25% 2mm steel spheres and 75% 2mm glass spheres.

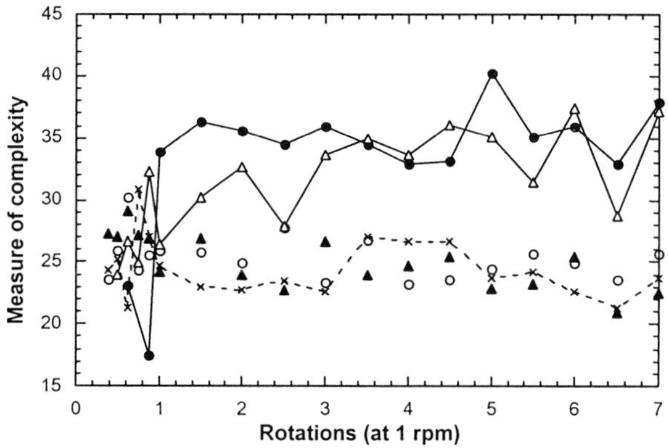

Fig. 7: Measurements taken for the same systems in Fig. 6. The key is the same as noted in that graph. The data were measured with a non-dimensional measure of the pattern complexity – p^2/a described in the text. Some differences between the more complicated structures in the ¼ and ½ filled square and the radial structure in the circle and other filling levels in the square are apparent in this measurement.

6.   Discussion

The experiments described above indicate that the previously observed radial segregation patterns are the exception rather than the rule when the mixer shapes are varied. The segregation patterns in rotating mixers are influenced both by segregation in the flowing layer and also by the underlying Poincaré sections. This hypothesis is supported by computational results obtained using a model that incorporates only diffusion and the differential flow in the flowing layer with various mixer shapes. As pointed out earlier, there are some discrepancies between the Poincaré sections and the equilibrium segregation patterns. One of the most prominent is the asymmetry in the segregation structures at the half-full level that is not present in the Poincaré section. This is due to that fact that the Poincaré sections are computed for a system of identical particles (i.e., a continuum description of the granular flow without segregation), while segregation experiments are necessarily done with systems of different particles. When mixtures of particles of different densities are used, the trajectory of the heavier particles deviates from this "base case" owing to their segregation velocity, and this deviation is always downward and backwards in the flowing layer: this leads to the asymmetry observed experimentally. Asymmetric structures are also obtained in the computational model including segregation.

The segregation model described in section 4.2 was developed for systems of equal sized, different density particles. Similar expressions can be obtained for the case of systems differing in size, at low solids volume fractions [8], though the system is clearly more complicated. The use of different-sized particles introduces additional important physical effects; unlike the case of identical or equal size particles, the flow in the layer depends on the concentration of the different particles present. For example, if the flowing layer consists entirely of small particles, it is thinner and moves faster; conversely larger particles form a deeper and slower flowing layer. In other words, the velocity field is now coupled to the composition of the particles in the layer.

A manifestation of the complexity introduced by the coupling between velocity and concentration is the radial streak structure shown in Fig. 5e. This instability appears to be independent of the underlying Poincaré section - it occurs in mixers of all shapes - and is left for future research.

One might ask to what extent these results apply to the full 3D case – a long circular "drum mixer" where axial segregation is typically observed. Previously, it has been suggested that radial segregation (observed in the circular mixer) is a necessary precursor to axial segregation. Since in the 2D square mixer we do not observe radial segregation for more than one revolution, this would suggest that either axial segregation should not exist in a 3D square mixer, or the two-arm structure shown in Figs. 2 and 4 should be present *only* in the 2D case. However, neither is true. We conducted experiments in 30" long drum mixers with a 5-inch *square* cross-section. When the degree of filling is ½, not only did axial segregation occur, but the axial segregated patterns show unmistakable signs of the two arm structure. Thus it appears that radial segregation is not a necessary condition for axial segregation. The full implications of the 2D patterns discovered here on mixing in long cylinders, however, await full investigation.

7.   Acknowledgments

This work was supported by the Department of Energy, Division of Basic Energy Sciences, and the National Science Foundation, Fluid, Particle, and Hydraulic Systems.

1.   Hill, K.M. and Kakalios, J.: Reversible axial segregation of rotating granular media, *Phys.Rev.E* **52** (1995) 4393-4400.
2.   Choo, K., Molteno, T.C.A, Morris, S.W.: Traveling granular segregation patterns in a long drum mixer, *Phys. Rev. Lett.* **79** (1997) 2975-2978.
3.   Hill, K.M., Caprihan, A., Kakalios, J.: Axial segregation of granular media rotated in a drum mixer: Pattern evolution, *Phys. Rev. E* **56** (1997) 4386-4393.
4.   Donald, M.B. and Roseman, B.: Mixing and demixing of solid particles. Part 1. Mechanisms in a horizontal drum mixer, *Br. Chem. Eng.* **7** (1962) 749.
5.   Das Gupta, S., Bhatia, S.K. and Khakhar, D.V.: Axial segregation of particles in a horizontal rotating cylinder, *Chem. Eng. Sci.* **46** (1991) 1513-1517.
6.   Cantelaub, F. and Bideau, D.: Radial segregation in a 2d drum: An experimental analysis, *Europhys. Lett.* **30** (1995) 133-138.
7.   Clement, E., Rajchenbach, J., Duran, J.: Mixing of a Granular Material in a Bidimensional Rotating Drum *Europhys. Lett.* **30** (1995) 7-12.
8.   Khakhar, D.V., McCarthy, J.J., and Ottino, J.M.: Radial segregation of granular mixtures in rotating cylinders, *Phys. Fluids* **9** (1997) 3600-3614.
9.   Hill, K.M., Gilchrist, J.F., Khakhar, D.V., McCarthy, J.J. and Ottino, J.M.: Segregation-Driven Organization in Chaotic Granular Flows, *Proc. Nat. Acad. Sci.* **96** (1999) 11701-11706.
10. Shinbrot, T., Alexander, A., and Muzzio, F.J.: Spontaneous chaotic granular mixing, *Nature* **397** (1999) 675-678.
11. Behringer, R.P.: The dynamics of flowing sand. *Nonlinear Sci. Today* **3** (1993) 1-15.
12. Metcalfe, G., Shinbrot, T., McCarthy, J.J., and Ottino, J.M.: Avalanche mixing of granular materials, *Nature* **374** (1995) 39-41.
13. Khakhar, D.V., McCarthy, J.J., Shinbrot, T. and Ottino, J.M.: Transverse flow and mixing of granular materials in a rotating cylinder, *Phys. Fluids* **9** (1997a) 31-43.
14. Ottino, J.M.: *The Kinematics of Mixing: Stretching, Chaos, and Transport*, Cambridge Univ. Press., Cambridge, U.K., 1989.
15. Savage, S.B.: Disorder, diffusion and structure formation in granular flow, in D. Bideau and A. Hansen (eds.), *Disorder and Granular Media*, Elsevier Science, Amsterdam, 1993, pp. 255-285.
16. Khakhar, D.V., McCarthy, J.J., Gilchrist, J.F. and Ottino, J.M.: Chaotic mixing of granular materials in 2D tumbling mixers, *Chaos* **9** (1999) 195-205.

SIZE SEGREGATION IN CYLINDRICAL HORIZONTAL COUETTE FLOWS OF PARTICLES: EXPERIMENTS AND SIMULATIONS

A. KARION and M.L. HUNT
Division of Engineering and Applied Science
California Institute of Technology
Pasadena, CA 91125

Abstract

The dynamics of a granular mixture can differ significantly from those of a same-size flow. For example, shearing of granular materials often leads to unwanted segregation of particles of different sizes. This research uses both experiments and discrete element computer simulations to study the segregation occurring in a cylindrical Couette flow. Both experiments and simulations show a complex segregation pattern in which the concentration of large particles varies with both the radial coordinate and the angular location in the cylinder. The large particles rise to the free surface at the top of the cylindrical container and also line the outer cylinder. The number of layers of large particles around the stationary outer cylinder varies with the angular coordinate.

1. Introduction

Granular mixtures of particles with a difference in grain size are used in many industrial applications. The study of mixtures is important in order to investigate the segregation that often occurs in flows of different-sized grains. The manner in which a flow will segregate is dependent on the flow geometry and its orientation with respect to gravity, because segregation is often gravity-induced. In this study, three-dimensional discrete element computer simulations are used in conjunction with experiments to investigate segregation in a horizontally oriented Couette flow.Computer simulations have been used in the past to study various bounded shear flows [1-3] and have also been shown to capture the qualitative effects of segregation in various geometries, including rotary kilns [4], vibrating beds [5], and chute flows [6]. Segregation has been a popular subject of experimental research as well. Segregation of particle mixtures have been studied experimentally in hoppers [7], horizontally rotating cylinders [8], and inclined chutes [9]. Vertically-oriented Couette flow experiments similar to the current study have been performed previously: Buggisch and Löffelmann [10] studied the self-diffusion

A.D. Rosato and D.L. Blackmore (eds.), IUTAM Symposium on Segregation in Granular Flows, 181–191.

of rods in a concentric cylinder Couette flow experiment; Tardos et al. [11] investigated stresses in a fluidized powder in a vertical Couette device; and Khosropour et al. [12] studied size segregation in a vertical Couette device of glass beads. In this last study, the trajectories of single 1, 2 and 3 mm spherical glass beads were observed as they rose to the top of a flow of 1 mm beads in a Couette device similar to that used in the current experiments. Unlike these three experimental Couette flow studies, the current Couette flow apparatus is oriented horizontally so that the cylinders are perpendicular to the direction of gravity.

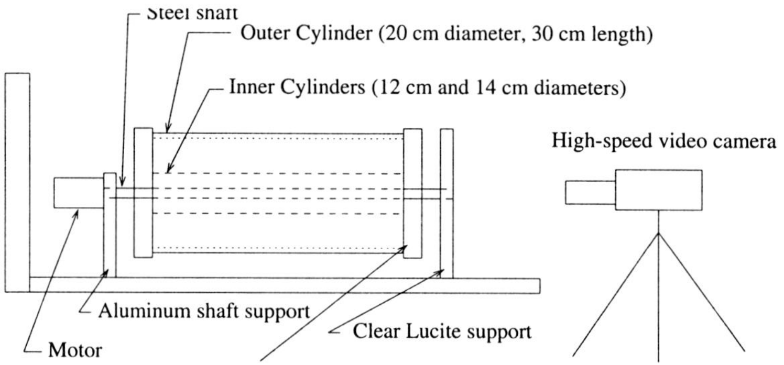

Figure 1. Diagram of experimental apparatus.

2. Experimental Apparatus

The experimental apparatus consists of a pair of concentric aluminum cylinders oriented horizontally as shown in the diagram in Fig. 1 and in Fig. 2. The inner cylinder rotates while the outer cylinder remains stationary. The surfaces of the cylinders are coated with sandpaper and the region between them is filled with spherical glass beads. The front of the cylinder is a glass window, so that the flow inside can be viewed. For the current experiments, a cardboard false bottom separator was installed 5 to 7 centimeters from the glass window, reducing the effective depth of the experiment. Only the region between the cardboard separator and the front window was filled with glass beads. The motor attached to the shaft that is coupled with the inner cylinder has variable rotation rates up to 42 rpm. The outer cylinder is held stationary by an external support. Mixtures of equal masses of 4 mm and 6 mm beads are sheared at various strain rates inside the apparatus. Various inner cylinder diameters can be used in the apparatus so that the ratio of gap width to particle size is varied. In the current experiments, the gap width, h, was either 3 or 4 cm, corresponding with $h/d_{small} = 7.5$ or 10, where d_{small} is the small particle diameter, or 4 mm. Segregation in the flow is visually observed through the glass window at the front. High-speed imaging and particle tracking is used to determine velocity profiles and number density profiles of each kind of particle at various angular locations.

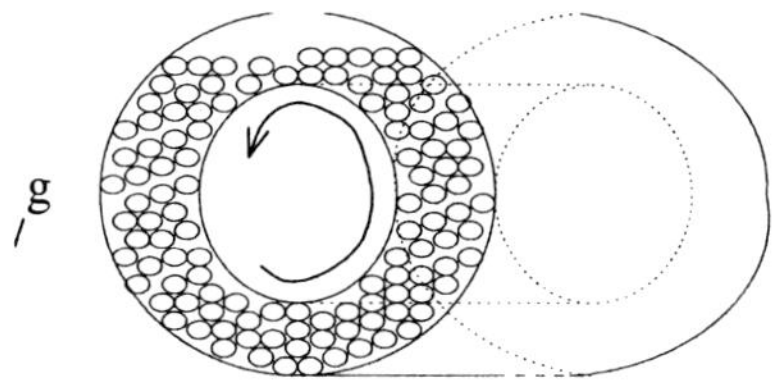

Figure 2. Schematic diagram of shear flow geometry.geometry

The flow is lit for the glass particles to reflect light. A high-speed camera records the flow at either 125 or 250 frames per second (depending on the flow velocity). The image frames are then digitized on a PC and the contrast is enhanced so that the light reflected off each particle appears as a white circle on a field of black. These white circles are then tracked using a commercial particle-tracking algorithm. Because of the short time lapse between individual frames, each white circle moves less than one particle diameter from frame to frame. This distance is measured by the software and the particle velocity can be deduced.

The number density of each kind of particle was measured by dyeing the 4 mm beads with black paint so that light did not reflect off of their surfaces. In this manner, only the 6 mm beads were tracked. At each frame, the total number of light reflections was counted by the digitizing program, as well as the location and velocity of each one. However, absolute numbers from each run could not be compared because of differences in lighting between experimental runs. Hence, not every 6 mm particle was counted, but great care was taken so that no 4 mm particles were counted. Care was also taken to light the flow so that there was no bias in the location of the lit particles. The same experiment was then repeated, but with the 6 mm beads dyed black and the 4 mm beads reflecting light. In this manner, the radial number distribution of each size was determined at a given angular location.

To determine the angular distribution of each particle size, the experiment was again run with the cardboard bottom installed. After running for between 10 and 20 minutes, the experiment was carefully rotated 90 degrees so that the glass window was facing up and the window was opened to allow access to the beads. Aluminum separators were inserted into the beads at 45 degree intervals. The particles from each section were removed and separated. The mass of each kind of particle in each sector was determined, resulting in a measurement of the angular dependence of segregation in this apparatus.

3. Computer Simulation

Computer simulations have been shown to be a powerful tool in the investigation of granular flows [1-3, 13, 14, 16]. Hard-particle models assume instantaneous particle collisions and have been used to study shear flows in the past [1-3, 13], but soft-particle models have proven to be more effective in simulating dense flows with enduring frictional contacts [16, 17], such as those presented here. This research utilizes a soft particle simulation to model granular shear flows with high solid fractions.

In this work, the three-dimensional discrete element simulation is an adaptation of that used by Wassgren [17] and is based on the contact model introduced by Cundall and Strack [18]. Normal contacts are modeled as linear springs with dashpots for dissipation. Tangential contacts are modeled by a tangential linear spring and a frictional slider. The interstitial fluid is not modeled in the simulations. Experiments of Couette flows in a vacuum and in air [12] have previously shown that there were no significant changes in the results due to the presence of air.

The flows studied are shear flows with gravity in the same geometry as the experimental flows. One difference is that the depth of the simulations is only 5 large particle diameters. The experimental apparatus, even with the cardboard false bottom, is approximately 5 to 7 cm, or 8 and 12 large diameters long. All the other dimensions (cylinder diameters and particle diameters) are the same as those of the experiment. The inner and outer cylinders have a coefficient of wall friction, $\mu_w = 0.7$, to drive the flow, while the inter-particle friction coefficient, μ_p, is 0.3. The front and rear boundaries are flat walls with a coefficient of friction of 0.1. The coefficient of restitution of inter-particle collisions, e_p, is 0.9, and the coefficient or restitution of all wall collisions, e_w, is 0.7. The density of the particles in the simulation is approximately that of glass, 2500 kg/m^3. As in the experiment, the inner cylinder rotates at a constant rate and the outer cylinder wall remains stationary. The nominal strain rate in the simulation, defined as $v_{wall}/h = \Omega R_{inner}/(R_{outer} - R_{inner})$, is 11 s^{-1} when $h/d_{small} = 7.5$ and 7s^{-1} when $h/d_{small} = 10$. These rates were chosen to coincide with the maximum shear rate of the experiments for each case. The overall solid fraction in three dimensions, ν, defined as the ratio of particle volume to total volume, is around 0.52, although it is much higher in most regions of the flow, since there is some empty space allowed at the top of the cylinder. As in the experiments, equal masses of large and small spherical particles are placed in the simulated container, and their diameter ratio is 1.5.

All simulations begin with the initial condition of randomly placed particles with small random velocities. They are allowed to settle under gravity and reach a steady state. A steady-state for the flow is generally defined as the state for which the kinetic energy is roughly constant. In the case of mixture flows, however, segregation is a time-dependent process and it is more difficult to determine the steady-state condition. The steady-state in the current work was determined by plotting the standard deviation of the concentration ratio of large to small particles at different angular locations against time. The simulation with $h/d_{small} = 7.5$ was run for 45 seconds of s imulation time. The standard deviation of the angular number density profile was roughly constant after 20 seconds elapsed. Radial velocity profiles and number distributions were averaged between 30 and 35 seconds and compared with the averages taken between 40 and 45 seconds, with no differences between the two results. The simulation with $h/d_{small} = 10$ was run for 40 seconds, and properties were averaged for the last 2.5 seconds of the run.

The sampling rate for all results was 20 times per second. Generally, the result at a typical location represents an average of between 200 and 1000 particles.

4. Results

Measurements in both computer simulations and experiments have been made at three angular locations of the flow. A snapshot of a mixture simulation running with $h/d_{small} = 7.5$ after 10 seconds is shown in Fig. 3, along with the three measurement locations, referred to as "left", "bottom", and "right".

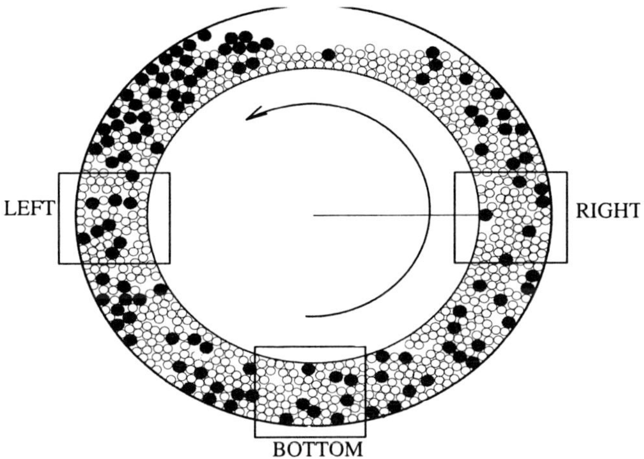

Figure 3. View of front of computer simulation with the three measurement locations. The larger particles are shaded, and the arrow illustrates the direction of rotation of the inner cylinder.

Angular velocity profiles and number density profiles are measured and compared at these three locations. In simulations, the measurements are made at each location over an arc of 45°. In experiments, the arc over which the measurements are made is usually between 15 and 30°.

4.1 VELOCITY PROFILES

Figures 4, 5 and 6 show the velocity of each kind of particle in the angular direction, v_θ, normalized by the velocity of the inner cylinder, $v_{wall} = \Omega R_{inner}$. Experiments performed at different cylinder rotation rates show that the velocity of the particles always scales with the wall velocity in this manner. The normalized velocity is plotted as a function of the radial coordinate normalized by the small particle diameter (which is 4 mm in this case). The position $R/d_{small} = 0$ corresponds to the inner moving cylinder, while $R/d = 7.5$ corresponds to the outer, stationary cylinder. The plots on the right side of each figure show the simulation results; these represent averages at the front surface of the simulations, to correspond with the glass window of the experiment. The velocity profiles (Figs. 4, 5 and 6) show no significant difference between the velocities of the two different kinds of particles. The experimental and simulation profiles agree very well for most cases.

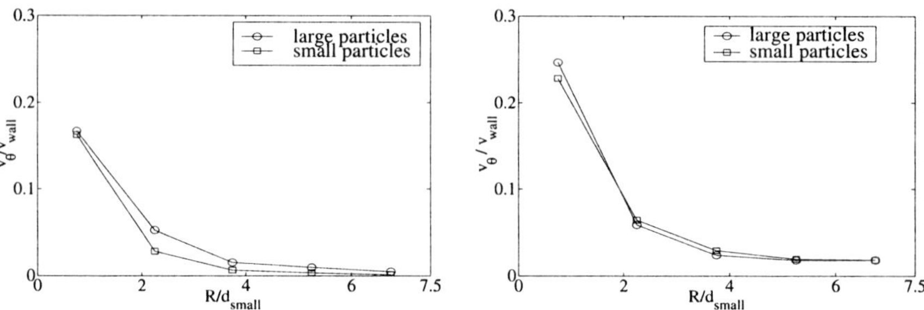

Figure 4. Velocity profiles for $h/d_{small} = 7.5$ at the left side of the circle for experiments (left figure) and simulations (right figure).

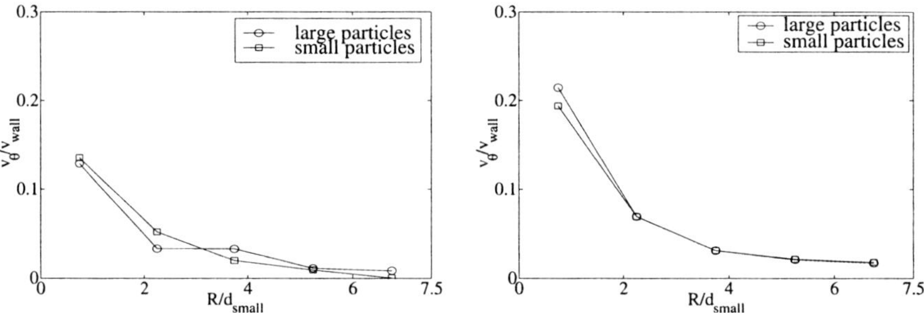

Figure 5. Velocity profiles for $h/d_{small} = 7.5$ at the bottom of the circle for experiments (left figure) and simulations (right figure).

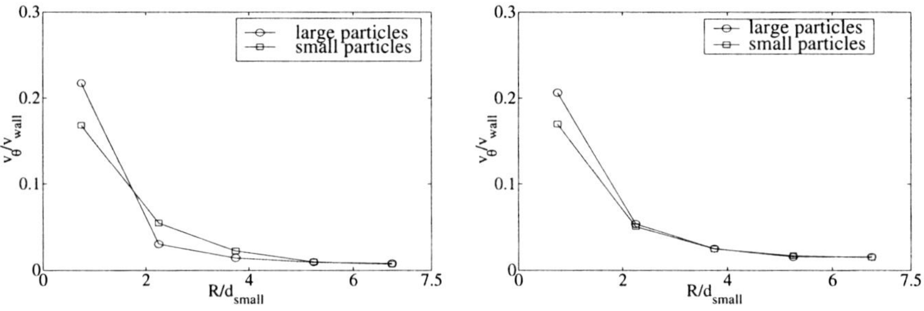

Figure 6. Velocity profiles for $h/d_{small} = 7.5$ at the right side of the circle for experiments (left figure) and simulations (right figure).

4.2 NUMBER DENSITY PROFILES

To determine angular as well as radial separation of large and small particles, measurements of number density of each kind of particle were made at each of three locations around the circle, as shown in Fig. 3. Due to changes in lighting among different experimental runs, the total number of particles counted using the digitizing method was not constant. It was thus necessary to normalize the number of particles of each type counted at a given radial location by the total number of particles of that size counted. To compare the number of large and small particles at each location, a quantity N has been defined as follows.

$$N_{small} = \frac{\frac{n_{small}}{n_{small,tot}}}{\frac{n_{small}}{n_{small,tot}} + \frac{n_{large}}{n_{large,tot}}} \qquad (1)$$

and, for the large particles,

$$N_{large} = \frac{\frac{n_{large}}{n_{large,tot}}}{\frac{n_{small}}{n_{small,tot}} + \frac{n_{large}}{n_{large,tot}}} \qquad (2)$$

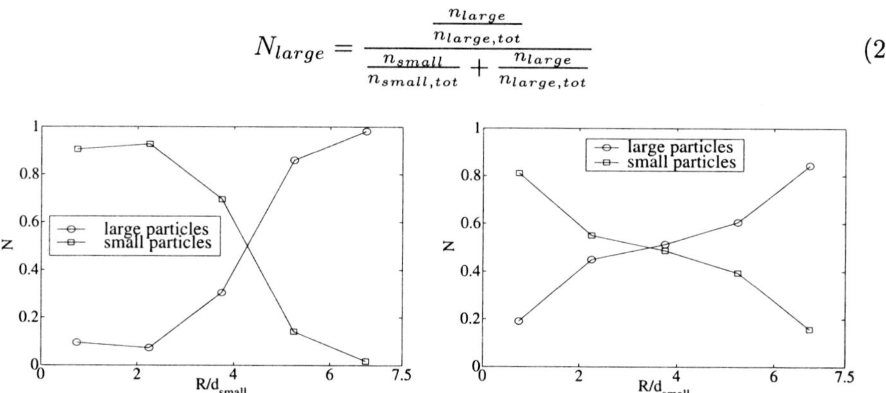

Figure 7. Results for $h/d_{small} = 7.5$ at the left side of the circle for both experiments (left figure) and simulations (right figure).

Using this formulation, $N_{small} + N_{large} = 1$ at every radial location. Figures 7, 8 and 9 show the results for both experiments and simulations of the same flow as a function of radial position R/d_{small}. Although more simple and accurate results are possible using the computer simulations, the same quantity has been plotted for comparison.

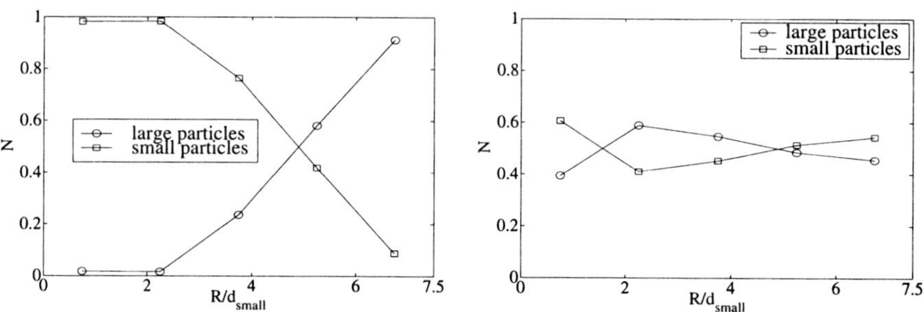

Figure 8. Results for $h/d_{small} = 7.5$ at the bottom of the circle for both experiments (left figure) and simulations (right figure).

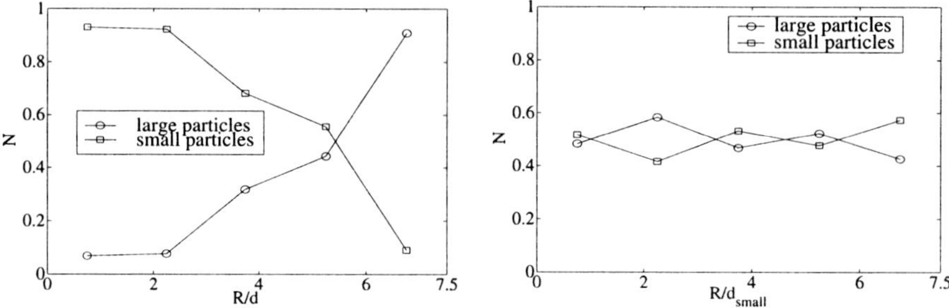

Figure 9. Results for h/d_{small} =7.5 at the right side of the circle for both experiments (left figure) and simulations (right figure).

Figures 7, 8 and 9 illustrate the pattern of segregation observed visually in the experiment. At all locations, large particles are concentrated heavily around the outer, stationary cylinder, while small particles are concentrated around the inner cylinder. But the ratio of the two kinds of particles in the central region varies depending on angular location. Toward the left side of the circle, there are more large particles in the flow, while on the right side there are more small particles. Thus, the radial segregation from the inner to the outer cylinder depends upon angular position as well. Although some segregation is observed in the simulations (notably in Fig. 7), significantly more radial segregation is observed in the experiment.

4.3   ANGULAR DEPENDENCE OF NUMBER DENSITY

To quantify the angular dependence of the number density, measurements of the mass of the large and small particles at eight different angular locations were made. In the experiment, eight aluminum separators were inserted into the granular material after the experiment had run for at least ten minutes and the amount of beads of each size in each sector were weighed. The same measurements were made in the computer simulations and experiments for the two different gap widths, $h/d_{small} = 7.5$ and 10. These measurements are made for the bulk material in the system, not just at the glass window.

Figures 10 and 11 show the mass fraction of small particles and large particles at each of eight angular locations in the flow. The figures show that the fraction of large particles is highest in the upper left of the circle. This result can be also be observed in Fig. 3, where the larger shaded particles are clustered on the left side of the free surface. The large particles segregate in the agitated free surface region, rising to the top. This gravitationally induced segregation is similar to that observed previously in chute flows [6, 9] and vertically vibrating flows [5]. In the more agitated region near the top of the circle, smaller particles have a higher probability of falling through an opening between other particles. Thus, there is a net flux of small particles downward toward the inner cylinder, while the larger particles are pushed up to the surface. A more complete discussion of this process can be found in the work of Savage and Lun [9] in chute flows. In the case of the horizontal Couette experiment, however, the mean motion to the left pushes the large particles to the left side of the free surface region, introducing an asymmetry in the general flow. This asymmetry can be seen in Figs. 10 and 11; if the flow were symmetric then the mass fraction graph would also be symmetric about the 180° mark.

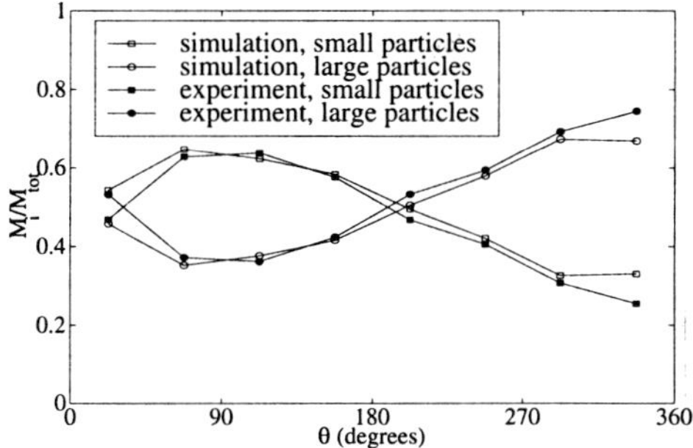

Figure 10. The fraction of each kind of particle as a function of angular location for both simulations and experiments with $h/d_{small} = 7.5$. The angle θ is measured clockwise from the top, so $\theta = 0$ corresponds to the top and $\theta = 90$ is on the right.

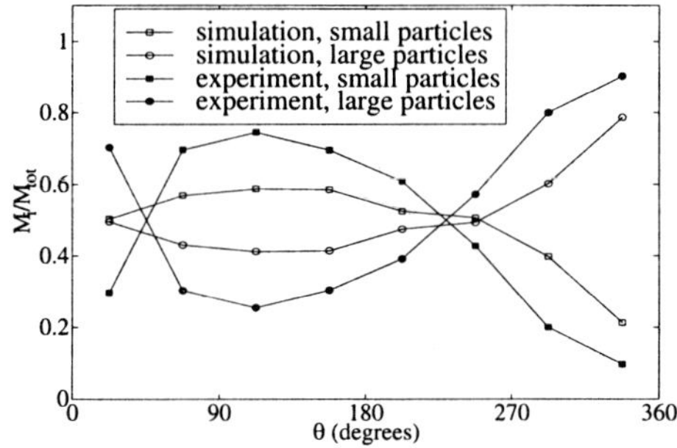

Figure 11. The fraction of each kind of particle as a function of angular location for both simulations and experiments with $h/d_{small} = 10.0$.

In the simulations (results are shown in the open symbols), the mass fraction measurements were made at the last sample taken of the simulation run. Simulations show the same qualitative behavior as experiments, and the results for $h/d_{small} = 7.5$ agree quantitatively as well (Fig. 10). In the simulations with $h/d_{small} = 10$, however, segregation in the simulations is significantly less than that in the experiment.

5. Conclusions

The results of this study illustrate a complex pattern of segregation in horizontally oriented Couette flows. In the flow studied here, particle segregation originates at the free surface of the flow, in the top area of the circle. In this area a high degree of particle agitation causes larger particles to migrate to the top of the free surface, just as has been observed by previous researchers in chute flows [6, 9] and vertically vibrating flows [5]. The upper portion of the cylinder, near the free surface, behaves very much like an inclined chute flow described and explained by Savage and Lun [9]. Smaller particles are able to fall through gaps between other particles more easily than larger ones, resulting in the percolation of small particles downward. Once the large particles are at the top, they remain close to the outer cylinder as they move left around the circle with the bulk flow. This radial segregation can be seen in the experimental results of Figs. 7-9. As the large particles travel around the circle to the bottom, some migrate to the faster-moving region closer to the inner cylinder and quickly travel around back to the top. Others remain near the outer boundary, traveling around at a slower pace. Thus, the number of layers of large particles adjacent to the outer surface changes with angular position. This angular segregation is caused by the horizontal orientation of the experiment with respect to gravity.

The current study also illustrates the degree of effectiveness of computer simulations when studying segregating flows. The simulations in this study captured the qualitative aspects of the segregation but not the degree to which the flow segregated. Several different values of the coefficient of friction and the coefficient of restitution were used in the simulations, in an attempt to find those that best fit the experimental results. The simulation results presented here are those that matched best, in addition to having realistic values for these parameters. In another attempt to match experimental data better, simulations were run with a latched spring normal contact model [19], similar to that used by Walton [16]. The results were nearly identical to those presented here, which use a linear spring and dashpot normal contact model.

Despite these attempts, segregation was not accurately predicted by the computer simulations. Radial number density profiles show much less segregation in the simulated flow than the experiment. Angular measurements of mass fraction also illustrate that the flow in the computer simulation with $h/d = 10$ does not segregate as much as the experimental flow, although in simulations with a smaller gap width the simulations predict the angular segregation very well. In a recent investigation, Cleary et al. [4] conclude that computer simulations capture the qualitative aspects of mixing and segregation but that quantitative measurements, such as the mixing rate, show a large sensitivity to parameters such as the particle coefficient of restitution and coefficient of friction. Similarly, the disagreement between the current simulations and experiments is probably due to a sensitivity of segregation on the particle and wall surface properties. In addition, although the implementation of a more physically accurate normal contact model did not alter results [19], perhaps the use of a more accurate tangential model would show better agreement with experiments.

6. Acknowledgements

Partial support for this work is provided by the Hitachi Research Laboratory, Hitachi, Ltd., and by the National Science Foundation under grant CTS-9530357.

7. References

1. Campbell, C.S., and Brennen, C.E.: Computer simulation of granular shear flows, *J. Fluid Mech.*, **151** (1985), 167–188.

2. Campbell, C.S.: Boundary interactions for two-dimensional granular flows. Part 1. Flat boundaries, asymmetric stresses and couple stresses, *J. Fluid Mech.*, **247** (1993), 111–136.

3. Savage, S.B., and Dai, R.: Studies of granular shear flows: Wall slip velocities, 'layering' and self-diffusion, *Mech. Mat.*, **16** (1993), 225–238.

4. Cleary, P.W., Metcalfe, G., and Liffman, K.: How well do discrete element granular flow models capture the essentials of mixing processes?, *App. Math. Modelling*, **22** (1998), 995–1008.

5. Gallas, J.A.C., Herrmann, H.J., Pöschel, T., and Sokolowski, S.: Molecular dynamics simulation of size segregation in three dimensions, *J. Stat. Phys.*, **82** (1996), 443–450.

6. Hirshfeld, D.,and Rapaport, D.C.: Molecular dynamics studies of grain segregation in sheared flow, *Phys. Rev. E*, **56** (1997), 2012–2018.

7. Arteaga, P., and Tüzün, U.: Flow of binary mixtures of equal-density granules in hoppers – size segregation, flowing density and discharge rates, *Chem. Eng. Sci.*, **45** (1990), 205–223.

8. Nakagawa, M.: Axial segregation of granular flows in a horizontal rotating cylinder, *Chem. Eng. Sci.*, **49** (1994), 2540-2544.

9. Savage, S.B., and Lun, C.K.K.: Particle size segregation in inclined chute flow of dry cohesionless granular solids, *J. Fluid Mech.*, **189** (1988), 311–335.

10. Buggisch, H. and Löffelmann, G.: Theoretical and experimental investigations into local granulate mixing mechanisms, *Chem. Eng. Process.*, **26** (1989), 193–200.

11. Tardos, G.I., Khan, M.I., and Schaeffer, D.G.: Forces on a slowly rotating, rough cylinder in a Couette device containing a dry, frictional powder, *Phys. Fluids*, **10** (1998), 335–341.

12. Khosropour, R., Zirinsky, J., Pak, H.K., and Behringer, R.P.: Convection and size segregation in a Couette flow of granular material, *Phys. Rev. E*, **56** (1997), 4467 -4473.

13. Campbell, C.S.: The stress tensor for simple shear flows of a granular material, *J. Fluid Mech.*, **203** (1989), 449–473.

14. Hunt, M.L.: Discrete element simulations for granular material flows: effective thermal conductivity and self - diffusivity, *Int. J. Heat Mass Transfer*, **40** (1997), 3059–3068.

15. Metcalfe, G., Shinbrot, T., McCarthy, J.J., and Ottino, J.M.: Avalanche mixing of granular solids, *Nature*, **374** (1995), 39–41.

16. Walton, O.R., and Braun, R.L.: Viscosity, granular – temperature, and stress calculations for shearing assemblies of inelastic, frictional disks, *J. Rheol.*, **30** (1986), 949–980.

17. Wassgren, C.R.: Vibration of Granular Materials, Ph.D. thesis, California Institute of Technology, Pasadena, CA, 1997.

18. Cundall, P.A., and Strack, O.D.L.: A discrete numerical model for granular assemblies, *Géotechnique*, **29** (1979), 47–65.

19. Karion, A.: Couette Flows of Granular Materials: Mixing, Rheology, and Energy Dissipation, Ph.D. thesis, California Institute of Technology, Pasadena, CA, 1999.

COUPLED SIZE AND DENSITY EFFECTS IN A 2D-ROTATING CYLINDER:

Two-Dimensional Radial Segregation

JON H. EGGERT[*] and DAVID T. WU[†]
*Department of Physics
†Department of Chemistry & Department of Chemical Engineering
Colorado School of Mines, Golden, CO 80401*

Abstract

Radial segregation of particulate mixtures in rotating cylinders has been the subject of much interest in recent years. We present a systematic study varying both density and size of a single (or a few) tracer disks in a two-dimensional cylinder half-filled with identical balls, and find a complex interaction between size and density effects. Our tracer size ratios ranged from 1 to 16 and our density ratios ranged from 0.02 to 9.6 times that of the balls. We slowly rotated the cylinder, measuring the radial and angular positions of the tracers each cycle, and found several unexpected results. First, size alone did not cause a tracer to be found primarily at the cylinder perimeter (as is observed in equal mixtures) indicating the role of collective effects in the segregation of general mixtures. Second, we found that due to differences in the width of the radial position distribution for large tracers, size and density cannot be made to completely compensate in segregation effects. Third, the boundary in density-size parameter space demarcating the transition from segregation towards the center and segregation towards the perimeter shows a surprising non-monotonic behavior. Finally, the angle of cylinder rotation required to cause a tracer to cycle once depends only on its radial position not on its size or density, consistent with the notion that segregation is driven solely by the tracer's depth within the avalanche.

1. Introduction

It has long been known that radial segregation in a vertically rotating cylinder occurs very rapidly in a binary granular mixture of different sizes or densities. [1-6] Although various phenomenological explanations have been proposed, several aspects of the fundamental nature of the mechanism are still incompletely understood. Some crucial questions that remain are: Can radial segregation be understood as a single-particle phenomenon or are collective effects necessary for a complete description? Do radial segregation due to size and to density behave in the same manner, or are the two

193

A.D. Rosato and D.L. Blackmore (eds.), IUTAM Symposium on Segregation in Granular Flows, 193–202.

mechanisms different? Can the size and density ratios of a binary system be tuned so that the radial segregation effects completely cancel and the material becomes truly well-mixed? Can we identify and understand the processes occurring within the avalanche region that lead to radial segregation? How well do discrete element and other simulations of radial segregation really agree with experiments? To address these and other questions we have performed extensive experiments in a slowly rotating two-dimensional cylinder with isolated tracer particles of widely varying size and density.

2. Experimental

We placed about 950 steel, aluminum, or plastic 3/32" (radius r_{ball} = 1.18 mm) balls, with areal mass densities ρ_{ball} = 1.247, 0.4355, and 0.1859 g/cm^2 respectively, in a half-filled cylinder of radius R = 56 mm. The cylinder end plate separation (< 2.6 mm) was just large enough to let the balls move freely. Motion of the balls was very nearly two-dimensional and we have treated the balls as disks throughout. The cylinder was constructed of Plexiglas and we found that by making the back plate of the cylinder conducting we could avoid the effects of static electricity; we used copper.

The circular tracer disks were machined out of lead, steel, copper, brass, aluminum, and plastic to cover a wide range of size ratios, $\Phi \equiv r_{tracer} / r_{ball}$ from 1 to 16, and density ratios, $\rho \equiv \rho_{tracer}/\rho_{ball}$ from 0.02 to 9.6. Fine manipulation of the density was obtained by drilling out the disk centers. By constructing tracers out of two types of material (e.g. brass and plastic), but with the same radius and density, we could test the influence of different moments of inertia. We found no evidence for any influence of the moment of inertia on our results. This is consistent with our qualitative observations that all tracers showed very little net rotation during the avalanches.

We followed the position of isolated tracers in an otherwise monodisperse medium of identical balls. In measurements on small tracers we often introduced several tracers in a single run and neglected any data when the disks appeared to interfere. The cylinder was rotated slowly by hand; when an avalanche occurred we immediately stopped rotation and only resumed rotation after the avalanche had stopped. This

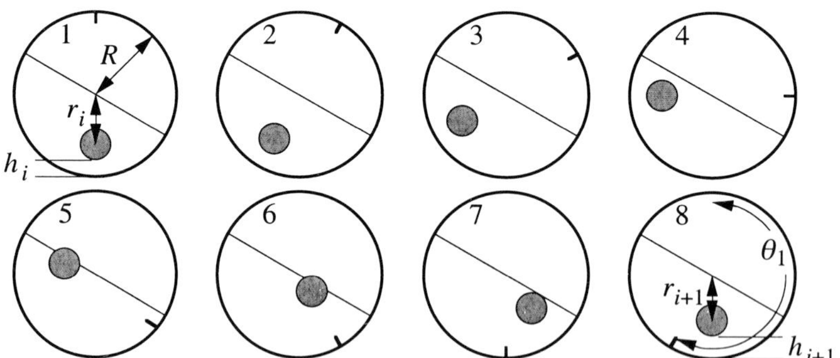

Figure 1. Experimental diagram. The tracer height, h_i, was determined each time the tracer passed below the cylinder center. Note that $h_i + r_i + r_{tracer} = R$. Also measured was the angular position of the tracer so that the rotation angle θ_1 for a single cycle could be measured and correlated to h_i and h_{i+1} for each cycle. (For the cycle shown, θ_1 = 210°.)

experiment is thus in the isolated avalanche regime of asymptotically slow rotation. [7] We had sufficient ball-cylinder friction that no sliding motion of the solid region within the cylinder was observed.

A tracer cycle was defined to begin when a tracer within the solid region passed directly below the center of the cylinder, and to end when the disk again passed below the center of the cylinder. At the beginning of each cycle we measured the height, h_i, from the perimeter of the cylinder to the *base* of the disk. Note that we use r_i to denote the distance from the center of the cylinder to the *center* of the tracer, and so we have

$$h_i + r_i + r_{tracer} = R. \tag{1}$$

We also recorded the position of the tracer with respect to the rotating cylinder to determine the rotation angle θ_1 for each cycle. These measurements are illustrated in figure 1. For each tracer, a typical run lasted between 50 and 100 cylinder rotations and contained roughly 70 to 150 tracer cycles.

3. Pure Density Segregation

In figure 2, we show histograms of the tracer height distribution $P(h)$ for size ratios near unity ($0.903 \leq \Phi \leq 1.038$) and a variety of density ratios. The distributions are normalized by the number of measurements and weighted by the radius r so that,

$$P(h) = \frac{N(h)}{r \sum N(h)} \tag{2}$$

where $N(h)$ is the number of times the tracer was observed within the histogram bin at height h. The weighting factor r is chosen so that $P(h)$ is proportional to the probability of measuring the tracer within a unit area at height h. Thus, for $\rho = 1.0$, $\Phi = 1.0$ (identical tracers), the histogram should be flat throughout the solid region, representing a broad unbiased distribution. As expected, for small ρ the tracer is nearly always found in contact with the perimeter, while for large ρ the tracer is found near the cylinder center.

As moments of these distributions, we have also calculated the mean height and standard deviation. Due to the r dependence of the cylinder area at height h discussed above for the histograms, we use weighted values for the mean height, $\bar{h}$, and standard deviation, σ_h, according to,

$$\bar{h} \equiv \frac{\sum h_i / r_i}{\sum 1/r_i} \tag{3a}$$

and

$$\sigma_h^2 \equiv \frac{\sum \left(h_i - \bar{h}\right)^2 / r_i}{\sum 1/r_i} \tag{3b}$$

respectively. Unless otherwise noted these definitions will be used throughout this report. Figure 3 displays the mean height and standard deviation of tracers with size ratios near unity. Since we expect an identical tracer with $\rho = 1.0$, $\Phi = 1.0$ (denoted by solid circles in all figures) to sample the maximum amount of radial space in the cylinder, it should also yield the largest standard deviation, as can be verified in figure 3b. In fact, this is another motivation for using the weighted values of equation (3); unweighted mean height and standard deviation calculations do not give σ_{max} at $\rho = 1.0$, $\Phi = 1.0$ since particles with small h dominate the solid region.

4. Pure Size Segregation

We studied segregation due to size differences only, using a set of disks with relative densities near unity ($0.955 \leq \rho \leq 1.000$) and size ratios from 1 to 16—a much larger range than previously reported. [3,4] Figure 4 shows the histograms for these data. As expected, large tracers segregate to the outside of the cylinder (small h). Note that the widths of the distributions narrow with Φ, and for our experiments on tracers with

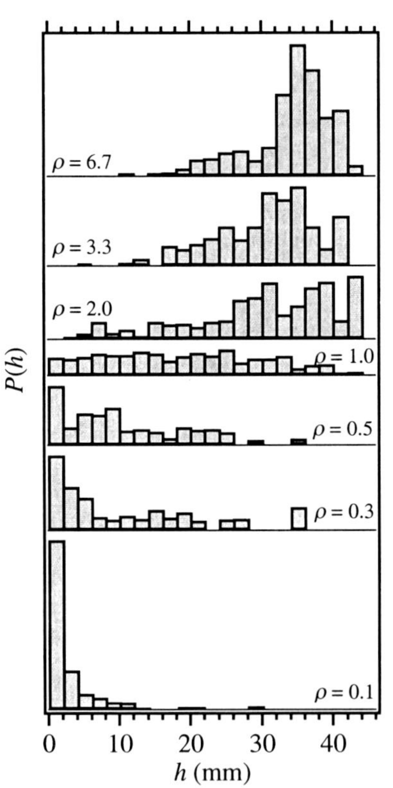

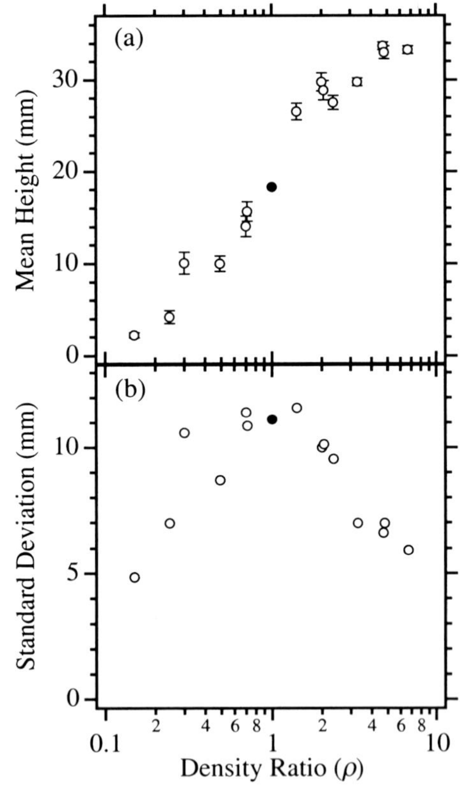

Figure 2. Histogram showing the probability $P(h)$ per unit area to find the tracer at height h. The size ratio, Φ, is near unity for these data.

Figure 3. a) Mean height of the tracer for size ratios near unity. b) Standard deviation for the same tracers. The solid circles denote identical ($\Phi = 1$, $\rho = 1$) tracers.

$\Phi \geq 5.5$, the tracer *never* touched the outer edge of the cylinder.   We have also performed experiments with the cylinder half-filled with equal areas of large disks and small balls, finding that the large disks do contact the perimeter in that case. Even in this system, the large disks were not observed to reach the cylinder perimeter when isolated; cooperative effects between the large disks--not their size alone--were needed to cause the disks to migrate completely to the cylinder wall. This contrasts with previous measurements using smaller size ratios that found no evidence for collective effects in radial segregation. [3,4]

The results for the mean height and the standard deviation are shown in figures 5a and 5b respectively. The mean height (figure 5a) approaches a constant value for relative diameters above about 2, while the standard deviation (figure 5b) decreases monotonically with the size ratio. As discussed above, even for very large diameters isolated tracers did not completely segregate to the cylinder perimeter ($\bar{h} = 0$) in our measurements. This result holds even for the largest tracers which have diameters in excess of the avalanche depth (which is $15 - 20$ mm deep), corresponding to $\Phi \gtrsim 7$.

The line in figure 5a is a fit of the data to the exponential function $h = Ae^{-\Phi/\Phi_0} + h_\infty$. The characteristic decay ratio is found to be $\Phi_0 = 0.37 \pm 0.03$ so that even very small

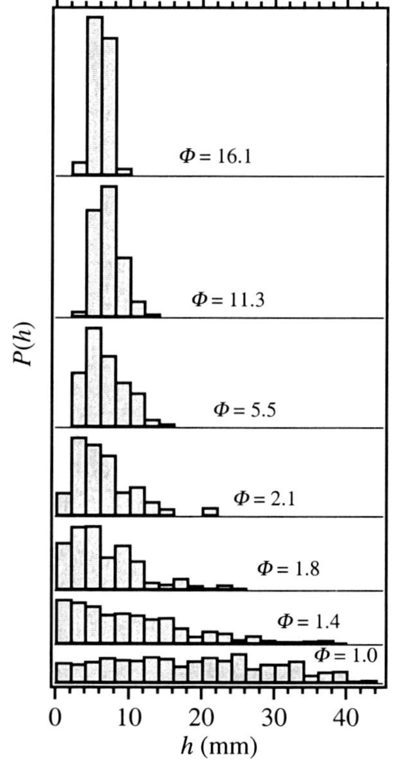

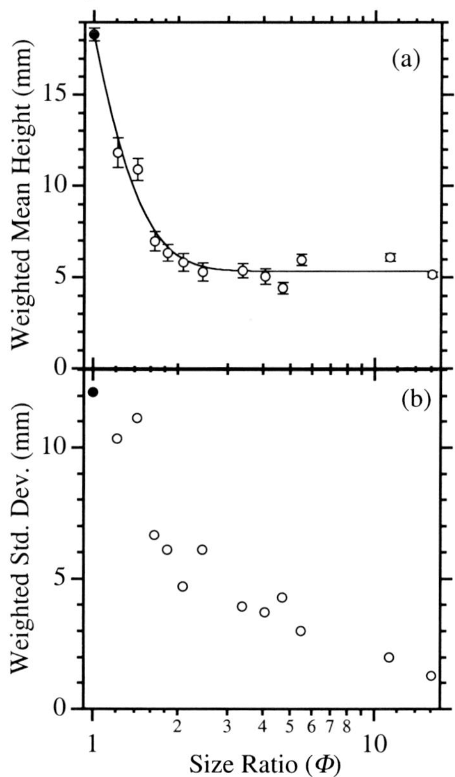

Figure 4. Histogram showing the probability $P(h)$ per unit area to find the tracer at height h. For these data $\rho \approx 1$.

Figure 5. a) Mean height of the tracer for density ratios near unity. b) Standard deviation for the same tracers. The solid circles denote identical ($\Phi = 1$, $\rho = 1$) tracers.

size differences lead to large segregation. We did not investigate the existence of a critical size ratio for the onset of segregation. The asymptotic height is $h_\infty = 5.33 \pm 0.10$ mm, which corresponds in our setup to about two ball diameters.

5. Coupled Size and Density Segregation

To gain an understanding of the competing effects of size and density segregation, we performed tracer studies for many density ratios and four different size ratios $\Phi = 1.00$, 1.57, 2.45, and 4.72. Our results are summarized in figure 6. For low density ratios the tracers approach complete segregation with $\overline{h} = 0$ and the tracers are nearly always in contact with the cylinder circumference.

In order to parameterize the segregation trends as a function of density, we fit the height versus density data. At low densities our measurements of a single size ratio are

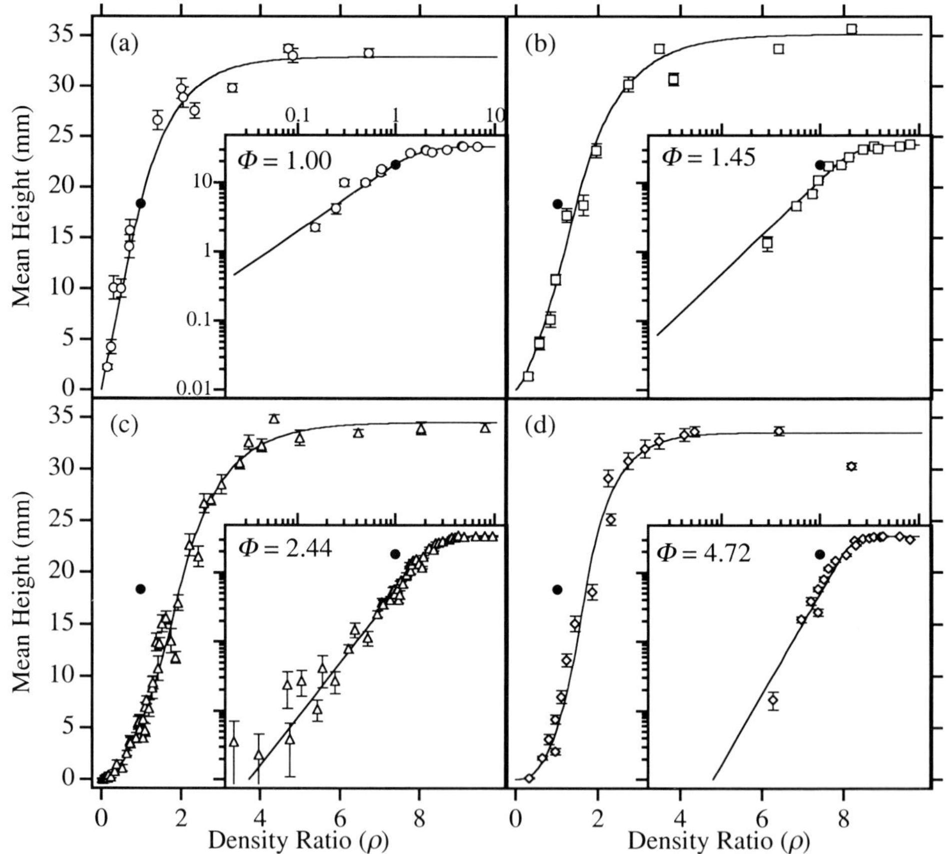

Figure 6. The mean height, $\overline{h}$, of the tracer base for various size ratios: a) $\Phi = 1.00$, b) $\Phi = 1.45$, c) $\Phi = 2.44$, and d) $\Phi = 4.72$. The insets show the same data plotted on a log-log scale to demonstrate the power law behavior for low densities at all size ratios. In all figures the solid circles denote identical ($\Phi = 1$, $\rho = 1$) tracers. The fits are discussed in the text.

well fit by a power law dependence on density, as shown in the insets to figure 6. Similarly, at high densities our measurements are consistent with an exponential approach to an asymptotic value. Thus, we fit our data to a power law at low density splined to an exponential at high density,

$$\bar{h} = \begin{cases} \dfrac{h_\infty}{2}\left(\dfrac{\rho}{\rho_B}\right)^n & \text{for } \rho \le \rho_B \\[2ex] h_\infty\left[1 - \dfrac{1}{2}\exp\left(-\dfrac{\rho - \rho_B}{\rho_0}\right)\right] & \text{for } \rho > \rho_B \end{cases} \tag{4}$$

where the spline break point, ρ_B, is chosen to correspond to $\bar{h} = h_\infty/2$, and also satisfies $\rho_B = n\rho_0$ to give a continuous first derivative. Fits to this function (varying the free parameters n, ρ_0, and h_∞) are shown as solid lines in figure 6. Figure 7 shows the fitted values for n and ρ_0 versus Φ. Note that the power law exponent n increases uniformly with the size ratio, while the characteristic decay density ρ_0 increases slightly before decreasing.

Ristow [1] performed a molecular dynamics simulation of density segregation for $\Phi = 1$ finding that the mean height versus density followed a pure exponential curve (roughly corresponding to $n = 1$) comparing well with our exponent of $n = 0.98 \pm 0.04$. To compare with his decay density ratio we also fit the

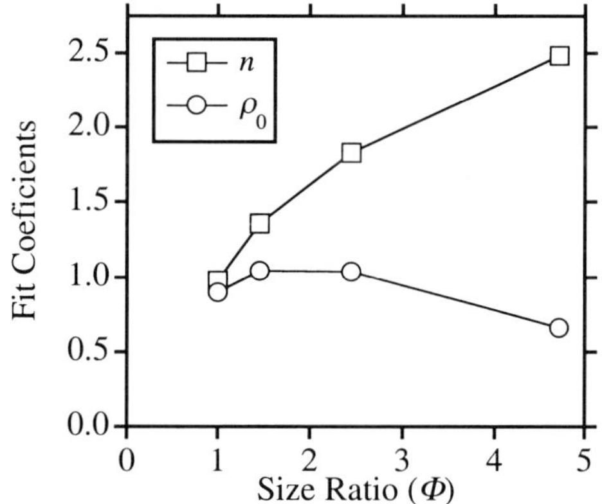

Figure 7. Parameters from equation (8) fit to the data shown in figure 6. Squares are the power law exponent n, and circles are the exponential decay density ρ_0.

$\Phi = 1$ height data (without weighting) to a pure exponential, finding $\rho_0 = 1.66 \pm 0.06$, in good agreement with Ristow's value of $\rho_0^{-1} = 0.6$ or $\rho_0 = 1.7$. In one run at $\Phi = 4.72$ and $\rho = 8.2$ we found that the disk sank below the avalanche and gave an anomalously low height value (figure 6d). This was also found by Ristow for $\Phi = 1$, $\rho > 10$. We neglected this datum in our fit.

Figure 8 shows the standard deviation versus the mean height for our four size ratios. The standard deviations as a function of the mean height are well fit by second-order polynomials. Not surprisingly, the standard deviation falls for very small and very large densities where segregation is strong and the tracer explores a very small region of the cylinder. Alternatively, where segregation is weak, the standard deviation reaches a maximum value implying that the tracer is sampling a large region of the cylinder. Thus, the region of maximum standard deviation corresponds to the best mixing for a given size ratio. This maximum decreases linearly with increasing Φ as seen in

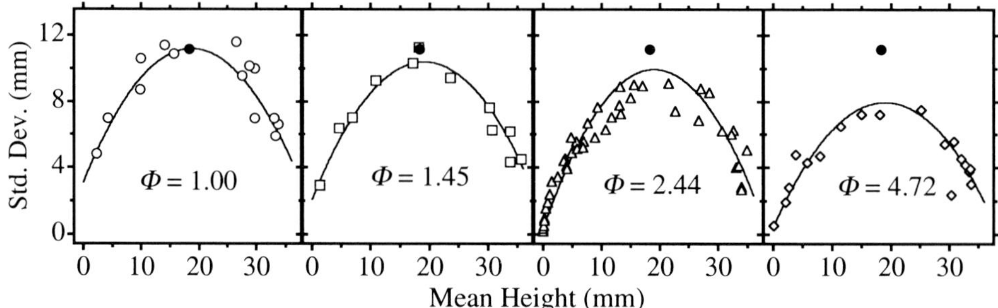

Figure 8. Standard deviation vs. the mean height for the size ratios from figure 6. The fits for the standard deviations are second-order polynomials. The solid circles denote identical ($\Phi = 1$, $\rho = 1$) tracers. Note that the identical tracer mean height corresponds very well to the mean height of the maximum standard deviation for each size ratio.

figure 8. Thus, even though adjusting the density can facilitate mixing of large tracers, for isolated tracers complete mixing is not possible with size ratios larger than one.

Figures 6 and 8 enable us to determine whether a specific tracer will tend to segregate inwards or outwards in the cylinder. Figure 9 shows this distinction in size - density space for three separate criteria: 1) the circles show the density ratio at which the standard deviation reaches its maximum value, 2) the diamonds show the density ratio at which the mean height is half the saturation height ($\rho_B = n\rho_0$ in equation (4)), and 3) the squares show the density ratio at which the mean height for a given size ratio equals the mean height for monodisperse balls. All three criteria produce nearly the same non-monotonic segregation boundary, providing a strong constraint on the plausibility of any proposed mechanism.

6. Rotation Angle and Buoyancy Mediated Segregation

The rotation angle θ_1 (as shown in figure 1) gives information regarding the amount of cylinder rotation during which each tracer participates in the avalanche. We find that there is a correlation between θ_1 and the average radial distance $r_{avg} = \frac{1}{2}(r_i + r_{i+1})$ from the tracer to the cylinder center. In figure 10 we plot the average rotation angle by binning the values of θ_1 according to r_{avg} for all densities and size ratios. The fact that there is no explicit dependence on size or density ratio strongly suggests that the influence of the avalanche experienced by a tracer is equivalent to that felt by other particles at

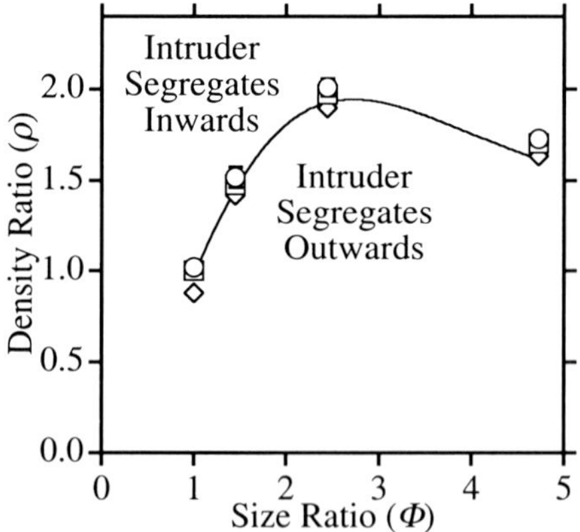

Figure 9. Transition in size-density space of the direction of segregation, as determined in three ways as discussed in the text (symbols). The solid line is a guide to the eye.

the same radius. This is in apparent contrast, for instance, to the notion that segregation occurs due to the rolling of large disks over small. Rather, it appears that the measured height, h, of any disk is determined entirely by the position of the tracer within the avalanche, regardless of density or size.

For disks near the cylinder circumference ($r > ~50$ mm), our experiments indicate a value of $\theta_1 \approx 225°$. Since the avalanche tends to be thin in this region, these tracers remain in the solid region for nearly 180°, implying that the tracer is entrained in avalanches for about 45° each cycle. Qualitatively, we observe that the disks near the circumference tend to participate in 3 to 5 avalanches every cycle. The radial dependence of θ_1 is nonlinear showing an apparent inflection point near the midpoint of the height distributions. This nonlinearity is very consistent for all size ratios and appears to be a property of the avalanche flow. Reproducing this curvature may serve as good test of potential segregation mechanisms.

For the density dependence of segregation, at least, it is easy to imagine that a denser tracer will be less buoyant and will tend to sink within the avalanche. [1,6] If the average shape of the avalanche-solid region is concave upward (like an upright bowl), then we might expect a less buoyant tracer to leave the avalanche nearer the cylinder center than would a more buoyant tracer. Indeed, based on qualitative observation, we believe that this interpretation is reasonable; tracers that flow deep within the avalanche segregate toward the center and tracers flowing near the surface of the avalanche segregate toward the circumference, for all sizes and densities. If Archimedes' principle were applicable, the buoyancy would depend only on density and not on size and we would not see pure size segregation as we do in figures 4 and 5. Obviously, to interpret radial segregation in terms of buoyancy, we need to understand how size affects the effective buoyancy.

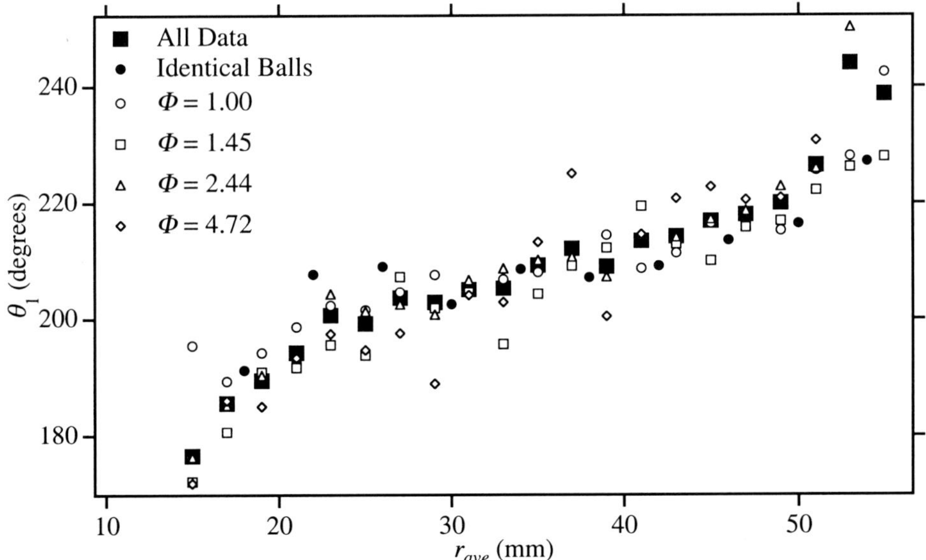

Figure 10. Rotation angle, θ_1, for a single cycle averaged over the average radial position, r_{ave}, of the tracer grouped into 2-mm bins. θ_1 is independent of Φ and ρ, depending only on r_{ave}.

7. Conclusions

By collecting statistically significant ensembles of individual tracer motion, we are able to draw several conclusions about radial segregation in a two-dimensional slowly rotating cylinder.   1) Within the range studied, size differences alone do not cause a tracer to be found primarily at the cylinder perimeter (as is observed in equal mixtures) implying that collective effects may play a role in the segregation of general mixtures. 2) Due to differences in the width of the radial position distribution for large tracers, size and density cannot be made to completely compensate in segregation effects. 3) The boundary in density-size parameter space demarcating the transition from segregation towards the center and segregation towards the perimeter shows a surprising non-monotonic behavior. 4) The angle of cylinder rotation required to cause a tracer to cycle once depends only on its radial position (and not on its size or density), consistent with the notion that segregation is driven solely by the tracer's depth within the avalanche, perhaps controlled by buoyancy.   These experimental results provide a quantitative criteria for assessing simulation or theoretical models for segregation.

Our experiments have extended the range of segregation studies in 2d-rotating cylinder to higher size ratios and to combinations of size and density effects. In doing so, we have discovered a new density regime where continuously increasing the size ratio results in segregation, first toward the center, then toward the perimeter, and finally toward the center again. Further regimes may be encountered at more extreme regions in the size-density parameter space. In particular, the observation that large tracers (larger than the avalanche thickness) are often excluded by a few layers of small balls at the cylinder perimeter might be understood as resulting from the partitioning of the avalanche balls into those behind and those ahead of the tracer (which cannot be easily evacuated). However, for tracers much larger than the avalanche thickness, one might expect the tracer to be able to contact the perimeter high above the avalanche layer. Another regime of behavior may also occur at very high densities, due to sinking of the tracer, as was noted in section 5. We are presently repeating all measurements reported here in a cylinder three times larger. This will enable us to determine the scaling behavior of our observations as well as to explore other possible segregation regimes.

8. References

1.   Ristow, G.H.: Particle mass segregation in a two-dimensional rotating drum, *Europhys. Lett.* **28** (1994), 97-101.
2.   Baumann, G., Janosi, I.M., and Wolf, D.E.: Surface properties and flow of granular material in a two-dimensional rotating-drum model, *Phys. Rev. E* **51** (1995), 1879-1888.
3.   Clement, E., Rajchenbach, J., and Duran, J.: Mixing of a granular material in a bidimensional rotating drum, *Europhys. Lett.* **30** (1995), 7-12.
4.   Cantelaube, F. and Bideau, D.: Radial segregation in a 2d drum: an experimental analysis, *Europhys. Lett.* **30** (1995), 133-8.
5.   Dury, C.M. and Ristow, G.H.: Radial segregation in a two-dim. rotating drum, *J. Phys. I* **7** (1997), 737-45.
6.   Khakhar, D.V., McCarthy, J.J., and Ottino, J.M.: Radial segregation of granular mixtures in rotating cylinders, *Phys. Fluids* **9** (1997), 3600-3614.
7.   We performed several runs with continuous rotation at 1 rpm (near the transition from isolated to continuous avalanches). There were no observable difference in these runs and we believe that our results are representative of the continuous avalanche regime as well as the isolated avalanche regime. However, we have found differences generated by rapid rotations when the particles adopt ballistic trajectories.

THE KINETICS OF SEGREGATION IN FLUIDISED BEDS

M.C. LEAPER
A.S. BURBIDGE
J.P.K. SEVILLE
School of Chemical Engineering,
University of Birmingham, Birmingham, B15 2TT, U.K.

G. BLAIR
BP-Amoco Oil
Sunbury on Thames, TW16 7LN, U.K.

Abstract

It is well known that particles in the fluidised state will segregate if their sizes or densities are sufficiently different. However, little research has been carried on the kinetics of this separation, nor on the separation of particles below 300μm; in addition, the effect of the distributor design on a segregating system has not been investigated. An ilmenite-glass ballotini system was studied where the particle sizes were between 80 and 300μm. By comparing the curves for fluidisation and defluidisation of an homogeneous mixture, it was shown that a fluidising curve of the system can be used to predict whether it will segregate or not. The type of gas distributor plate also had an effect on the segregation behaviour, with differences occurring between porous plate and multi-orifice distributors at an identical gas flow rate. Where segregation occurred concentration profiles under various conditions were plotted against time to determine the behaviour of the bed. This was achieved by removing the material layer by layer, separating magnetically and analysing the mass ratio of the contents. Within the top 35% layer, separation showed an exponential approach to steady state, enabling a "rate constant" for separation to be obtained. This constant was shown to depend on the excess gas velocity.

1. Introduction

Segregation is a common feature of fluidised beds. Extensive studies of particles in Geldart's group B were performed by Rowe *et al.* [1-3], and, more recently, Hoffmann *et al.* [4], but these are limited to the investigation of equilibrium segregation and do not address the kinetics of the process. It is generally agreed that the mechanisms of segregation [5] are:

- preferential entrainment of fine particles within the channels or bubbles at velocities which exceed their terminal velocities.
- bulk movement of fines in the wake/drift material dragged upwards by bubbles.
- stripping of fines through voids between particles in the dense phase by percolation.

A.D. Rosato and D.L. Blackmore (eds.), IUTAM Symposium on Segregation in Granular Flows, 203–210.
© 2000 *Kluwer Academic Publishers. Printed in the Netherlands.*

- the downward movement of all particles in the bulk in areas where there are no bubbles.

The aim of this study is to investigate the segregation of a glass ballotini-ilmenite system within the group A fluidisation regime, with the intention of using the data to design a continuous separator. The potential applications of such a system are numerous, allowing multi-particulate systems to be separated on a large scale with a much lower air requirement per tonne of material processed than lean-phase processes. Ilmenite was chosen because it is a paramagnetic material, which allowed it to be separated magnetically from the soda glass for analysis purposes; it is also a much denser material (4500 kgm^{-3} compared with 2500 kgm^{-3} for soda glass) and its dark colour contrasts sharply with that of the glass, allowing easy visual inspection.

The study focused on four main areas:
- how the distributor affected the segregation process.
- the relationship between the equilibrium fluidising curve and the extent of segregation.
- the equilibrium state attained as a function of fluidising velocity.
- the kinetics of segregation.

2.   Experimental Materials

The experimental apparatus consisted of a 15 cm perspex fluidising column, with air supplied by a rotameter board; two gas distributors were used – a plastic sintered porous plate of thickness 3 mm (sinter size 50 μm) and a steel perforated plate of thickness 5mm with 0.5mm holes on triangular pitch of 7mm; filter paper was placed under the perforated plate to prevent material from falling through.

The materials used were soda glass, which had a density of 2500 kgm^{-3}, and ilmenite with a particle density of 4500 kgm^{-3}. The initial mixture in all experiments was of a 50/50 composition. The cumulative undersize distribution is shown in Figure 2. Initial analysis showed that the U$_{mf}$ values of the glass and ilmenite were 0.011 and 0.038 m/s, respectively.

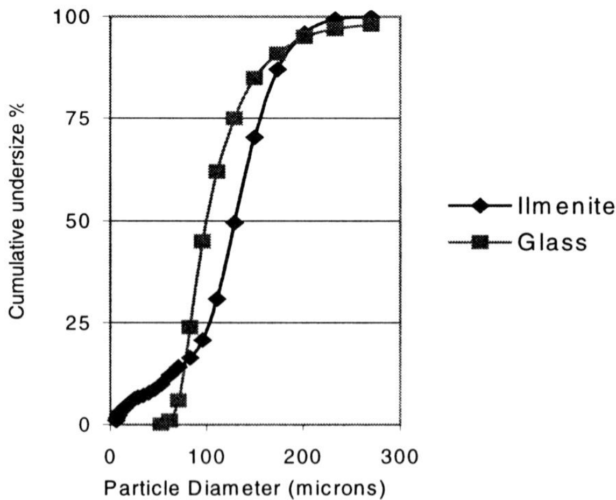

Figure 1. Cumulative undersize distribution for ilmenite and glass ballotini

3. Experimental Method

Sampling was performed by suction of particles, using an ejector connected to a compressed air line, with the particles being disengaged from the resulting air flow in a sampling bottle with a glass filter.

Two kg. of each material was used for every experiment, which gave a fluidised bed height of 12-13 cm and a defluidised bed height of 11 cm, resulting in an aspect ratio of about 0.8. To ensure that each experiment began under identical homogeneous conditions, the materials were first fluidised at a velocity in excess of 0.12 m/s (which would give a mixing index close to 1) for at least 5 minutes. After removing the fluidising air, homogeneity was confirmed by dividing the defluidised bed into layers and vacuuming off each layer for separate analysis. The analysis was performed by passing each mixture through a rotating drum magnetic separator of diameter 75mm with a field strength of between 1 and 1.5 Tesla.

3.1 Investigating the Effect of Distributor Choice on Segregation by Measuring Equilibrium Concentration Profiles

According to Geldart and Baeyens [6], porous distributor plates promote segregation because the small bubbles formed have insufficient energy to transport jetsam particles the whole height of the bed; this implies that the distributor plate, which governs the initial bubble size, may have an effect on segregation. The aim of this part of the study was to examine this and to ascertain whether other types of distributor plate were capable of creating a segregating bed of similar characteristics. A 50/50 by mass mixture of glass and ilmenite was prepared in the fashion described in the previous section and fluidised at 0.017 m/s for 40 minutes using a porous plate. This was repeated using a fluidising velocity of 0.034 m/s and again using the perforated plate. Samples were then taken at various heights as described in the previous section. The results are shown in Figure 3, expressed in terms of the mass fraction of glass (x-axis) and the bed height (y-axis); the sample height is taken as the average of each layer.

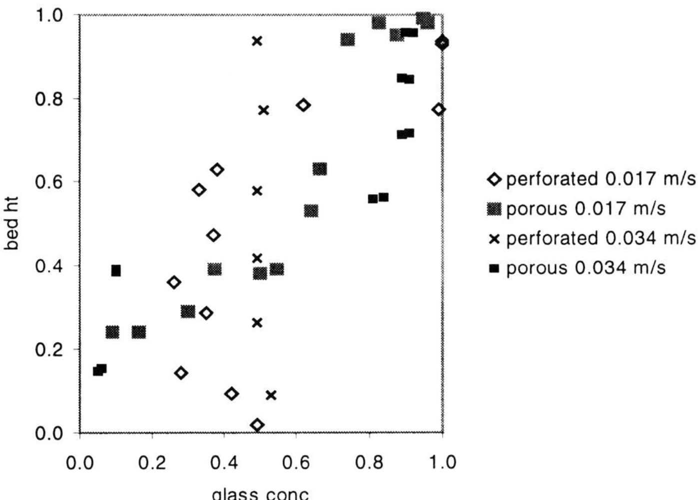

Figure 2. Bed concentration profiles using perforated and porous plate distributors

The results show that the perforated plate gives a segregating system over a smaller range of fluidising velocities than the porous plate, which gives a more effective total separation and is most suitable for concentrating the ilmenite. Because the upper fluidised layer is smaller when a perforated plate is used, the defluidised layer is less rich in ilmenite and so the separation is less efficient. It is also significant that near the distributor the glass concentration tends towards the 50/50 state, suggesting a defluidised layer in this region.

3.2 EXAMINING THE FLUIDISING CURVE OF THE BINARY MIXTURE

The fluidising curve was plotted for both distributor plates in turn by measuring the pressure drop across the bed at equilibrium for a known fluidising velocity. Initially the mixture was prepared in the way described in the experimental method (i.e. the starting point was an homogeneous mixture). The equilibrium pressure drop was then measured for each increase in fluidising velocity to a value well above U_{mf}. The fluidising velocity was then progressively reduced. Figures 4(a) and 4(b) show the variation of bed pressure drop over bed weight per unit area against fluidising velocity for both fluidisation and defluidisation. If the bed is kept in the segregated state and fluidised again, the plot follows the *defluidisation* curve; this is consistent with previous observations [7].

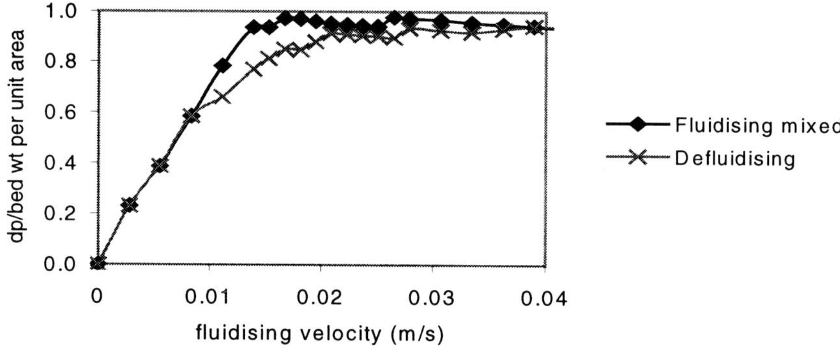

Figure 3(a). Fluidising/defluidising curves for porous plate

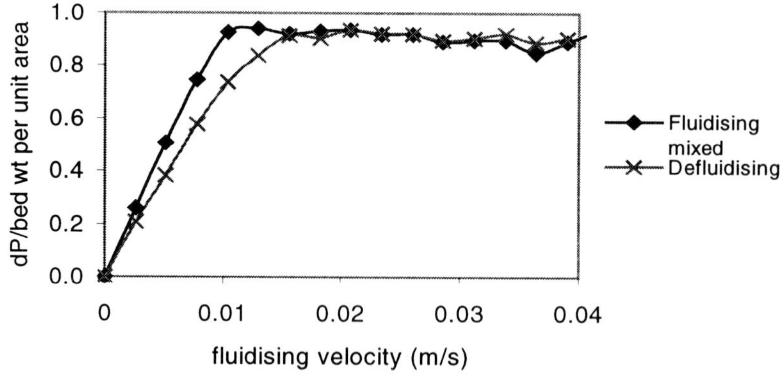

Figure 3(b). Fluidising/defluidising curves for multi-orifice perforated plate

There was a difference between the fluidising and defluidising curves at velocities where segregation occurred and mixing where they were coincident; the plot demonstrates that the operating window for segregation is much larger with the porous plate (0.012 to 0.038 m/s as compared with 0.01 to 0.02 m/s).

3.3   EXAMINING THE EFFECT OF VELOCITY ON THE KINETICS OF SEGREGATION

Having investigated the effect of the distributor on the fluidising curve and equilibrium concentration profile, the kinetics of the process were examined in greater detail. The equilibrium bed concentration profiles for the two selected fluidising velocities were determined by fluidising the material for 4 hours; the bed concentration profiles for intermediate times were also measured and are shown in Figures 5-7.

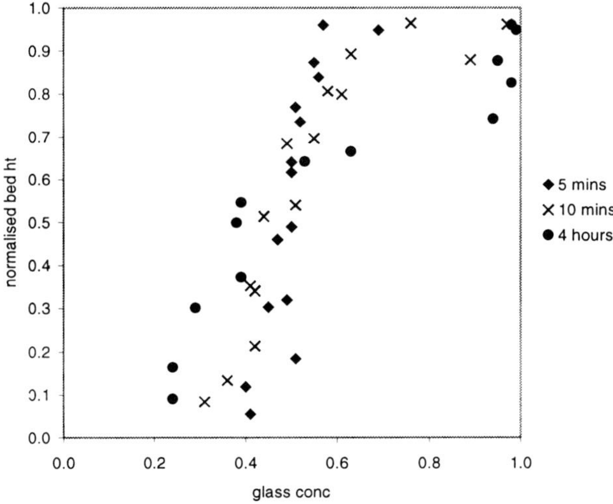

Figure 4. Glass concentration profile of the bed at 0.017 m/s for porous plate system

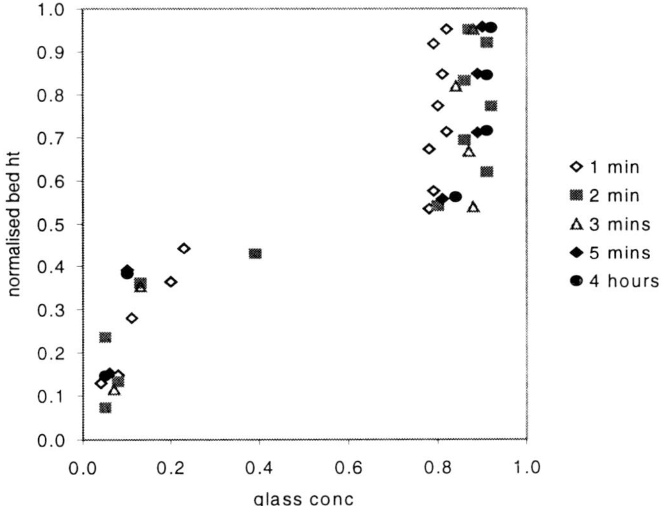

Figure 5. Glass Concentration Profile at 0.034 m/s for Porous Plate System

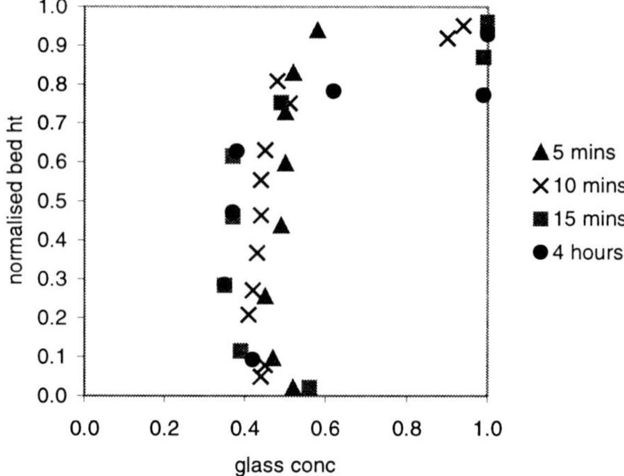

Figure 6. Glass concentration profile at 0.017 m/s for perforated plate system

During all of the experiments, a significant amount of the ilmenite fell to the bottom of the bed, forming a defluidised layer; glass particles were seen to be trapped within this layer, forming discrete areas of localised concentration.

The simplest assumption about the kinetics of segregation is that the concentration at each point in the bed makes an exponential approach to a steady state value. For example, if the glass concentration in the top 35% of the bed is considered then the plot of glass concentration against time might be expected to give an exponential curve of the form:

$$(C_\infty - C) = (C_\infty - C_0) \cdot e^{-kt} \tag{4}$$

where $C = f(t)$ and C_∞ and C_0 are the equilibrium and initial (i.e. 50% in this case) glass concentrations respectively. The segregation "rate constant" k can then be found thus:

$$-\ln \frac{(C_\infty - C)}{(C_\infty - C_o)} = kt \tag{5}$$

An example of one of these plots is shown in Figure 8:

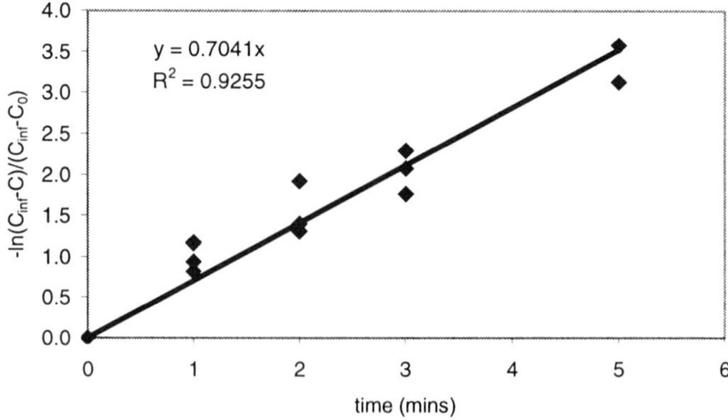

Figure 8. Semi-log plot for u = 0.034 m/s, porous plate

The predictions of equation (4) are compared with experimental results in Figure 9, using the fitted values of k and C_∞. The results are summarised in Table 1.

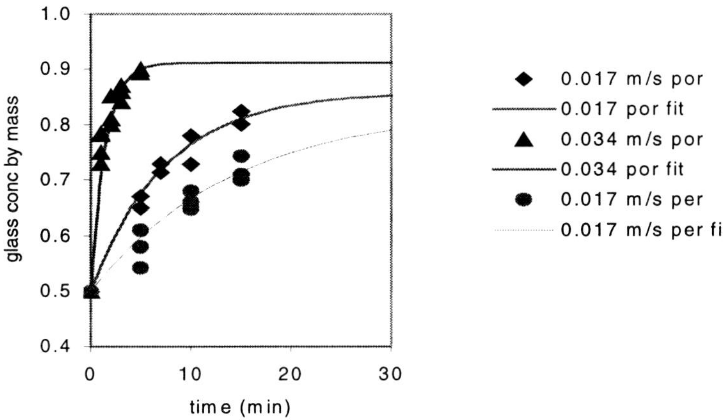

Figure 9. Comparing results and fit for top 35% of the bed

TABLE 1. Comparing the equilibrium concentration and rate constant for the top 35% of the bed

Fluidising Velocity (m/s)	Equilibrium Concentration	k (s^{-1})
0.017 porous	0.86	8
0.017 perforated	0.83	4
0.034 porous	0.91	42

4. Discussion

Identical mixtures behave very differently when fluidised with different distributors, indicating that bubble characteristics close to the bottom of the bed play a significant part in determining whether a binary mixture will segregate. By measuring the fluidising curve of a binary mixture under a known set of conditions, it is possible to predict whether it will segregate or not; this is done by obtaining the pressure drop across the bed for each velocity at an equilibrium state, first by fluidising an homogeneous mixture progressively to a fully fluidised state, and then by reversing the process, producing a defluidising curve. If the two curves are coincident, then this shows evidence of a system that will mix perfectly at all stages of the fluidising curve. In a segregating system, particles within a fully fluidised bed will re-arrange to have the lowest possible energy state as the fluidising velocity is reduced, accounting for the lower pressure drop on the segregating part of the defluidising curve as compared with the equivalent points at the same velocities on the fluidising curve of an homogeneous mixture. If a segregated mixture is then re-fluidised, it will follow the defluidising curve described above.

Because of the relatively wide spacing between the holes in the multi-orifice plate, it is possible that below a certain velocity there is a defluidised layer close to the distributor from the outset, and that this increases the tendency of the perforated plate to behave more like a porous plate, producing small bubbles rather than the discrete jets that appear at higher velocities; this may be supported by the fact that the glass concentration *increases* near the bottom of the column towards the original 50/50 concentration at a fluidising velocity of 0.017 m/s.

These results show that with regard to the separation of both materials, the porous plate is more effective than the perforated plate because at the same velocity it produces a larger glass-enriched, fluidised layer, increasing its ability to concentrate the denser material. From the kinetic experiments it also shows that it approaches an equilibrium at a faster rate. In addition, a greater fluidising velocity using the porous plate causes equilibrium to be reached more quickly. The increased amount of bubble movement at the higher gas velocities has two effects: it transports the lighter material more quickly to the surface and it disturbs the "dense phase" of the fluidised material, so enhancing percolation of the denser material towards the bottom of the bed , a key mechanism in segregation [5]. One might expect that the rate of segregation would be related to the bed "turnover time" [5], the average time taken for a particle to circulate around the bed. This is proportional to $H_{mf}/(U-U_{mf})$, where H_{mf} is the height of the bed at minimum fluidisation. Two factors make the turnover time difficult to calculate – the uncertainty in the U_{mf} definition of the mixture and the presence of a defluidised layer, which reduces the effective value of H_{mf}.. However, in the case of the porous plate, the turnover frequency increases as the gas velocity is increased from 0.017 to 0.034 m/s, by approximately the same factor as the rate constant. Further studies of this kind are planned, including direct measurement of particle trajectories and turnover time using Positron Emission Particle Tracking [8].

5. Conclusions and Recommendations

This study has shown that the type of distributor used has a large effect on the behaviour of a fluidised binary system in the size range 80-300µm, providing some indication of the choice to be made where perfect mixing or complete segregation are required. It also demonstrates that separation of the particles is more efficient at higher velocities because the size of the defluidised layer is reduced and the structure disturbed to such an extent that the majority of flotsam particles are entrained in the upper fluidised layer. In order to develop this work further, it is necessary to investigate the effect of aspect ratio and other distributors on the process and to assess how particles of a larger size difference would behave.

6.   References

1.   Rowe, P.N., Nienow, A.W. and Agbim, A.J.: A Preliminary Quantitative Study of Particle Segregation in Gas Fluidised Beds – Binary Systems of Near Spherical Particles, *Trans.Instn. Chem. Engrs.* 50 (1972), 324-333.
2.   Rowe, P.N., Nienow and A.W.: Particle Mixing and Segregation in Gas Fluidised Beds. A Review., *Powder Technology* **15** (1976), 141-147.
3.   Rowe, P.N., Nienow, A.W. and Agbim, A.J.: The Mechanisms by Which Particles Segregate in Gas Fluidised Beds – Binary Systems of Near-Spherical Particles, *Trans.Instn. Chem. Engrs.* 50 (1972), 310-323.
4.   Hoffmann, A.C. and Romp, E.J.: Segregation in a Fluidised Powder of a Continuous Size Distribution, *Powder Technology* **66** (1991), 119-126.
5.   Geldart, D.: Gas Fluidisation Technology, John Wiley and Son (1986) ISBN 0 471 90806 1.
6.   Geldart, D., Baeyens, J.: The Design of Distributors for Gas-Fluidized Beds, *Powder Technology* **42** (1985), 67-78.
7.   Chiba, S., Chiba, T., Nienow, A.W. and Kobayashi, H.: The Minimum Fluidisation Velocity, Bed Expansion and Pressure-Drop Profile of Binary Particle Mixtures, *Powder Technology* **22** (1979), 255-269.
8.   Stein, M., Seville, J.P.K., Benton, D., McNeil, P. and Parker D.J.: Particle Movement and Segregation in Gas Fluidised Beds, *I.Chem.E Research Event* (1996), 1060-1062.

USING GAS FLOWS TO PROBE SEGREGATION IN MOVING GRANULAR BEDS

W. R. PATERSON, D. L. MOPPETT, L. P. B. COELHO,
G. E. SOUTHERN, J. E. FITZGERALD
Department of Chemical Engineering
University of Cambridge
United Kingdom

D. M. SCOTT
Department of Chemical Engineering
Monash University
Victoria, Australia

Abstract

We study the segregation of spherical particles of different diameters in moving beds *i.e.* steady-state dense-phase flows of granular solids. The transients typical of, for instance, hopper discharge are absent. Such flows often occur in vessels in which a fluid flows, either co-currently or counter-currently to the particles, which themselves move vertically downwards under gravity. They are widely used in industry in applications such as continuous catalyst regeneration reactors and blast furnaces.

We have used three different "gas probes". The first is the pressure drop measured when known steady flows of air are passed through moving beds. This lets us identify how to arrest the granular flow while preserving the structure of the bed, permitting most subsequent experiments to be conducted more conveniently on such "frozen" beds. Our second consists of injecting a pulse of CO_2 tracer into a steady inlet flow of air; the CO_2 concentration in the outlet air is measured using a mass spectrometer. We thence infer gas residence time distributions (RTDs). Our third "gas probe" consists of injecting the CO_2 tracer at a particular position within the bed: the response at a position vertically above the injection point is measured. Radial traverses then reveal the radial profile of tracer axial velocity and so the radial variation in resistance to gas flow.

The probes were proven by experiments using monosized glass spheres, or "ballotini", of diameter 1, 3 and 5 mm, revealing significant maldistribution of gas flow, dependent on the frictional properties of the vessel wall. By adopting walls of the type which cause little maldistribution when enclosing mono-sized particles, we have been able to focus on the dramatic maldistribution caused by segregation in binary mixtures. For example, with a mixture of 40% (by mass) 1 mm spheres with 3 mm spheres, the RTD may display two peaks separated by a factor of 4 in time, consistent with a variation over the cross section of the gas axial velocity by a factor of four.

The particle segregation which imposes the radial variation in resistance to gas flow and so causes the maldistribution is found to be strongly influenced by the

A.D. Rosato and D.L. Blackmore (eds.), IUTAM Symposium on Segregation in Granular Flows, 211–221.

"free fall gap" *i.e.* the distance between a control orifice in the particle inlet tube and the base of the free surface of the moving bed. Under suitable conditions, the smaller particles accumulate preferentially near the central axis; under other conditions they accumulate instead in an outer annulus near the vessel wall; under yet further conditions they spread approximately uniformly over the cross-section. Visual observation discloses the segregation mechanisms at work.

1. Three "gas probes" applied to beds of monosized particles.

1.1 PRESSURE DROP, OUR FIRST GAS PROBE

After three bed volumes of particles had passed through a column, we measured the pressure drop when known steady flows of air were passed through the moving bed and through such moving beds brought to rest, *i.e.* "frozen beds", thus identifying how to arrest the granular flow while preserving the structure of the bed (Paterson *et al.* [1], Happel [2]). This let later experiments be conducted conveniently on such "frozen" beds. Later we supported this approach, Crawshaw *et al.* [3], by measurement of:-

1.2 RESIDENCE TIME DISTRIBUTION (RTD), A SECOND GAS PROBE

A pulse of tracer gas is injected into a steady flow of air entering a bed and the transient tracer concentration in the outlet air is measured, yielding the RTD. A typical arrangement is:-

1.2.1 Experimental apparatus

A schematic diagram of the apparatus is given in Fig.1. Glass ballotini of *e.g.* 3 mm diameter were fed from a hopper, at *ca.* 1-3 kg/min, to a perspex column of internal diameter 140 mm and length 0.5 m, 1 m, 1.1 m or 1.5 m. The solids flowed out via a centrally located pipe of diameter 22, 27 or 40 mm. A tracer gas, CO_2, was injected so that the shape of the tracer concentration-time curve at the bed inlet was an adequate approximation to an ideal Dirac delta function. The main gas flow through the bed, low pressure compressed air, was introduced via the solids outlet pipe. The range of Reynolds numbers (based on the particle diameter and gas superficial velocity) covered was 5.0 to 34. Either sample point could be connected to a mass spectrometer. Three types of beds were studied: (i) a well compacted fixed bed, (ii) a fully developed moving bed, and (iii) a frozen bed.

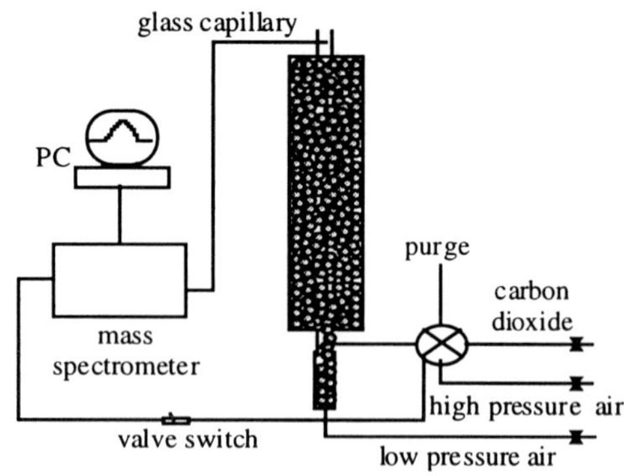

Figure 1. Apparatus for measuring gas RTD.

1.3 GAS TIME-OF-FLIGHT, A THIRD GAS PROBE.

We injected the CO_2 tracer directly into the bed rather than into the inlet gas: the response at a position vertically above the injection point was measured. Radial traverses then revealed the radial profile of tracer axial velocity, indicating the radial variation in resistance to gas flow.

1.3.1 Experimental apparatus

An injection tube was pushed into the bed through a small port in the wall of the column, so that it protruded into the bed, and was sealed in position to prevent gas leaks. The detection tube was installed in a similar fashion, with its end 0.5 m vertically above the injection tube's. Preliminary experiments proved that the tubes did not substantially disturb the bed. Ports were located in various different axial positions.

1.4 RESULTS AND DISCUSSION

1.4.1 RTD

The quantity which is measured in the RTD experiments is the concentration of tracer as a function of time, $C(t)$. This is usually normalised by the total quantity of tracer injected, which may conveniently be calculated as the integral of the outlet concentration curve. When the tracer injection is a delta function this gives $E(t)$, usually called the "external residence time distribution" (but strictly a probability density function)

$$E(t) = C(t) / \int_0^\infty C(t')dt'; \quad E_\theta \ [or \ E(\theta)] = \tau E(t) \qquad (1)$$

where the mean residence time, τ, and the dimensionless time, θ, are

$$\tau = \int_0^\infty tE(t)dt; \quad \theta = t / \tau \qquad (2)$$

The technique's first success was in finding that the E-curves for the moving and frozen beds are almost identical, thus vindicating our focus on frozen beds (Fig. 2).

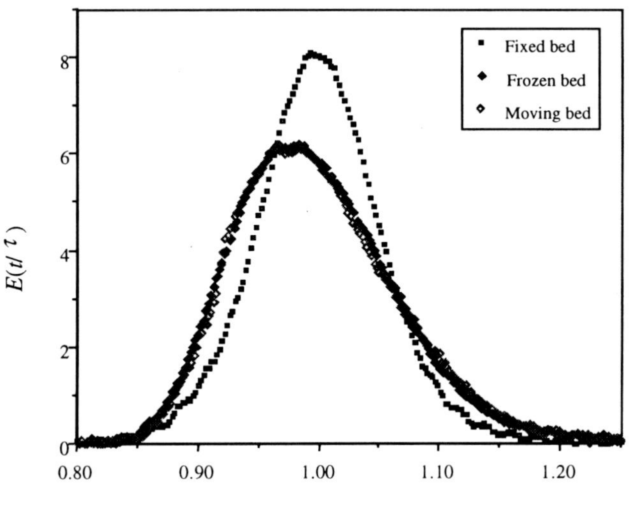

Figure 2. Comparison of E curves for three types of bed.

Qualitative interpretation of such RTDs is aided by mathematical transformation (Paterson *et al.* [4]). Quantitative interpretation requires us to allow for end effects near the top and bottom of the bed, caused both by atypical bed structure and also by the atypical flow of the gas, as it diverges from, or converges to, a narrow inlet/outlet pipe. We have developed two techniques:-

(a) we "subtract" the end effects using FFT deconvolution applied to data from beds of two different lengths. The response of the bed to a pulse input is represented as the convolution of the responses of (i) the entrance and exit zones of the bed and (ii) the zone remote from the ends. It is this latter zone which constitutes a characteristic or typical section of bed. The validity of the deconvolution technique has been demonstrated by applying it to well-consolidated fixed beds (Crawshaw *et al.* [5]).

(b) We use a mathematical model (Moppett [6]) which represents the end zones as well-stirred "tanks" of length L_{cst}, not because we believe that there is substantial real backmixing, but because of the wide spread of residence times which might reasonably be supposed to approximate the atypical flow in the end zones. We allow each "tank" to have part of its volume, fraction *df*, inaccessible to the flowing gas, a "dead volume". The central zone of the bed, remote from the ends, is of length L_c and is supposed to support axially dispersed plug flow, described using a Peclet number. The good fit of this model to the data, with parameter values that are physically reasonable, is taken as evidence that the model components are well chosen, including the representation of the end effects, see Table 1 (Paterson *et al.* [7]).

We have then found that the extent of maldistribution depends crucially on the frictional properties of the vessel wall. When it is smooth, the RTD has the Gaussian shape characteristic of axially dispersed plug flow; when the wall of the vessel is rough, the RTD is wide and asymmetric, evidence of substantial flow maldistribution - see Fig. 3. Even the intermediate, perspex-walled case shows, after deconvolution, a bi-modal RTD, unmistakable evidence of channelling, Fig. 4.

TABLE 1. Parameter values for the modelling of gas RTD in frozen beds constructed using walls of different frictional properties.

	PTFE	Unlined	Emery paper
L_c (m)	1.33	1.4	1.2
L_{cst} (m)	0.085	0.05	0.15
df(%)	0.59	0.56	0.72

1.4.2 Time of flight

We again validated our probe by reproducing familiar fixed bed results. Then we found that our traverses, within the typical zone of the frozen beds, explain maldistribution by finding "channelling" and by identifying the location of the "channels". When the wall of the vessel is smooth, the radial profile of tracer gas velocity is fairly flat, as shown in Fig. 5 - there is little channelling, consistent with the RTD of Fig. 4 By contrast, the rough wall causes pronounced channelling, with a low gas velocity near the centre-line and higher velocities near the wall, again consistent with the RTD of Fig. 4. For the walls without a PTFE lining, the gas velocity profiles for 1, 3 and 5 mm spheres have qualitatively the same shape, with minima at 7-10 particle diameters from the wall, coinciding approximately with the position of the edge of the wall shear zone measured in flowing granular materials.

Toyama [8] reports the thickness of the shear zone in cylindrical beds to be 4-8 particle diameters, and Nedderman and Laohakul [9] report the thickness of the shear zone in parallel sided bins to be 5-6 particle diameters; see also Takahashi and Yanai [10]. There would seem to be a broad consistency between the results of our gas probe studies and the direct study of the solids flow. We seek to develop further the experimental detail of our injection and detection tubes and their mode of operation, and our technique for interpreting their results.

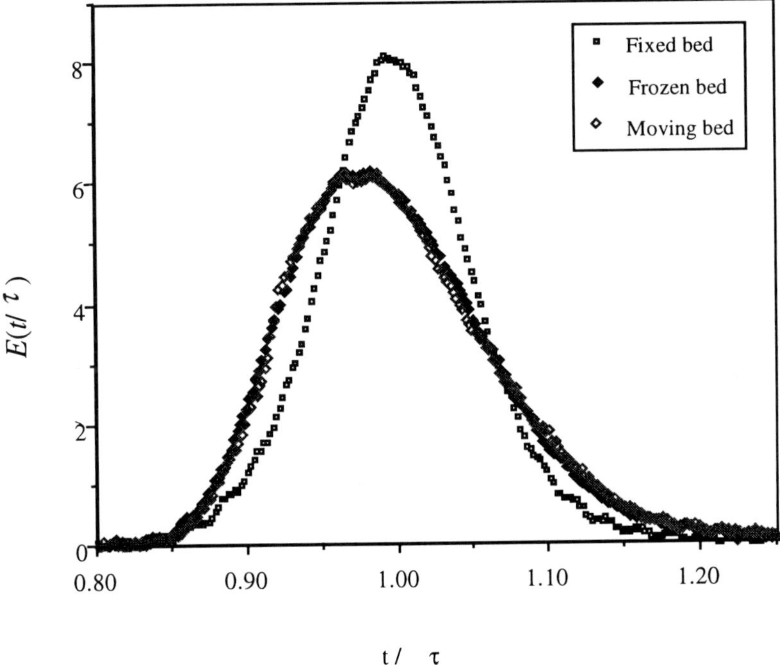

Figure 3. Comparison of frozen bed E curves for three types of wall.

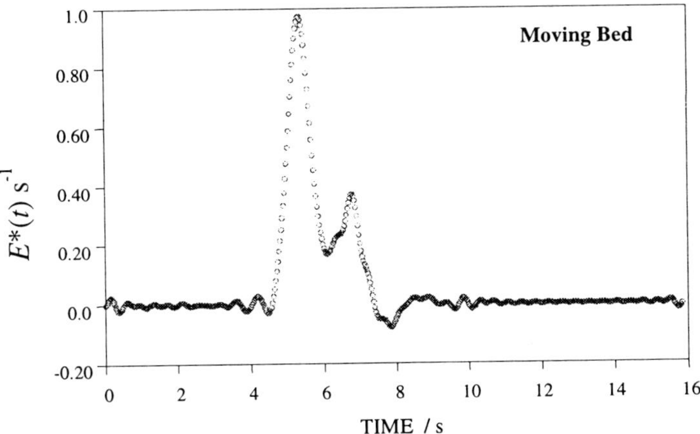

Figure 4. Deconvoluted RTD for a bed of effective length 0.5m.

1.5 CONCLUSIONS

The three gas probes have been proved on fixed, moving and frozen beds of monosized particles. Our methods of coping with end effects have been tested succesfully and we have discovered how to operate moving beds with minimal gas maldistribution due to wall friction. We are therefore well equipped to use these probes on binary mixtures in such a way that maldistibution may be interpreted undistorted by end effects, and unambiguously attributed to segregation rather than to wall friction.

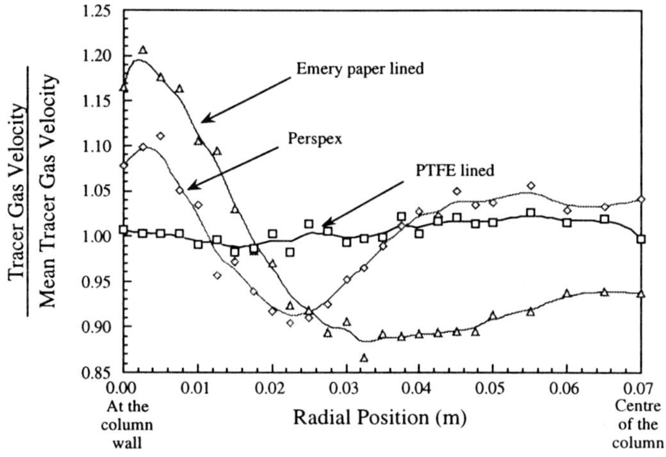

Figure 5. Normalised interstitial velocity: effect of wall friction

2. First applications to segregation in binary mixtures

2.1 FORMING THE BEDS

Mixtures of 1 mm and 3 mm particles flowed in the range 1–3 kg/min through a 1.5 m column. Care was taken to ensure that the inflowing particles were reasonably well mixed: the particles were introduced to the hopper above the column in small batches (approximately 1 litre) which had been mixed by hand. The bed walls were perspex: We found it possible to further reduce the friction at perspex walls by introducing a slow countercurrent stream of humid air into the particle inlet tube during construction of the frozen beds. This helped to discharge the static electric charge which otherwise accumulated and increased the attractive forces between the bed wall and the ballotini. The location of the free surface of the bed within the vessel was controlled by manipulation of the outflow rate of the ballotini, such that the peak of the free surface is in contact with the top plate and particle entry tube. Thus the incoming ballotini are in a dense flow regime and do not fall freely before striking the bed surface.

2.2 SOLIDS PROBES

To permit comparison with the inferences from the gas probe results, we determined the particle spatial distributions in a column of length 1.10 m, with two diametrically opposed holes situated 0.58 m from the its top. Through these a particle sampling device was introduced radially. This "sampler" was a horizontal cylinder which had 10 cells at different positions along its length. When open, the sampler allowed approximately free flow of particles; the flow pattern of the particles above the device was negligibly perturbed by the presence of the sampler. The sampler could be closed, trapping the particles that were passing through the cells at that instant. The particles trapped in each cell were sorted by size, yielding the radial distribution of particles in the bed (Paterson *et al.* [11]).

The results were confirmed qualitatively in another apparatus, a perspex column which could be separated on a central vertical plane. The particle flow was suitably stopped, and the column was flooded with gelatine. When the gelatine had cooled and set, the bed was cut in half from top to bottom, and a thin slice of bed was cut from one of the exposed faces. The particle distribution was measured by melting the gelatine over a square grid, in which the particles were trapped. The particles were then sorted by size. The results were consistent with those from the sampler.

2.3 RESULTS

The key feature of the RTD results, Fig. 6, is that a tracer, introduced as a single pulse into the gas flowing into the bed, may emerge from the other end of the bed as two distinct pulses, with peak times separated by up to factor four. This evidence of maldistribution is broadly consistent with the corresponding radial profiles of axial gas velocity ploted in Fig. 7 (wherein the different profiles differ according to the algorithm adopted to interpret the tracer concentration curve from the detection tube). The cause of this gross maldistribution is found in the particle distributions, the smaller particles being concentrated near the bed centre-line and the larger near the walls, leading to two regions in the bed - a central core of high resistance to gas flow, and an outer annular region of lower resistance to gas flow.

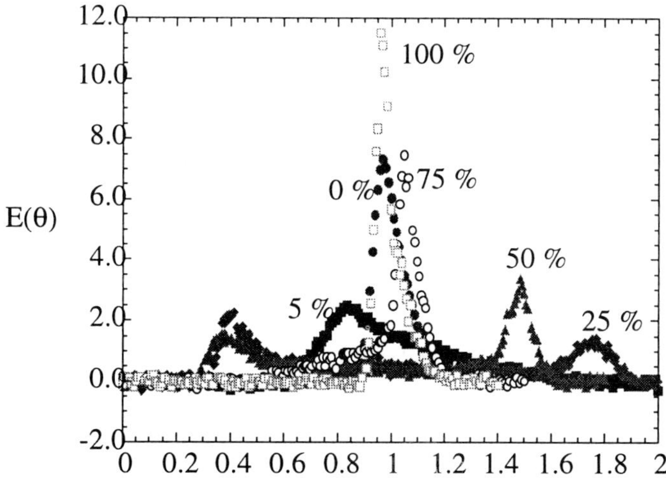

Figure 6. Gas RTDs for frozen beds of binary particle mixtures (mass % 1mm with 3mm diameter)

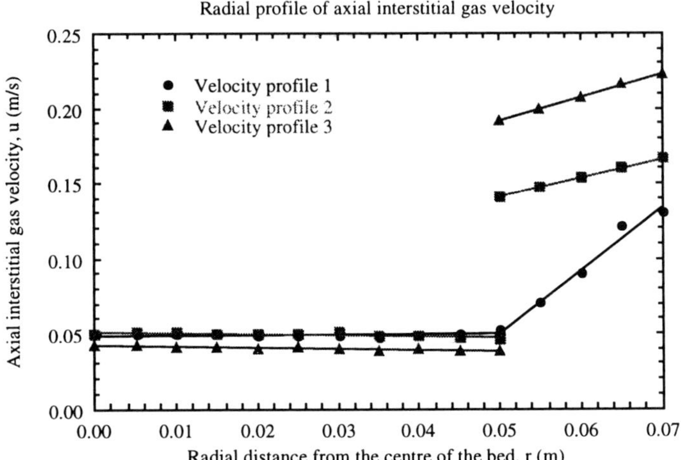

Figure 7. Velocity profiles in a frozen bed (40 mass % 1 mm with 3 mm):
three different interpretations of the experiments

2.4 MATHEMATICAL MODELLING: CONSISTENCY OF PROBES

The qualitative consistency is clear: mathematical modelling can determine whether quantitative consistency is achievable. We have concentrated our studies on the particularly complex case of a gas RTD with two well defined peaks - implying extreme variations of gas velocity over the bed cross section - and specifically on the results for a frozen bed of composition 40% 1 mm particles, with 3 mm particles, supporting an air flow of 30 litres per minute.

2.4.1 Use of time-of-flight measurements

The intermediate, characteristic zone of the moving bed is described as supporting, at each radial position, advection of gas plus axial dispersion, with interstitial gas velocity

$$
\begin{aligned}
v &= 0.04054; & 0.00 \leq r \leq 0.05 \\
v &= -0.4831 + 10.4730r; & 0.05 \leq r \leq 0.07
\end{aligned}
\tag{3}
$$

We claim that this profile is essentially consistent with the different data sets of Fig. 7 because plausible physical arguments imply that where the gas velocity is low, $r <$ 0.05 m, the velocity must be essentially that of "profile 1", whereas where the velocity is high, $r >$ 0.05 m, the velocity must rise continuously to a value approximately equalling that of "profile 3" adjacent to the wall. At an intermediate stage in the calculation, the radial voidage profile is "backed out" from the interstitial gas velocity profile, the spatial distribution of particles and the measured mean voidage characteristic of the bed. The crucial fitting parameter for the RTD prediction is then the length of the intermediate section of the bed. The axial Peclet number is assumed uniform at all radial positions of the moving bed and is estimated using a fixed-bed correlation applied using a local mean particle diameter inferred from the particle spatial distributions. It proved unnecessary to assume non-zero dead zones in the identical CSTs used to model the end effects.

2.4.3 Comparison of model predictions with measured RTD

The model prediction of the gas RTD is shown in Fig. 8, together with the experimental points: the agreement is good (once the degree of ambiguity and corresponding necessity for judgement in the interpretation of the time-of-flight results is acknowledged), the position and width of the peaks being practically coincident [12]. The "good fit" length of the intermediate section of the bed is 1.40 m. The consistency of the two gas probes used is demonstrated. Since the channelling feature being modelled for segregated beds is much less subtle than for mono-sized beds, a simpler representation of the end effects proved adequate. There remains scope to exploit the pressure drop probe for binary mixtures.

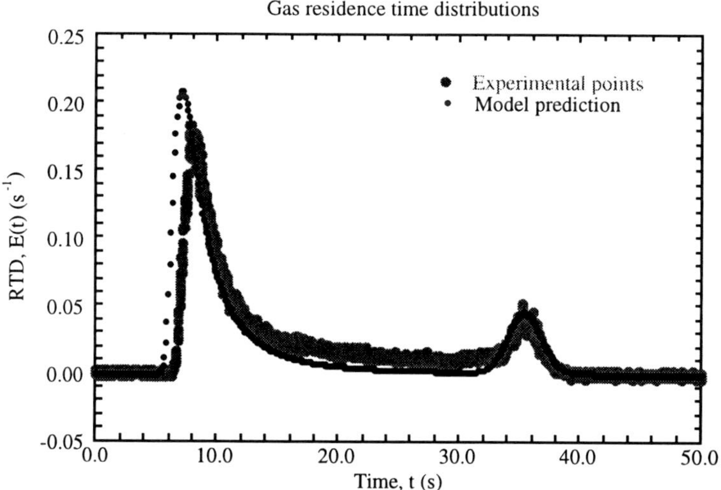

Figure 8. Model prediction vs. experimental gas RTD for a frozen bed of 40% mass fraction 1mm ballotini with 3mm ballotini.

3. Further studies of segregation in binary mixtures

3.1 "FREE-FALL HEIGHT" OR "ENTRY GAP"

It has proved possible to control the particle inflow and outflow rates so that there is a constant, non-zero "free-fall height" or "entry gap", with the particles falling from the inlet tube in a dilute flow regime before striking the bed surface. We have studied mixtures of 1 mm and 3 mm particles flowing at 3.5 kg/s, with fines fractions of 5% and 50%, and entry gaps of up to 30 cm.

3.1.1 RTDs

We form our beds in a two-section column of height 1.5 m and then push a horizontal metal plate through the frozen bed to allow us to remove the top 0.5 m of column, allowing us then to measure the RTDs on a bed of height 1.0 m. The lessons are most intriguing for the mixtures with 50% fines. For 20 cm and 30 cm entry gaps, the RTD is wide, although less so than for the "no gap" case, and bimodal. However, a 10 cm gap results in a much narrower and almost unimodal RTD.

3.2 VISUAL OBSERVATIONS AND MECHANISMS

3.2.1 Experimental procedures and observations of cross-sections

Dividing the frozen bed with a horizontal metal plate has allowed us to photograph the bed cross-section at a "typical" location, well below the zone where the bed structure is undergoing its formation. The results are striking and are supported by the evidence available by visual inspection through the bed wall. The 20 cm case has a "reversed" structure: the fines dominate the outer annulus and the 3 mm spheres dominate the core. The 10 cm case has a near-homogeneous structure over the whole cross-section. For the 30 cm case, there is evidence of a three-zone structure: an outer annulus of fines, an inner anulus dominated by 3 mm spheres and a core of binary mixture. Moreover, observation of the behaviour of the stream of particles as it strikes the bed surface permits us to sketch the likely mechanisms of segregation.

3.2.2 Discussion

The "no gap" case is readily explained. The 3 mm spheres tend to roll down the free surface of the bed and come to rest at the wall, dominating that region, whereas the 1 mm fines tend to drop through voids in the (presumably rather open-structured) surface and become absorbed into the central core of the bed. They do not tend to reach the wall. With a finite entry gap, the fines tend to bounce on colliding with 3 mm particles on the bed surface, giving them a means of reaching the wall in high proportions. The fines seem not to bounce on striking other fines or a void. It is the relative importance of the three mechanisms of bouncing, rolling and absorbing that determines the bed structure: thus the intermediate entry gap results in enough bouncing of fines to give a fairly uniform radial distribution whereas the larger entry gaps give longer bounces of fines, resulting in concentration of the fines at the wall.

3.3 CONCLUSIONS

Gas probes have real merit as a useful part of an experimental tool kit, complementary to other tools. Our gas probes yielded results consistent with each other, with direct studies of bed structure and with various visual observations. In particular, the gas probes have proved their worth as means of identifying cases worthy of further, detailed, investigation.

4. Nomenclature

C　　tracer concentration (kgm^{-3})

df　　fraction of one end zone occupied by a "dead volume" (dimensionless)

$E(t)$　exit residence time distribution density function (s^{-1})

E_θ　E expressed as a function of θ (dimensionless)

$E(\theta)$　E expressed as a function of θ (dimensionless)

L_c　　length of the central zone *i.e.* characteristic section of the bed (m)

L_{cst}　length of one of the end zones (m)

t　　time (s)

u　　superficial fluid velocity (ms^{-1})

v interstitial fluid velocity (ms^{-1})

τ mean residence time, s

θ reduced time (dimensionless)

5. References

1. Paterson, W.R., Crawshaw, J.P., Hart, G., Parker, S.R., Scott, D.M.and Young, J.P.: Pressure drop in countercurrent gas flow through moving particulate beds, *Chem. Eng. Res. Des.* **70(A)** (1992), 252-253.
2. Happel J.:Pressure drop due to vapour flow through moving beds, *Ind. Eng. Chem.* **41(6)** (1949), 1161-1174.
3. Crawshaw J.P., Paterson W.R., and Scott D.M.: Gas flow maldistribution in fixed and moving beds. *Chem. Eng. Res. Des.* **71, Part A**, (1993), 643-648.
4. Paterson W.R., Crawshaw J.P. and Scott D.M.: Gas flow maldistribution in fixed and moving beds. *Proceedings of 20th Australian Chemical Engineering Conference (CHEMECA 92)*, Canberra, Australia, Vol. 1, (1992), pg. 230.
5. Crawshaw J.P., Paterson W.R., and Scott D.M.: Gas channelling and heat transfer in moving beds of spherical particles. *Chem. Eng. Res. Des.* (1998), in press.
6. Moppett, D.L.: Investigations of gas flow maldistribution in granular moving beds. Ph.D. Dissertation, University of Cambridge, 1996, UK
7. Paterson W.R., Berresford E.L., Moppett D.L., Scott D.M., Simmons V.K. and Thorpe R.B.: Gas flow maldistribution in moving beds of monosized particles. *Chem. Eng. Sci.* (1999), in press.
8. Toyama, S.:The flow of granular material in moving beds. *Powder Technology* **4** (1971), 214.
9. Nedderman, R.M. and Laohakul, C: The thickness of the shear zone of flowing granular materials. *Powder Technology* **25** (1980), 91-100.
10. Takahashi, H. and Yanai, H.: Flow profile and void fraction of granular solids in a moving bed. *Powder Technology* **7** (1973), 205-214.
11. Paterson W.R., Coelho L.P.B., Scott D.M. and Thorpe R. B.: Segregation in moving beds of particulate solids. *World Congress on Particle Technology 3*, 1998, Brighton, UK (ISBN 0-85295-401-8, paper 328), pp. 3514-3524.
12. Coelho, L.P.B.: Gas percolation through segregated particulate beds. Ph.D. dissertation, University of Cambridge, 1998, UK.

6. Acknowlegements

We gratefully acknowledge SERC (UK) for support of this work, grant Transport Properties in Moving Beds, GR/D 18292. LPBC would like to acknowledge the Portuguese institution JNICT for their scholarship through the programme PRAXIS XXI. DLM thanks EPSRC, UK for his studentship. DMS thanks Professor Tam Sridhar for his hospitality at Monash University during a sabbatical leave from the University of Cambridge.

PARTICLE SEGREGATION IN COLLISIONAL SHEARING FLOWS

B. Ö. ARNARSON and J. T. JENKINS
Department of Theoretical and Applied Mechanics
Cornell University, Ithaca, NY 14853

M. Y. LOUGE
Sibley School of Mechanical and Aerospace Engineering
Cornell University, Ithaca, NY 14853

Abstract

We first indicate the place that collision flows have in the phenomena of particle segregation. Then we outline the development of a kinetic theory for particle segregation in a collisional shearing flow of a binary mixture of nearly elastic spheres. Finally, we indicate a means of testing the predictions of the theory against the results of computer simulation using only local values of data measured in the simulation.

1. Introduction

Most experimental studies of particle segregation are carried out on flows involving a free surface that are condensed by gravity into layers of grains in relative motion [1]. In each layer, grains interact with at least several of their neighbors through rubbing and bumping. Phenomenological models of segregation in such flows that take into account their significant features often provide a satisfactory description [1]. Because the interactions of a single particle with its neighbors are so complicated and because the statistical characterizations of the interactions certainly involve strong correlations between the positions and velocities of the particles, it is likely that a more detailed treatment is not possible.

Careful observations of particle segregation in less condensed free surface flows are less common. Such flows involve interactions between particles that are more violent than rubbing and bumping; these interactions can often be realistically modeled as binary collisions. Also, in these more agitated flows, the particle interactions are more random and, as a consequence, their statistical description is simpler.

Collisional flows can most easily be maintained in a shear cell. Typically, in such a cell, the flow takes place in a cylindrical annulus in which the bottom and sides are rotated and the top is held fixed [2]. The weight of the top is supported by collisions with the agitated grains of the shearing flow. This weight can be adjusted in order to maintain a fixed flow thickness or to vary the thickness in a controlled way. Knowledge of how the cell has been filled and the thickness of the flow permit the calculation of the average volume fraction. A measurement of the force necessary to keep the top from

223

A.D. Rosato and D.L. Blackmore (eds.), IUTAM Symposium on Segregation in Granular Flows, 223–229.

rotating permits the determination of the shear stress required to maintain the relative velocity of the bottom and top.

At high enough rates of shear, the weight of the grains in the cell is often a small fraction of the weight of the top of the cell. In this event, the influence of gravity on the flow may be neglected. However, the centripetal acceleration is always important; at the highest shear rates, it is typically more than five times the gravitational acceleration. Also, at the high shear rates necessary to render gravity negligible, collisions between the particles and between the particles and the boundaries may become so violent as to cause significant damage.

Theories for the prediction of particle segregation in collisional shearing flows make use of the analogy between the molecules of a dense gas and the agitated macroscopic grains that exists provided that the collisions between grains do not dissipate too much energy [3]. The most refined theory for segregation in dense molecular gases is a kinetic theory for mixtures of elastic spheres based on the assumption of molecular chaos, and the correct extension to mixtures of Enskog's characterization of the influence of the finite volume of the particles on their frequency of collision [4, 5]. The derivation of the theory employs the Chapman-Enskog expansion to obtain the velocity distributions to first order in spatial gradients. The Enskog approximation is then employed in order to get explicit expressions for the velocity distribution functions. The results of the theory are consistent with irreversible thermodynamics, have been presented graphically, and are available numerically.

In our application of this theory to mixtures of inelastic grains, we have characterized the amount of permitted dissipation and we have obtained analytical results valid to second order in the Enskog approximation for binary mixtures [3]. Analytical expressions permit the rational approximation of the coefficient governing transport, segregation, and dissipation over a range of mixture and species volume fractions and particle size and mass ratios.

The presence of dissipation in collisions permits collisional shearing flows of inelastic grains to achieve a steady balance, in which the rate at which the kinetic energy of the velocity fluctuations increases due to collisions driven by gradients in the mixture mean velocity, is equal to the rate at which it decreases due to the energy lost in each collision, and the rate at which it is transported due to inhomogeneities in the flow. The experimental study of shearing flows permits various aspects of the phenomena of segregation to be studied separately and helps in developing intuition about the behavior of more complicated mixtures.

In order to carry out the analysis of such experiments, we have extended existing theory for shearing flows of identical spheres in several ways. We have accounted for friction in collisions between particles in the flow by incorporating the additional energy loss into an effective coefficient of restitution [6]. We have extended existing boundary conditions for a single species interacting with either a bumpy or a frictional boundary to apply to a binary mixture interacting with a boundary that is both bumpy and frictional [7]. We have generalized existing theory for a dense shearing flow of a single species to apply to the prediction of the profiles of mixture mean velocity and mixture mean kinetic energy in dense shearing flows of a binary mixture between bumpy, frictional boundaries.

In order to incorporate friction into the kinetic theory, a simple but realistic model of a frictional collision must be employed. The simplest such model distinguishes between collisions in which the relative velocity of the points of contact is momentarily zero during a collision and those in which it is not. The former are called sticking collisions, the latter are called sliding collisions. The model employs three parameters: a coefficient of normal restitution for both sticking and sliding collisions, a coefficient of friction for sliding collisions, and a coefficient of tangential restitution for sticking collisions [8]. We have employed the model to interpret the results of experiments on binary collisions between identical spheres and collisions between a single sphere and a flat or bumpy boundary. The parameters determined in this way provide an excellent fit to the date [9, 10].

Computer simulations have played important roles in the development of granular mechanics [11, 12]. Such simulations of dissipative but frictionless disks and spheres have, in some instances, supported simple predictions of the kinetic theory and, in other instances, shown that collisional granular flows of frictionless inelastic particles could behave in ways far more complicated that predicted. A corresponding contribution from numerical simulations for frictional particles was not possible until a simple, physically realistic model of the frictional interaction between two colliding spheres could be verified in physical experiments. Using the simple model, numerical simulations have contributed to an understanding of the flow of inelastic frictional spheres down inclines, and in the determination of boundary conditions for frictional, inelastic spheres interacting with a flat, frictional elastic wall [13, 14].

2. Theoretical Framework

We sketch the structure of the theory for segregation in a binary mixture of spheres in a rectilinear, collisional shearing flow, first with gravity acting perpendicular to the flow and then in the absence of gravity. The detailed forms of the terms not given explicitly may be found in [3] and [15].

2.1 MOMENTUM BALANCE

In a simple unsteady, rectilinear flow, the horizontal and vertical components of the balance of momentum for the mixture as a whole may be written as

$$\rho\dot{u} = S' \quad \text{and} \quad 0 = -P' - \rho g$$

Respectively, where ρ is the mixture mass density, u is the horizontal component of the mass-center mixture velocity, S and P are the mixture shear stress and mixture pressure, g is the gravitational acceleration, and overdots and primes indicate derivatives with respect to time and vertical coordinates, respectively. The mixture shear stress and the mixture pressure are the components of the rates of momentum transfer per unit vertical area due to particle translation and particle collisions in directions parallel and perpendicular to the flow.

The mixture pressure P is proportional to the mixture fluctuation energy T:

$$P = \sum_{i=A,B} \sum_{j=A,B} K_{ij} T$$

where K_{ij} are complicated but known analytic functions of the diameters σ_A and σ_B of the two species and the number densities n_A and n_B, the number of particles of each species per unit volume. The mixture fluctuation energy T is two-thirds the sum of the average kinetic energy associated with the velocity of each species relative to the mixture velocity, weighted by the number fractions n_A/n and n_B/n of that species, where $n \equiv n_A + n_B$ is the mixture number density.

The shear stress S is proportional to u' :

$$S = \eta u'$$

where the shear viscosity η is

$$\eta \equiv \frac{1}{2} \sum_{i=A,B} b_{i0} \left(n_i + \frac{4}{5} \sum_{j=A,B} K_{ij} \frac{m_j}{m_{ij}} \right) T + \frac{1}{5} \sum_{i=A,B} \sum_{j=A,B} \left(\frac{2}{\pi} \frac{m_i m_j}{m_{ij}} T \right)^{1/2} K_{ij} \sigma_{ij}$$

where m_{ij} is the sum of mass m_i and m_j, σ_{ij} is half the sum of the diameters and the b_{i0}, associated with the perturbation to the Maxwellian velocity distribution function, are known analytic functions of the number densities, diameters, and masses of the two species that are given by the Enskog approximation at first order.

2.2 ENERGY BALANCE

Time dependent segregation of the spheres is described by an expression related to an approximate form of the difference between the balances of momentum for each species:

$$v_A - v_B = -\frac{n^2}{n_A n_B} D_{AB} (d_A + K_T T') \tag{1}$$

where v_A and v_B are each the vertical component of a diffusion velocity, defined as the average velocity of a species relative to the mixture velocity, D_{AB} and K_T are, respectively, the coefficients of ordinary and "thermal" diffusion, and d_A is the vertical component of the diffusion force. These have the form:

$$D_{AB} \equiv \frac{n_A n_B}{n} \frac{\sigma_{AB}}{K_{AB}} \left(\frac{\pi}{8} \frac{m_{AB}}{m_A m_B} T \right)^{1/2} ,$$

$$K_T \equiv \frac{2}{3} \sqrt{\pi} \frac{n_A n_B}{n} \sigma_{AB}^2 \left[\left(\frac{m_A}{m_B} \right)^{3/2} a_{A1} - \left(\frac{m_A}{m_B} \right)^{3/2} a_{B1} \right],$$

and

$$d_A \equiv -\frac{\rho_A}{\rho} \frac{1}{nT} P' + \frac{1}{nT} \left(n_A + K_{AA} + 2 \frac{m_A}{m_{AB}} K_{AB} \right) T' + \frac{n_A}{nT} \left(\frac{\partial \mu_A}{\partial n_A} n_A' + \frac{\partial \mu_A}{\partial n_B} n_B' \right),$$

where μ_A is the chemical potential of species A, with μ_A/T a known analytic function of the number densities and the diameters of the two species.

When $v_A - v_B$ is different from zero, segregation is taking place; when $v_A - v_B$ vanishes, a steady balance between the gradients of fluctuation energy and gradients of species number density is attained. Segregation in influenced by gravity through the presence of P' in d_A, with

$$P' = \frac{\partial P}{\partial n_A} n_A' + \frac{\partial P}{\partial n_B} n_B' + \frac{\partial P}{\partial T} T' = -\rho g. \tag{2}$$

Arnarson and Willits [15] follow Kincaid, Cohen, and Lopez de Haro [5] and characterize steady, fully developed segregation in the presence of gravity in terms of a segregation coefficient α_{AB} defined by

$$\alpha_{AB} \frac{T'}{T} \equiv \frac{n_A'}{n_A} - \frac{n_B'}{n_B}. \tag{3}$$

In this, n_A' and n_B' are expressed in terms T' using (2) and the steady form of (1):

$$0 = \frac{\rho_A}{nT} g + \frac{1}{nT}\left(n_A + K_{AA} + 2\frac{m_A}{m_{AB}} K_{AB} \right) T' + \frac{n_A}{nT}\left(\frac{\partial \mu_A}{\partial n_A} n_A' + \frac{\partial \mu_A}{\partial n_B} n_B' \right) + K_T T'.$$

(4)

Then, with the constitutive relations obtained at second order in the Enskog approximation, α_{AB} may be expressed as an explicit analytic function of five parameters: the total volume fraction v; the ratio ρ_A^s/ρ_B^s of the mass densities of the material of the spheres; the number fraction n_A/n; the ratio σ_A/σ_B of the diameters; and a parameter β that characterizes the strength of gravity relative to that of the temperature gradient, $\beta \equiv \rho g/nT'$. Arnarson and Willits [15] show contours of α_{AB} in the plane of density ratio and diameter ratio for various values of the total volume fraction, the number fraction, and gravitational strength.

In a steady fully developed state maintained in the absence of gravity, Eq. (4) becomes,

$$0 = \frac{1}{nT}\left(n_A + K_{AA} + 2\frac{m_A}{m_{AB}} K_{AB} \right) T' + \frac{n_A}{nT}\left(\frac{\partial \mu_A}{\partial n_A} n_A' + \frac{\partial \mu_A}{\partial n_B} n_B' \right) + K_T T'$$

This is the balance between gradients of number density and gradient of fluctuation energy that will be interrogated in the shear cell in a microgravity environment [16].

In Figure 1, we show a comparison between values of α_{AB} calculated in two different ways using the measured values of the number densities and mixture fluctuation energy in a computer simulation of a steady, fully-developed, collisional shearing flow carried out with no gravity. The spheres were made of the same material,

$\sigma_A/\sigma_B = 4/3$, the effective coefficient of restitution was 0.9, the mean mole fraction of species A was 0.2, and the mean mixture volume fraction in the simulation was 0.27. First, the data was differentiated in order to obtain local values of n_A'/n_A , n_B'/n_B , and T'/T ; then a numerical value of α_{AB} was determined using the definition given in Eq. (3). Second, α_{AB} was determined using the explicit function of v and n_A/n provided by the kinetic theory.

As indicated in the figure, over a range of mixture volume fractions from 0.15 to 0.21, the two values of α_{AB} differ by as much as twenty percent. However, this is not too bad for a comparison that involves differentiating the data. Direct comparisons of the fields of v, μ T, and n_A predicted by the kinetic theory and those measured in the computer simulations are possible, but require the solution of a boundary-value problem. Such comparisons will be discussed elsewhere.

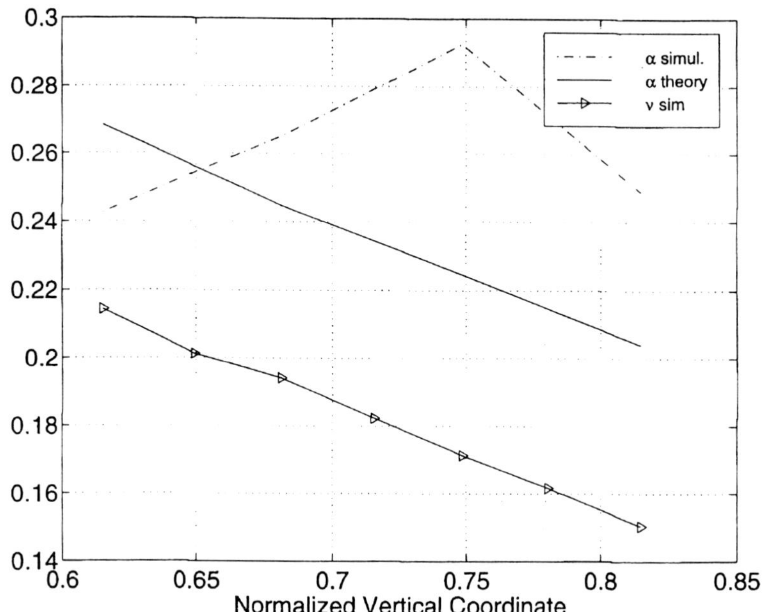

3. Acknowledgement

A significant part of this research has been supported by NASA's Microgravity Science and Applications Division under contract NCC3-468.

4. References

1. S. B. Savage and C.K.K. Lun: Particle size segregation in inclined chute flow of dry cohesionless granular solids, *J. Fluid Mech.* **189** (1988), 311-335.
2. K. Craig, R.H. Buckholtz, and G. Domoto: An experimental study of rapid flow of dry cohesionless metal powders. *J. Appl. Mech.* **53** (1986), 935-942.

3.   J. T, Jenkins and F. Mancini: Kinetic theory for smooth, nearly elastic spheres, *Phys. Fluids* **A1** (1989), 2050-2057.

4.   M. López de Haro, E.G.D. Cohen, and J .M. Kincaid: The Enskog theory for multicomponent mixtures. I. Linear transport theory. *J. Chem. Phys.* **86** (1987), 2746-2759.

5.   J.M. Kincaid, E.G.D. Cohen and M. López de Haro: The Enskog theory for multicomponent mixtures. IV. Thermal diffusion. *J. Chem. Phys.* **86** (1987), 963-975.

6.   J. T. Jenkins and C. Zhang: Kinetic theory for identical, slightly frictional, nearly elastic spheres. *Phys. Fluids*, under review (1997).

7.   J. T. Jenkins: Boundary conditions for collisional grain flow at a bumpy, frictional wall. *In preparation* (1997).

8.   O. R. Walton: Numerical simulation of inelastic, frictional particle interactions, in M. C. Roco (ed.), *Particulate Two-Phase Flow*, Boston, 1992, Butterworth-Heinemann.

9.   S. F. Foerster, M. Y. Louge, H. Chang, and K. Allia: Measurements of the collision properties of small spheres, *Phys. Fluids* **6** (1994), 1108-1115.

10. A. Lorenz, C. Tuozzolo, and M. Y. Louge: Measurements of impact properties of small, nearly spherical particles, *Experimental Mechanics* **37** (1997), 292-298.

11. C. S. Campbell: The stress tensor for simple shear flow of a granular material, *J. Fluid Mech.* **203** (1989), 449-473.

12. O. R. Walton and R. L. Braun: Stress calculations for assemblies of inelastic spheres in uniform shear, *Acta Mech.* **63** (1986), 73-86.

13. O. R. Walton, R. L. Braun, R. G. Mallon, and D. M. Cervelli: Particle-dynamics calculations of gravity flow of inelastic, frictional spheres, in M. Satake and J. T. Jenkins (eds.), *Micromechanics of Granular Materials – New Models and Constitutive Relations*, Elsevier, 1989, Amsterdam, pp. 153-162.

14. M. Y. Louge: Computer simulations of rapid granular flow of spheres interacting with a flat frictional boundary, *Phys. Fluids* **6** (1994), 2253-2269.

15. B. O Arnarson and J. T. Willits: Thermal diffusion in binary mixtures of smooth, nearly elastic spheres without gravity, *Phys. Fluids* **10** (1998), 1324-1328.

16. M. Y. Louge, J. T. Jenkins, A. Reeves, and S. Keast: Microgravity segregation in collisional granular shearing flows, in A. D. Rosato and D. Blackmore (eds.), *Segregation in Granular Flows*, Kluwer, 2000, Dordrecht.

ORIGINS OF SOME SIZE SEGREGATION PHENOMENA

"Why the Brazil Nuts are on Top" (in a nutshell!)

LESLIE V. WOODCOCK
Department of Chemical Engineering
University of Bradford
BRADFORD BD7 1DP
W. Yorkshire, U.K.

Abstract

Segregation of granular mixtures when fluidised by shaking or shearing under gravity, whereupon larger particles migrate upwards, may be partly explained by corresponding states relationships between a thermodynamic fluid and fluidised granular media at steady state. The predominant underlying effect is dilatancy. Particulate systems fluidised by shearing motions of the container tend towards the equipartition-of-energy. Then, scaling shows that larger particles exhibit higher dilatancy under steady shear. In the presence of gravity, the least dilatant size fractions, i.e. smaller particles, migrate downwards, and *vice versa*, to minimise the gravitational potential during shearing motions.

1. Introduction

The phenomenon of size segregation is a common phenomenon in everyday life, but there appears to be little or no formal scientific investigations of these effects prior to those of Williams in the early 1960's [1,2]. These early experimental papers were essentially designed to elucidate and describe the mechanisms for size segregation rather than to explain in a formal theory the origins of the effects. The paper of Bridgwater [3] greatly extended our knowledge and mechanistic description of size segregation. Despite much reported experimental research since, mainly in the chemical engineering literature, however, there is presently no generally accepted explanation as to "why" the large particles migrate upwards on shearing motions, rather than "how". The lack of an understanding of the underlying physical science appears to have been first brought to the attention of the physics community by a

A.D. Rosato and D.L. Blackmore (eds.), IUTAM Symposium on Segregation in Granular Flows, 231–240.
© 2000 *Kluwer Academic Publishers. Printed in the Netherlands.*

significant publication in *Physical Review Letters* of a paper with the title, "Why the Brazil Nuts are on Top" [4]. This paper essentially showed that controlled Monte Carlo simulation of idealised model disks also leads to size segregation phenomenon akin to our real life 3D observations. Further computer simulations have been reported since [5, 6], but a comprehensive explanation of size-segregation phenomena is still lacking. Controversy and speculation as to the origins of these size-segregation effects persists, and has extended more recently to disputes in the literature [7, 8].

Consider a mixture of large and small hard-spheres, which could be fluidised by vibration of its container under the influence of gravity. At rest, the packing densities of equal spheres are independent of the sphere size. There is no potential energy preference, therefore, for either particular sized sphere; the mechanics of systems at rest can provide no driving force for size segregation phenomena. The answer appears to rest in the study of steady fluidised states that do show size effects, with the dilatancy (expansion) due to shaking, or shear, varying as particle size.

Computational means have recently been developed [9, 10] to test, and subsequently enable some of the tenets of statistical thermodynamics to be applied to granular fluids. A program has been developed to simulate the steady-state fluidisation of granular inelastic spheres, by vibration [9]. The simulation monitors the granular pressure and the kinetic energies of the particles (sometimes referred to as the "granular temperature" and thereby tests applicability of the principle of equipartition-of-energy at steady state.

It has been found [10] that there is a fundamental driving force for dynamical granular systems to obey the equipartition-of-energy principle of equilibrium statistical mechanics. This requires a dynamical system of particles to continuously redistribute the kinetic energies of the particles evenly between all the available degrees-of-freedom, insofar as is possible within the constraints of the system. The particles continually receive their energy from dynamical driving surfaces, or external forces, dissipate the energy either by inelastic collisions, or viscous flow if there is a hydrodynamic medium present, and, at the same time, redistribute the kinetic energy according to the principle of equipartition.

2. Corresponding states

The general theory of corresponding states is applicable to all fluids that have interparticle potentials of the same functional form. If the interparticle potential contains an energy scale, e, and a distance scale, s, then simple dimensional analysis shows that the equation of state, or the pressure caused by a system of such particles can be expressed as,

$$P^* = f(T^*, V^*) \tag{1}$$

where the reduced variables P^*, T^* and V^* are defined by $P^* = P\sigma^3/\varepsilon$, $T^* = k_B T/\varepsilon$ (k_B is Boltzmann's constant), and $V^* = V/(N\sigma^3)$. The equation of state, when expressed in this reduced form, is the same for all fluids having the same form of potential. The values of P, T and V are determined by the parameters ε and σ, which will be different for different fluids. This, in effect, is a statement of the theory of corresponding states: at equivalent values of reduced temperature, the P-V isotherms of different fluids only differ in the scaling along the P and V axes. If plotted in terms of the reduced units P and V, the isotherms would coincide.

The hard-sphere model has no characteristic scale of energy, only one of length. An energy scale may be introduced by choosing an arbitrary value for the system kinetic energy, $E_0 = m<v^2>$, where m is the particle mass and $<v^2>$ is the mean-squared velocity. Different systems of hard spheres at the same density but at different energies are however equivalent, by a simple scaling of the times over which the different systems evolve. The characteristic unit of time for such a hard-sphere system is in fact determined by the arbitrary energy, E_0. There is no significance in the value of E_0 chosen for the energy scale; an increased value of E_0 is equivalent to a faster system evolution, in the form of a smaller unit of time. When the system experiences no explicit time-dependent perturbations (i.e. it is at equilibrium), all different energies, also temperatures and pressures, may be simply scaled from any other energy state at the same density. The determination of a single isotherm T(V), is therefore, a complete characterisation of the equilibrium P-V-T behaviour of the hard-sphere fluid.

This corresponding states theory may be utilised for systems of particles not strictly in equilibrium, i.e. undergoing deformation or heat flow, provided that the system is not perturbed too far from equilibrium. This deliberately vague statement may be made more precise by observation of the results of computer simulation experiments of the steady shear of hard spheres. A region of Newtonian flow behaviour is found at low shear rates, within which the corresponding states theory may be applied. Non-linear non-Newtonian flow behaviour is found when the mean flow velocity and the mean thermal velocity are approximately equal. In terms of the reduced units of the thermalised hard-sphere fluid (where the unit of mass is m, that of length is the particle diameter s, and that of time is $(m\sigma^2/kT)^{0.5}$ these are called 'isokinetic' units, and are denoted by the superscript †), nonlinearity occurs when the reduced shear rate is approximately equal to one:

$$\dot{\gamma}^{\dagger} \equiv \dot{\gamma}\sqrt{\frac{m\sigma^2}{E_O}} = \frac{\sigma\dot{\gamma}}{\sqrt{E_O/m}} \approx 1 \quad . \tag{2}$$

Once this non-linear flow behaviour is reached, the corresponding states theory becomes inapplicable, as the mean shear flow disrupts the equilibrium structure.

The use of dimensional analysis and reduced variables enables one to establish complicated scaling relationships between seemingly disparate systems. The flow behaviour of inelastic or rough powder particles can be directly equated with that of solid particles in suspension, and also with the artificially thermalized hard-sphere fluid. It is now well known that simple fluid models such as soft and hard spheres exhibit high flow rate behaviour, such as shear thinning and shear thickening, characteristic of more complex fluids [11]. Here it is shown that the linear, low-shear flow rate behaviour of the hard-sphere fluid, together with its scaling relation to the simplest granular-sphere model, allows for the analytic prediction of the non-linear behaviour of this latter system.

3. Rapid granular flow

The ideal rapid granular flow system is defined to be a large number of monodisperse frictionless spheres undergoing laminar homogeneous shear flow. The stipulation of laminar homogeneous shear flow is the simplest flow geometry possible, which also excludes the effects of any surfaces or boundaries on the behaviour of the system. Such a system may be characterised by the following variables: describing the particles, the mass (m), diameter (σ), and coefficient of inelasticity (K); the system state is specified by the packing/solid fraction (y) and the shear rate (γ).

The behaviour of the idealised granular system is an unknown function of all these system variables. An element of the stress or pressure tensor, Pij, for example can be expressed

$$Pij = f\,(m, \sigma, K, y, \gamma) \tag{3}$$

By applying Buckingham's Pi Theorem, the six variables in this equation can be reduced to three dimensionless groups to give:

$$P^{\#}{}_{ij} = f\,(K^{\#}, y) \tag{4}$$

where $K^{\#} = K/m$ for a velocity-independent coefficient-of-restitution e (defined below). Since the real stress has dimension t^{-2}, this approach immediately generates the relationship between stresses and the square of the shear rate, originally obtained by Bagnold [12] using an approximate argument, for rapid granular flow. The result is thus seen as a consequence of describing the granular system in its natural units (where the unit of mass is m, that of length is σ, and that of time is the inverse of the shear rate; these are called 'isoshear' units, and are denoted by the superscript (#)). The stresses generated, when expressed in the reduced form above, depend only upon an unknown function of the particle packing fraction and the particle inelasticity.

For a typical 'head-on' collision, the relative velocity after a collision will be less than the relative velocity before the collision. Any collision may be characterised by the ratio of final to initial relative velocities, which is known by the misnomer coefficient- of-restitution (the restitution ratio would be more appropriate)

$$e = -\frac{v'_i - v'_j}{v_i - v_j} \tag{5}$$

where v_i and v_j are the velocities before the collision, and v'_i and v'_j are the velocities after the collision.

This definition was introduced by Isaac Newton in the seventeenth century in the early origins of mechanics. Although, e is now known not to be a material constant, Newton's simple approximation is widely retained in the study of collision dynamics of granular materials.

Using the principle of momentum conservation the energy lost in such a collision is given by:

$$\Delta E = \frac{m}{4}(1 - e^2)v_0^2 \tag{6}$$

where v_o is the impact velocity. The exact form of e for a collision between two bodies is not precisely known, but is usually accurately represented as a decreasing function of the impact velocity v_o. For the generalisation to arbitrary collisions, the impact velocity is defined as the component of the relative velocity taken along the normal to the surface at the point of impact. For spherical particles this is equivalent to the component of relative velocity along the line of particle centres:

$$v_0 = (\mathbf{v}_i - \mathbf{v}_j) \bullet \mathbf{k} \tag{7}$$

where $\mathbf{k} = (\mathbf{r}_j - \mathbf{r}_i)/\left|\mathbf{r}_j - \mathbf{r}\right|$.

If e is taken to have a constant value, the energy loss in a collision (ΔE) is proportional to the square of v_o:

$$\Delta E = Kv_0^2 \tag{8}$$

with $e = \sqrt{1 - 4K/m}$.

This particular case is of special interest, because it is the commonest approximation made for the coefficient-of-restitution in previous theoretical treatments. Its importance in the present analysis is that it introduces no new unit of time into the model granular system, so that the natural unit of time remains the inverse of the shear rate. Using isoshear units to scale all physical quantities ensures that these scaled quantities have no dependence on the shear rate. More simply put, the reduced shear rate in isoshear units is simply $\gamma^\# = 1$.

It is straightforward to relate quantities of interest expressed in isoshear units to those same quantities expressed in isokinetic units. The relevant scaling relationship for the pressure tensor is [13]:

$$P^\#(y,e) = P^\dagger(y,\dot{\gamma}^\dagger)/(\dot{\gamma}^\dagger)^2 \tag{9}$$

When a granular system is subjected to a shear deformation, the action of the shear field tends to increase the individual particles' kinetic energies. Energy is removed, in the form of heat, by inelastic collisions, but there is no particular significance in the actual method of energy removal or thermosetting mechanism used to reach steady state conditions. Alternate sources of dissipation, including viscous drag due to an interstitial fluid, or even artificial thermalisation (e.g. a periodic rescaling of the velocities) also give rise to steady states. Two different nonequilibrium systems correspond when they have equal rates of energy dissipation (which must also be equal to the rate of energy being input, by virtue of the steady-state condition). This energy balance condition provides an explicit relation between the particle inelasticity and the shear rate; when used in conjunction with scaling relationships such as (9), this allows for a completely interchangeable description of the corresponding systems.

For a system of N perfectly elastic, frictionless, spherical particles in a volume V, undergoing a steady shear deformation, with a rate specified by the velocity gradient dv_x/dy, the rate of shear work causes an energy input per unit time of

$$Q_{in} = P_{xy}\dot{\gamma}V \tag{10}$$

Using the definition of an effective viscosity,

$$\eta = P_{xy}/\dot{\gamma} \tag{11}$$

And substituting in Equation (9) gives,

$$Q_{in} = \eta\,\dot{\gamma}^2\,V \tag{12}$$

For a system particles undergoing homogeneous steady-state shear, this power input must be balanced by the energy dissipation due to inelastic collisions. This may be expressed as:

$$Q_{out} = N\nu\langle\Delta E\rangle \tag{13}$$

where <ΔE> is the average kinetic energy lost in an inelastic collision, and ν is the average collision frequency per particle and is known exactly given the equation of state for a thermalised hard-sphere fluid,

$$\nu^\dagger = \nu\left(\frac{m\sigma^2}{E_o}\right) = \frac{3}{\sqrt{\pi}}\left(\frac{PV}{NE_o}-1\right) \tag{14}$$

where P is the system pressure and E_o replaces the usual thermodynamic term ($k_B T$). By equating (12) and (13), and making use of the exact result [13],

$$\left\langle v_o^{\dagger 2}\right\rangle = m\left\langle v_o^2\right\rangle / E_o = 4 \tag{15}$$

then

$$1-e^2 = \frac{\sqrt{\pi}\,\eta^\dagger\dot{\gamma}^{\dagger 2}}{12\rho^\dagger(Z-1)} \tag{16}$$

The left-hand side gives the value of the coefficient of inelasticity required to thermalise the granular system sheared at the reduced shear rate $\dot{\gamma}^\dagger$. The principle of corresponding states predicts that each value of the coefficient of restitution in an inelastic system is mapped onto an isokinetic system with a different value of the isokinetic shear rate.

Equation (16) may be simply understood via the following thought experiment; consider imposing a steady shear on a granular system which is perfectly elastic, e = 1. The shear will cause the granular kinetic energy of the system to continually rise as a function of time, and the system will never achieve a steady state. This granular system is equivalent to an isokinetic system where the shear rate is a steadily decreasing function of time ($\dot{\gamma}^\dagger = \dot{\gamma}\sqrt{m\sigma^2/E_0}$). Eventually, the isokinetic shear rate will drop to a low enough value so that the isokinetic system is in its Newtonian region, after which point in time the quantities $\eta^\dagger$, $\rho^\dagger$ and Z appearing in Equation (16) are equal to their equilibrium values. The Newtonian-equilibrium region is thus seen to be the region where kinetic energy gradients are dissipated on a time scale, which is short, compared to the inverse of the shear rate. For granular systems which have values of e close to 1, repeating the thought experiment shows that different values of e (and so the complete flow curve, by virtue of Equation (4)) are effectively described by the same equilibrium hard-sphere quantities $\eta^\dagger$ and Z for any $\rho^\dagger$.

4. Rheology predictions

These scaling relationships, Equations (9) and (16), may be used to make quantitative predictions of non-Newtonian rapid granular flow behaviour without recourse to any numerical simulations of the granular system. By making use of our knowledge of the equilibrium hard-sphere fluid [14,15] we may obtain values of the quantities $\eta^\dagger$ and Z as a function of $\rho^\dagger$ or y (the volume fraction $y=\pi\rho^\dagger/6$) . The mean normal pressure of the hard-sphere fluid may be found from an accurate empirical equation of state. For dense amorphous systems the following simple free volume equation suffices,

$$\frac{PV}{NE_0} = \frac{3y/y_a}{1-y/y_a} \tag{17}$$

where y_a is the maximum amorphous close-packing volume fraction for random-packed spheres (approximately $2/\pi$), y_0 the maximum crystal packing fraction ($\pi/6$). The equilibrium hard-sphere fluid viscosity may also be accurately expressed over the entire density range up to a glass transition by a similar empirical relationship we have found empirically

$$\eta^\dagger = \frac{\eta\sigma}{\sqrt{mE_0}} = \eta_0^\dagger\left[1+3\pi\left(\frac{y}{1-(y/y_a)}\right)^{3/2}\right] \tag{18}$$

where $\eta_0^\dagger = \dfrac{5}{16\sqrt{\pi}}$ is the kinetic theory low-density result, and y_a is again the volume fraction at amorphous close packing. Expressions (17) and (18) allow calculation of $\eta^\dagger$, $\rho^\dagger$ and Z for any packing fraction as input.

More importantly, the scaling relationships Equations (9) and (16) provide qualitative predictions of non-Newtonian behaviour for the granular system in the region of rapid granular flow. Using Equation (9) we may express the system mean normal pressure as,

$$P = \frac{m\dot{\gamma}^2}{\sigma}P^{\#}(e,y) = \frac{m\dot{\gamma}^2 P^{\dagger}(\dot{\gamma}^{\dagger^2},y)}{\sigma\dot{\gamma}^{\dagger^2}} \qquad (19)$$

We make now use of Equation (16) to express $\dot{\gamma}^{\dagger}$ as a function of e; choosing a given value of e fixes $\dot{\gamma}^{\dagger}$ as a function of y, leaving for the pressure only the explicit dependence on $\dot{\gamma}$ and, for e close to 1, the equilibrium of the pressure, a known function of y:

$$P = \frac{m\,P^{\dagger}(y)}{\sigma(\dot{\gamma}^{\dagger}(y))^2}\dot{\gamma}^2 = \text{Constant}(y)\sigma^2\dot{\gamma}^2 \qquad (20)$$

Equation (20) is simply the statement that the granular material is dilatant, i.e. its mean normal pressure increases with shear rate squared. At ambient pressure a sheared granular material in rapid flow expands. Dilatancy is a fundamentally non-Newtonian phenomenon.

The predicted constitutive rheology of these idealised granular materials is everywhere shear thickening, and everywhere shear dilatant, in the rapid granular flow limit. This result represents a remarkable analytic prediction of a complex, non-linear rheological flow phenomena from the equilibrium behavior of the classical hard-sphere fluid and its linear Newtonian viscosity, <u>without introducing any new physics.</u>

The collisional constitutive behaviour is analytically predictable by these scaling laws over the entire range of shear rate. The dependence of the stresses on the square of the shear rate is independent of the extent of inelasticity and represents exact confirmation the earlier predictions of this result by Bagnold [12] from empirical observations and *ad hoc* approximations.

The same principles can be applied to much more complex material behaviour, including a more realistic dependence of the coefficient of restitution on impact velocity, the effects of surface friction and the influence of an interstitial fluid. In fact, the particle model chosen could contain all these effects and the same scaling arguments applied. However, the more complex the model, the more parameters there are to be determined and their effects on the overall behaviour must be elucidated.

Since the normal pressure is a constant up to the onset of nonlinearity, and the stress is proportional to the rate of shear, this information can be used in corresponding-states scaling treatments to predict the corresponding shear flow curve for a system of inelastic spheres of any prescribed degree of inelasticity for rapid granular flow. If the coefficient of restitution (e) is taken to be a constant, scaling predicts that the effective steady-state homogeneous pressure tensor of a dry idealised fluid of frictionless monodisperse hard spheres is shear thickening over it's entire range of existence, and has the simple form (equation 19 of reference 13),

$$p^{\neq}(y,e) = \frac{p^{\dagger}(y,\gamma^{\dagger})}{\gamma^{\dagger^2}} \qquad (21)$$

in which the right hand side is determined entirely from the known viscosity (η) and pressure equation-of-state $p(V,T)$ of the equilibrium hard-sphere fluid at the same density. The corresponding state scaling equation (4) becomes analytically exact in the limit of high shear rate, i.e. rapid granular flow.

5. Shear thickening and Dilatancy

The result for the shear stress as a function of shear rate is

$$p_{xy}(y,\sigma,\gamma) = \frac{m\gamma^2\eta^\dagger}{\sigma\gamma^{\dagger 2}} = \text{Const.}\times\sigma^2\gamma^2 \tag{22}$$

and the shear induced increase in normal pressure, i.e. the trace of **p** is

$$Tr|\mathbf{p}| = \frac{1}{3}(p_x + p_y + p_z) = \frac{m\dot\gamma^2 z^\dagger\rho^\dagger}{\sigma\dot\gamma^\dagger} = \text{Const}\times\sigma^2\dot\gamma^2 \tag{23}$$

Equations (22) and (23) show that in the region of rapid granular flow, the hard-sphere fluid is everywhere shear thickening (i.e. increasing effective viscosity with shear rate), and everywhere shear dilatant. This follows because, under only the materials own gravitational weight, the granular pressures generated by shear are manifested as dilatancy. The degree of dilatancy, moreover, for steady shear under a prescribed density increases as sphere diameter squared. In the presence of a sufficient external field such as gravity, where the potential energies are dominant, these steady shearing motions will also, therefore, give rise to the segregation effect with the least dilatant size fractions (smaller particles) moving in the direction of gravity, and vice versa, in its attempt to minimize the potential energy under shear motion.

6. Conclusion

It is suggested that the present scaling relationships for idealised fluidised steady - states provide the underlying driving force, and hence reason, for many, but not all, hitherto inexplicable size segregation phenomena. For most macroscopic granular materials, this steady-state condition is difficult to attain. Under some conditions of processing, for example, real fluidised granular materials are clearly far from the steady state purported here. Nevertheless, all evolutionary time-dependent processes, undergoing localized motion at an instant in time, will correspond to some steady state. The steady-state scaling effects for spheres are so strong, as to suggest that they must always be present, if not dominant, in complex granular media with polydispersity of size, shape, density and other properties. Thus whilst the scaling effect reported here may not be the whole story of the diversity of size segregation phenomena, it must always be a significant driving force for size segregation in systems which are being sheared by shaking.

7. References

1.   Williams, J. C.: The Segregation of powders and granular materials, *Fuel. Soc. J* (Sheffield) **14**, (1963), 29-34.
2.   Williams, J. C.: The segregation of particulate materials. A review, *J. Powder Tech.* **15** (1976) 245-251.

3.   Bridgwater, J.: Fundamental powder mixing mechanisms, *Powder Tech.* **15** (1976), 215-236.

4.   A. D. Rosato, A. D., Strandburg, K. J., Prinz, F., Swendsen, R. H.: Why the brazil nuts are on top: size segregation of particulate matter by shaking, *Phys. Rev. Lett.* **58** (1987), 1038-1042.

5.   Jullien R., Meakin, P. and Pavlovitch A.: Three-dimensional model for particle-size segregation by shaking, *Phys. Rev. Lett.* **69** (1992), 640-643.

6.   Fitt, A.D. and Wilmott, P.: Cellular-automaton model for segregation of a two-species granular flow, *Phys. Rev. A* **45** (1992), 2383-2388.

7.   Maddox, J.: Why pebbles float to the surface, *Nature* **358** (1992), 535.

8.   Barker, G.C. and Mehta, A. : Size segregation in powders, *Nature* **361** (1993), 308.

9.   Woodcock, L.V.: Steady states in granular dynamics, in C. Thornton (ed.) *Powders and Grains,* Balkema, Rotterdam 1993, pp. 261-264.

10.   Knight, T.A., and Woodcock, L.V.: Test of the equipartition principle for granular spheres in a saw-tooth shaker, *J. Phys A Math. Gen.* **29 (1996),** 4365.

11.   Hoover, W.G , Computer simulation of many-body dynamics, *Physics Today* **37** (1984), 44-50.

12.   Bagnold, R.A.: Experiments on a gravity-free dispersion of large solid spheres in a Newtonian fluid under shear, *Proc. Roy. Soc.* **A225** (1954) 49-63.

13.   Turner, M. C. and Woodcock, L.V., Scaling laws for rapid granular flow, *J. Powder Tech.* **60** (1990), 47-60.

14.   Alder, B.J. & Wainwright T.E.: Studies in molecular dynamics II: behavior of a small number of elastic spheres, *J. Chem. Phys.* **33** (1960), 1439-1451.

15.   Alder, B.J., Gass, D.M. & Wainwright, T.E.: Studies in molecular dynamics VIII: the transport coefficients for a hard-sphere fluid, *J. Chem. Phys.* **53**, (1970), 3813-3825.

THE MECHANICS OF PARTICLE-FILLED FLOWS AT HIGH SOLIDS VOLUME FRACTION

E. BRUCE PITMAN
Department of Mathematics
State University of New York
Buffalo, NY

Abstract

Classical modeling predicts instability when a bed of powder is fluidized by gas. Older experiments by Rietema and colleagues, and more recent ones by Tsinontides and Jackson, present a compelling argument that, at least for Geldart Type-A powders, there exists an interval of stable fluidization. Moreover, these experiments indicate that constitutive models for the particle phase must account for plastic, solid-like, stresses, including tensile strength. We propose a simple one-dimensional model accounting for some of the observed behavior. We also present preliminary results from a numerical model for particle-fluid interaction.

1. Introduction

Conventional fluidized beds models postulate mass and momentum conservation equations for solid and fluid. There are two sources of difficulty that are encountered in the modeling. First is the constitutive relation for the particles. This is usually modeled by a fluid-like relation, with viscosity and pressure depending on the solids volume fraction. The second difficulty is the particle-fluid interaction. This is usually modeled by a drag term, with the drag coefficient dependent on the solids volume fraction. Although there have been some successes with these theories, including recent progress on the development of instabilities, there are many issues still not resolved. The paper [3] gives a perspective of these developments.

When powders are fluidized by gas, the conventional theory predicts linear instability. That is, once the fluid flow rate is greater than the mass of the powder bed (plus any tensile strength), the powder should bubble or a channel form. However some very fine powders, classified by Geldart as Type A, do appear stable for an interval of gas flow rates during fluidization. The origin of this stable behavior is not clear.

Over a decade ago, Rietema and colleagues studied fine powders by fluidizing a bed of material and tipping it [14]. He noted that the upper free surface of the bed did not remain horizontal, as a fluid would, but rather it inclined during tipping, up to some yield threshold. These experiments convinced Rietema that a solids stress was active in

A.D. Rosato and D.L. Blackmore (eds.), IUTAM Symposium on Segregation in Granular Flows, 241–254.
© 2000 *Kluwer Academic Publishers. Printed in the Netherlands.*

the powder, and he postulated an elastic origin of this stress to explain the experimental findings.

More recently, Tsinontides and Jackson [18] have looked carefully at the cycle of fluidizing and de-fluidizing. They discovered several interesting facts:

-   a hysteresis in the pressure drop through the bed, as gas flow rate cycles;

-   history dependence of bed height at zero gas flow, after fluidization-defluidization cycling;

-   striations in particle volume fraction, as gas is flowing.

Taken together, these experiments provide clear evidence that in a bed of granular-material – at least Type-A powders – modeling of stress must include a solid-like elastic or plastic component; see [1, 5]. This paper contains a modeling effort that at least partially accounts for the observed phenomenon. The principle contribution here is a one-dimensional, two-phase model that is similar to classical Coulomb friction models; this model is introduced in the next section. Linearized stability is examined, and dependence on model parameters is explored in Section 3. Finally, preliminary results from a two-dimensional computation of fluid-particle flow are presented. Experimental powders range from 100μ diameter cracking catalyst [18] to 10μ xerographic toners [20], and the specifics of the powder influence parameter choices later.

2. Two-Phase Modeling

We being with standard equations for two-phase mass and momentum balance laws [5]

$$\frac{\partial}{\partial t}\phi + \nabla\cdot(\phi\mathbf{v}) = 0 \tag{1}$$

$$\frac{\partial}{\partial t}(1-\phi) + \nabla\cdot((1-\phi)\,\mathbf{u}) = 0 \tag{2}$$

$$\rho_s\phi[\frac{\partial}{\partial t}\mathbf{v}+\mathbf{v}\cdot\nabla\mathbf{v}]+\nabla\cdot T_s = -\phi\nabla P+(1-\phi)\beta(\mathbf{u}-\mathbf{v})+\phi\rho_s\mathbf{g} \tag{3}$$

$$\rho_f(1-\phi)[\frac{\partial}{\partial t}\mathbf{u}+\mathbf{u}\cdot\nabla\mathbf{u}]+\nabla\cdot T_s = -(1-\phi)[\nabla P+\beta(\mathbf{u}-\mathbf{v})+\rho_f\mathbf{g}] \tag{4}$$

Here ϕ is the solids volume fraction, composed of particles of density ρ_s and moving with velocity $\mathbf{v}$ in a fluid of density ρ_f moving with a velocity $\mathbf{u}$. The solid stress is denoted T_s, and the gas pressure, the only fluid stress considered, is P, and $\mathbf{g}$ is gravity. The drag $\beta = \beta(\phi) = (\phi\rho_s g)/(v_T(1-\phi)^{n-1})$, where v_T is the terminal particle free-fall velocity (about 10 cm/sec for the catalyst, and about 0.3 cm/sec for toner powder in air) and the exponent n is about 4.

It is useful to scale these equations, to determine the most important terms. However in doing so, choices must be made for characteristic velocity and length scales; it is not at all clear what are the proper choices for these scales. We will use a length scale set by a typical experimental vessel (on the order of $1 - 10$ cm in diameter, or up to about 5 - 40 cm in total bed height); a velocity scale set by v_T. For cracking catalyst, the density is about 1.5 g/cm^3; xerographic toner powders have a density just a little above 1 gm/cm^3. There is a three order-of-magnitude difference in the densities of particle and fluid; we drop all terms of $O\!\left(\rho_f/\rho_s\right)$. This non-dimensionalization leads to the following system:

$$\frac{\partial}{\partial t}\phi + \nabla \cdot (\phi \mathbf{v}) = 0 \tag{5}$$

$$\frac{\partial}{\partial t}(1-\phi) + \nabla \cdot \left((1-\phi)\mathbf{u}\right) = 0 \tag{6}$$

$$\phi[\frac{\partial}{\partial t}\mathbf{v} + \mathbf{v}\cdot\nabla\mathbf{v}] = \frac{-1}{Fr}[\nabla \cdot T_s + \nabla P + \phi \mathbf{e}_z] \tag{7}$$

$$\nabla P = -\beta(\mathbf{u}-\mathbf{v}) \tag{8}$$

The parameter $Fr = U^2/gL$ is the Froude number, and is about 2 for glass beads, and about 10^{-5} for toner powder. In writing the solids momentum equation, the gas momentum equation has been used to simplify terms. The term $\mathbf{e}_z$ is a unit vector in the z-direction – this is the non-dimensional form of the gravitational body force.

It is sometimes useful to recast the equations for gas pressure and the sum of the mass balance laws to write:

$$\frac{\partial}{\partial t}\phi + \nabla \cdot (\phi \mathbf{v}) = 0 \tag{9}$$

$$\phi[\frac{\partial}{\partial t}\mathbf{v} + \mathbf{v}\cdot\nabla\mathbf{v}] = \frac{-1}{Fr}[\nabla \cdot T_s + \nabla P + \phi \mathbf{e}_z] \tag{10}$$

$$\nabla \cdot \frac{1-\phi}{\beta}\nabla P = \nabla \cdot \mathbf{v} \tag{11}$$

The gas velocity can be recovered as $\mathbf{u} = \mathbf{v} - \frac{1}{\beta}\nabla P$.

At this stage, classical particle-fluid modeling assumes the particle stress is fluid-like, and can be written

$$T_s = P_s + \mu_s (\nabla \mathbf{v} + \nabla \mathbf{v}^T)$$

where P_s, μ_s are functions of ϕ. It is precisely the weaknesses of this hydrodynamic model that we hope to overcome by including solids frictional constitutive relations.

Before proceeding, it is worth recalling facts about single phase (i.e, particle only) frictionally-based constitutive laws. Schaeffer [15] showed that using an incompressible Mohr-Coulomb constitutive model, the governing PDEs are always ill-posed. That is, the equations linearized about a constant solution look like a mixed forward-backward equation $\partial_t w = \partial_{xx} w - \partial_{yy} w$. Such a model is not useful. Later, [12] showed that using a critical state soil mechanics model greatly mollified the linearized blow-up; two dimensional flows are ill-posed on the expansion side of the yield locus, but well-posed on the consolidation side, and three dimensional flows are well-posed. These results give us warning that any model system we choose for particle-fluid flow may contain instabilities due to the solids-only constitutive relation.

Two further reductions should be mentioned before specific stress relations are considered. First, consider the limit $Fr \to 0$. The inertial terms in the solids momentum drop out, leaving

$$\nabla \cdot T_s + \nabla P = -\phi \mathbf{e}_z$$

and we have but one PDE with explicit time-dependence. (We mention that a study of a similarly reduced system for solids only shows the equations are well-posed; see [16].)

Second, consider a model in 1 space dimension. Because of the reduction to one dimension, wall friction must be taken into account (in other words, average the momentum equations in the directions orthogonal to the vertical, and incorporate the lateral stress boundary conditions into the remaining momentum equation). We have

$$\frac{\partial}{\partial t}\phi + \frac{\partial}{\partial z}(\phi v) = 0 \tag{12}$$

$$\phi[\frac{\partial}{\partial t} v + v\frac{\partial}{\partial z} v] = -\frac{1}{Fr}[\frac{\partial}{\partial z} T - \alpha T + \frac{\partial}{\partial z} P + \phi] \tag{13}$$

$$\frac{\partial}{\partial z}(\frac{1-\phi}{\beta}\frac{\partial}{\partial z} P) = \frac{\partial}{\partial z} v \tag{14}$$

Again, $u = v - \dfrac{1}{\beta}\dfrac{\partial}{\partial z} P$. Note that a wall stress factor has been introduced. In the $Fr \to 0$ limit, the momentum equation becomes a Janssen equation, generalized to include the gas pressure gradient.

Now turn attention to the particle stress constitutive model, written only for the 1-dimensional model systems. Perhaps the simplest relation would be to assume a volume-fraction dependency, not unlike both the critical state and the classical hydrodynamic models. A simple fit for compressive stresses might look like

$$T = T(\phi) = \frac{\phi^2}{(\phi_o - \phi)^2}$$

where ϕ_o is the maximum solids packing. A similar kind of relation might hold for tensile stresses. However transiting between compressive and tensile stress relations is difficult without further information about the deformation. Moreover, a relation like this assumes plastic deformation everywhere at all times, a condition not likely met in experiments. To overcome this shortcoming, some form of elastic-plastic constitutive relation must be postulated. In a simple elastic-plastic model, the plastic yield locus causes a singularity in the stress relation. We assume instead an elastic-visco-plastic relaxation relation,

$$\frac{\partial}{\partial t}T + v\frac{\partial}{\partial z}T + E^2\frac{\partial}{\partial z}v = -\frac{1}{\tau}\langle T - Y\rangle$$

The first two terms on the left are the usual convective derivative, the third term is an elastic response, with modulus E^2. Note the elastic response is positive in compression and negative in tension, which is opposite from the usual convention in continuum mechanics, but consistent with that in soil mechanics. These three terms are, in effect, a rate form of the linear elastic stress-strain (Hooke's) law. On the right, $Y = Y(\phi)$ is the yield locus, the compressive and tensile relations given above, and τ is the relaxation time. The brackets are defined as

$$\langle T - Y\rangle = \begin{cases} T - Y & |T| > |Y| \\ 0 & \text{otherwise} \end{cases}$$

This equation says that the stress responds elastically to the strain rate $\dfrac{\partial v}{\partial z}$ and relaxes back to the appropriate yield locus in a characteristic time τ. Finally, it should be noted that, when non-dimensionalized like the balance laws, the form of this stress equation remains the same, with the elastic modulus scaled by the stress scaling $\rho_s gL$.

3. Analysis of Model Equations

In this section, we present a brief analysis of the time independent model equations in one dimension. Consider only plastic deformation, with a plastic yield function as above

$$T = T(\phi) = \frac{\phi^2}{(\phi_o - \phi)^2}; \quad \text{inverting,} \; \phi = \phi_o\frac{\sqrt{T}}{1+\sqrt{T}}.$$ We consider $v = 0$. Then in a coordinate system with $z = 0$ at the base and $z = H$ at the upper free surface of the granular bed,

$$T(z) = T(0) + \int_0^z -\alpha T + \phi + \beta \frac{u(0)(1 - \phi(0))}{1 - \phi} dz$$

Using the stress-volume fraction relation, this is an integral relation for T, dependent on the parameters u and ϕ at the base:

$$T(z) = \int_0^z F(T; u(0), \phi(0)) dz.$$

We now turn to a linear stability analysis, in the zero Froude number limit system. We think of the volume fraction as a function of stress. There is an approximate solution deep in a bin, much like a Janssen solution, in which the solids fraction, the stress, and the gas velocity are constant, and the solids velocity is zero; the gas pressure gradient, also constant, counteracts the gravitational body force. Linearize around this solution. (In the following, a "0" subscript implies evaluation at the steady solution). We separate the elastic and plastic cases, to illustrate the role of the modeling assumptions.

$$\frac{\partial}{\partial t}\phi + \phi_o \frac{\partial}{\partial z} v = 0 \tag{15}$$

$$\frac{\partial}{\partial z}T - \alpha T + \frac{\partial}{\partial z} P = -\phi \tag{16}$$

$$B_o \frac{\partial}{\partial z}\frac{\partial}{\partial z} P = \frac{\partial}{\partial z} v \tag{17}$$

$$\text{either } \frac{\partial}{\partial t}T = -E^2 \frac{\partial}{\partial z} v \quad \text{or} \quad T = Y(\phi) \tag{18}$$

Here, B_o is the D'Arcy drag coefficient at the constant solids fraction ϕ_o. In the last equation, the two cases correspond to elastic and rigid-plastic deformation. After algebraic manipulation, we find

-Elastic Case

$$\frac{\partial}{\partial t}\phi - B_o (\phi_o + \alpha E^2) \frac{\partial}{\partial z}\phi = B_o E^2 \frac{\partial}{\partial z}\frac{\partial}{\partial z}\phi$$

The solution is stable.

-Plastic Case

$$\frac{\partial}{\partial t}\phi - B_o \phi_o (\alpha Y_o' - 1) \frac{\partial}{\partial z}\phi = B_o \phi_o Y_o' \frac{\partial}{\partial z}\frac{\partial}{\partial z}\phi$$

The solution is stable, if $Y_o' > 0$ - that is, for compressive stresses. In tension, $Y_o' < 0$ and the solution is unstable.

It is worth noting that the plastic instability is of the form of a backward heat equation with advection – so a smooth solution "roughens", and the linearized growth rate is $O(\xi^2)$, where ξ is the wave number.

As the gas pressure increases to the point of balancing the bed weight, there is a question whether this constant solution remains valid. Plots like Figure 22 of Tsinontides and Jackson suggest that the assumption may not be without basis; a more complete analysis is needed.

4. A Numerical Model

A weakness in current models of fluid-particle mechanics is the description of particle-particle and particle-fluid momentum transfer. Conventional two-phase theories as outlined above assume a fluid-like constitutive relation for particles, and drag, virtual mass and buoyancy terms are postulated to model inter-phase momentum transfer. In this section we propose a numerical model for particle-fluid motion in two dimensions that overcomes the difficulty of phenomenological modeling.

Our approach is to balance three competing goals: fast computations, accurate fluid-particle momentum transfer, and reasonable flow field detail. In light of these considerations, we propose extending the idea of Peskins's Immersed Boundary Technique (IBT) to particle-fluid flows; see [2]. The essential idea is to model collisional (and cohesive) forces based on particle location, and to project these forces onto a fixed computational grid. A solution of the fluid equations provides a local velocity for each particle, whose position and velocity are updated. In addition to simulating specific flow configurations, this methodology will enable quantifiable comparison between phenomenological and 'first principles' approaches to the problem of particle-fluid flow.

Related ideas can be traced to Fogelson [2] who used the Immersed Boundary methodology for Stokes flow. He "built" macroscopic Stokes particles as collections of marker points and springs. The mass of any macroscopic particle is included in the momentum equations only as a body force, in the spirit of a Boussinesq approximation. Fourier methods are used to solve the fluid equations. Similar ideas have been employed by Sulsky [17]. In related work, Maxey [11] uses a force coupling model and pseudo-spectral methods to study sedimentation of small particles at finite Reynolds number. Joseph and colleagues [8] solve fluid equations (both Newtonian and non-Newtonian) on a finite element mesh that is the complement of all particles. No slip conditions are enforced on the boundaries (i.e., at the surface of all particles). As particles move, the mesh must be updated. In this algorithm, particles cannot collide; a fluid region must be present between every pair of particles. Similar ideas are used by Tezduyar [7]. Glowinski et al. [4] use a "fictitious domain" method for Navier-Stokes simulations. Here, the fluid equations are solved on a rectangular grid. Particles are represented by surface marker points and the mesh nodes interior to this surface, and these distinguished nodes (that is, the virtual particle) are constrained to move together as a rigid body. A Lagrange multiplier is incorporated into the momentum equations to enforce this constraint. Again, particles are not allowed to collide during the simulation. Tsuji [19] has simulated particle flows in gas and liquid. In both cases, particle-particle interaction is modeled using a discrete element method. Gas flows are computed by a Monte Carlo technique, and liquid flows by finite elements. Phenomenological interaction terms are postulated for the particle-fluid momentum transfer. A similar approach was used by Schwarzer [9] for liquid flows. Ladd [10] uses a lattice Boltzmann method for solving the fluid equations, and models fluid-particle interaction by momentum exchange between representative fluid particles and marker nodes representing the particle surface.

In our adaption of the IBT, we assume the fluid phase is incompressible. If accelerations are small, this assumption is not overly restrictive, even if air is the interstitial fluid. Particles, thought of as solid disks, are free to move throughout the computational domain. To be clear, we will represent a macroscopic particle by a single numerical particle, in the spirit of MD (a.k.a. discrete element) calculations. If particles touch, there is an elastic interaction force proportional to the distance between particle centers (as in so-called 'soft particle' Molecular Dynamics). These interaction forces are interpolated to the fixed computational grid by approximate δ-functions. On the grid, the Navier-Stokes equations are solved, with the interpolated force acting as a source of momentum. The newly computed velocities are interpolated back to the particles, whose positions are then updated based on these velocities. To model heavy particles, the numerical "particles" are given extra mass (that is, a density greater than the fluid), which must be included in the inertial terms. To model solid particles, the numerical "particles"are given extra viscosity (that is, a viscosity greater than the fluid). Thus a variable density, variable viscosity Navier-Stokes solver is required [13]. To correctly capture fluid-particle momentum transfer affected through boundary layers, the particle diameter must be several times the typical grid spacing.

In order to clarify our proposed method, let us sketch the algorithm for non-cohesive particles. Assume N identical particles of radius r_k and density ρ_k are located at $\theta_k(t)$ and moving with velocity $\xi_k(t)$, in a fluid of density ρ_{fluid}. For $k = 1, \cdots, N$, compute the local particle-particle force using a molecular dynamics model,

$$f_{kl}^N = \begin{cases} E^2(2r - |\theta_k - \theta_l|) & |\theta_k - \theta_l| < 2r \\ 0 & otherwise \end{cases}$$

$$f_{kl}^T = v f_{kl}^N \, sgn\left((\xi_k - \xi_l)^\perp\right)$$

$$f_k = \sum_{l \neq k} f_{kl}$$

Here, f_{kl}^N, f_{kl}^T are normal and tangential particle-particle forces, which must be projected into x, y-forces. Now interpolate these forces and the particle mass to a fixed Eulerian grid

$$F(x) = \int f_k \delta(x - \theta_k) d\theta$$

$$\rho(x) = \rho_{fluid} + \int (\rho_k - \rho_{fluid})\delta(x - \theta_k)d\theta$$

$$\mu(x) = \mu_{fluid} + \int (\mu_k - \mu_{fluid})\delta(x - \theta_k)d\theta$$

The integrals are over the region occupied by particles. Here, $\delta(z)$ is an approximation to the Dirac delta function; in this paper, δ is a piecewise-linear, scaled "hat" function. This force F is now used as a source term in the Navier-Stokes equations

$$\rho(u_t + u \cdot \nabla u) = -\nabla P + \nabla \cdot \mu(\nabla u + \nabla u^t) - \rho g + F$$

$$\nabla \cdot u = 0$$

Particle Location

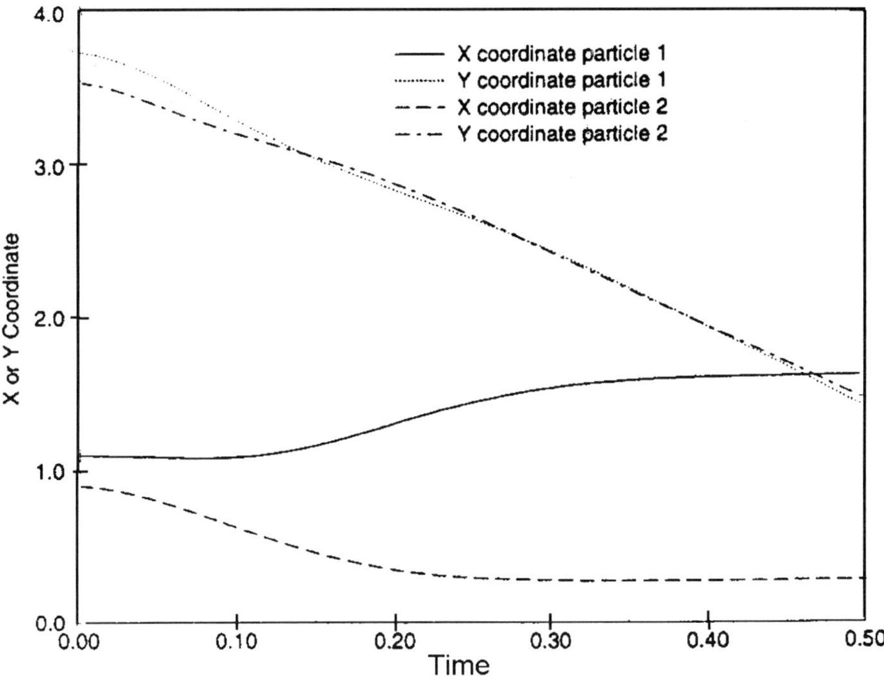

Figure 1. Particle location as a function of time. The two particles start one above and to the side of the other. They approach each other briefly, then tumble apart and fall next to each other, a stable configuration.

The solution gives a velocity field defined throughout the computational domain. Update the particle locations by interpolation (this is a re-interpretation of the no-slip condition for particles)

$$\xi_k = \int u(x)\delta(x - \theta_k)dx$$

$$\theta_k(t + \Delta t) = \theta_k(t) + \Delta t \xi_k$$

The force, density, and viscosity distribution are recomputed based on these new locations, and the procedure repeated. To be clear, the density field is recomputed at every time step based on particle location and interpolation; there is no continuum density advection (as in the multi-fluid calculations of [13]). The extra "particle viscosity" provides some resistance to deformation during one time step. By construction, the particle moves at the locally-averaged fluid velocity. In this way, there is no accumulated deformation of the fluid "drop" that represents the particle.

If one ignored the fluid dynamics, the algorithm computes discrete element forces from particle interactions, projects them onto a fixed grid, and updates the average particle velocity and positions from a grid calculation. This is reminiscent of particle-particle/particle-mesh schemes in astrophysics.

In the method outlined, the elastic (i.e., force calculation) and the fluid equations are solved at every time step, and that time step is set by the elastic interaction, an updating strategy used in other IBT applications[1]. The important issue for this application, and for IBT applications by others, is to adequately resolve the disparate timescales of the problem. Particles are necessarily stiff, with a large coefficient of elasticity (i.e., E^2 is large); the elastic force generated by particle interaction imposes a severe time step requirement for numerical stability. The fluid dynamics evolves on a much slower timescale. In the fluid solver, significant computational work is done in the iteration loop required in the Poisson solver of the projection step, a calculation performed at every time step. For the small time step set by the elastic stiffness, few iterations are required in this Poisson solver; for larger time steps (for example, as set by the fluid dynamics if only one particle is present) many iterations are required.

Cohesion based on van der Waals force can be incorporated easily into the framework just outlined. Although the van de Waals attraction is long-range, it is sufficiently weak beyond one or two particle radii that we can truncate the potential at finite length, making the particle-particle force calculation local once again.

The method will not fully resolve all the details of the flow field around every particle, at least not without prohibitive computational costs on present day readily available hardware. Our goal, however, is to develop an algorithm capable of adequately simulating particle-fluid flows at high particle volume fraction, with acceptable computational costs. This, we believe, can be accomplished. Preliminary tests show the method can simulate expected behavior of a few particles in low Reynolds number flows with $O(1-4)$ solid-fluid density ratio (essentially sand in

---

[1] As in MD simulations, in our scheme typical particle-particle collisions take about 50 time steps.

water). We want to simulate systems with $O(1000)$ density ratios (sand in air), but still small Reynolds numbers. In following [13] for the fluid solver, we were influenced by the demonstrated capabilities of the method in solving moderate Reynolds number flows with density ratios of 1000.

Here we report on some initial test cases with few particles sedimenting under gravity. These computations take just under three hours on a single R10000 processor of an SGI Origin 2000.

Figure 1 shows the x and y-position of two particles of density 2 (fluid density 1) falling, at a Reynolds number of 1 and a Froude number of 10^{-2}. Figure 2 shows the entire velocity field at three times, early, mid, and late, and the location of the particles. The grid is 20 x 40, and the particle radius is one grid spacing; furthermore, the approximate delta function projection operator – which is a constant over the entire particle and falls to zero linearly over one additional grid spacing – makes the "effective" particle size large still, and about twice the size of the particle symbol in the graph. These figures show the long-side configuration losing stability to a broadside configuration, as others have demonstrated experimentally.

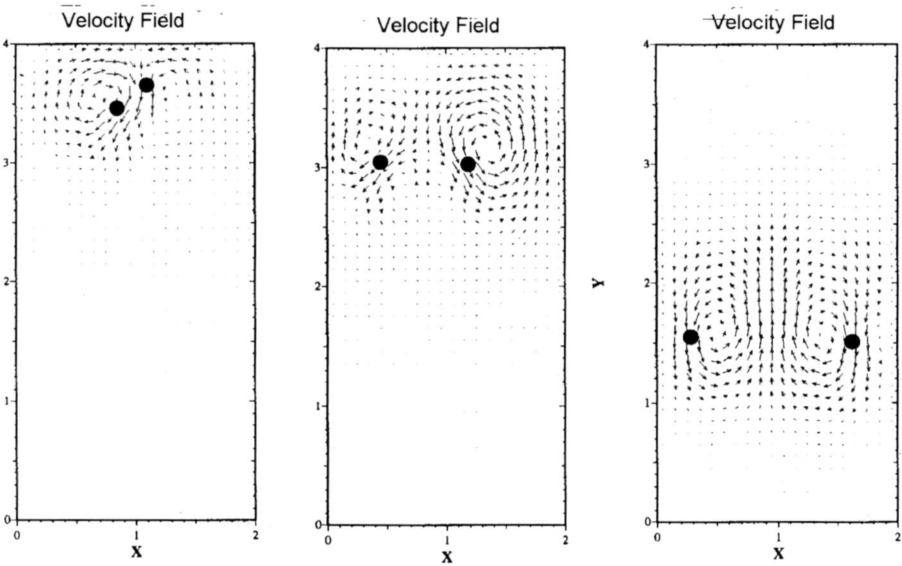

Figure 2. The velocity field and markers of particles at three instances in time. Early on, the particles tumble apart, and establish a broadside-on configuration for the remainder of the simulation.

Figures 3 shows five particles sedimenting. The two plots show a snapshot of the particles at three different times, early, mid-time and late. The diamond markers show the particles at the middle time.

In both plots, the velocity field shown corresponds to the middle time, corresponding to the particles located at the diamond markers. The difference in these plots is cohesion. On the left, the particles are non-cohesive. On the right, a cohesive force (about 2.5% of the elastic interaction force) is added to particles if they are closer than 1.05 times a particle diameter (the elastic interaction is zero until particles are less than one diameter apart). This square well is a little weaker than estimates of van de Waals interaction, but keeps its attractive well a little further in its reach. Nonetheless, the cohesive attraction alters the subsequent flow, suggesting significant segregation and solids non-uniformity. Note that flow breaks the cohesion bond later in the flow, but the effects are significant nonetheless.

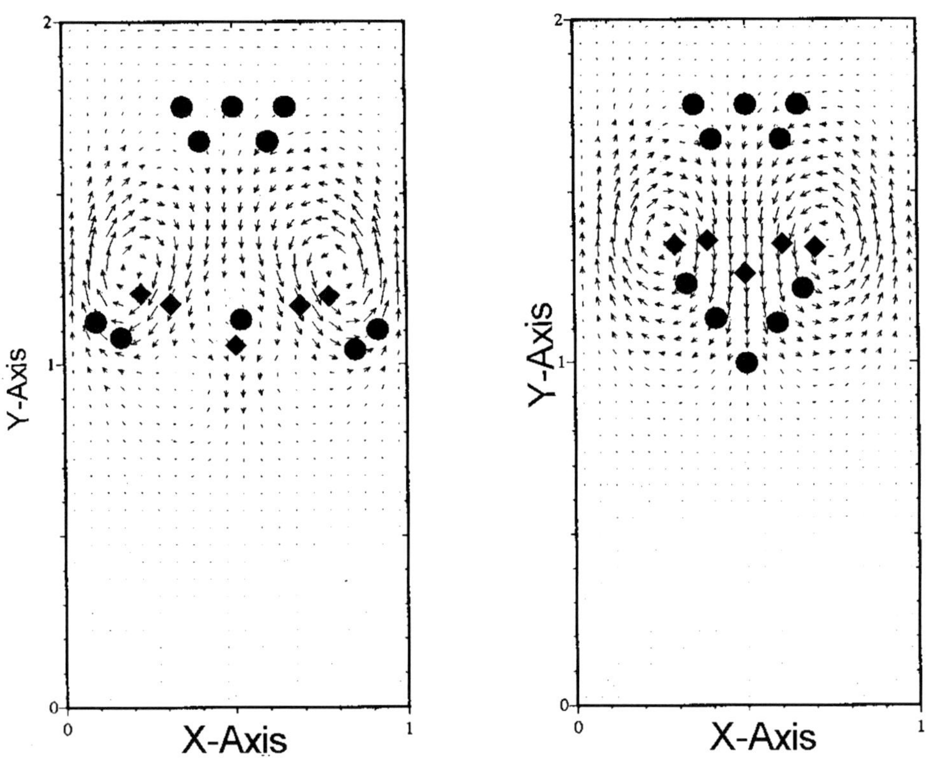

Figure 3. A snapshot of five particles falling under gravity at three times, and the velocity field corresponding to the middle time. The plot on the left shows non-cohesive particles. In the plot on the right, a cohesive force has been introduced; that force is active for the middle part of the simulation, but the particles separate later in the flow.

If one ignored the fluid dynamics, the numerical procedure outline above would project MD-based forces onto a fixed grid, on which velocities would be updated. Another projection gives velocities at the particles, updating their positions. Now imagine using the force calculation to mimic the particle-particle constitutive relation. After the projection, instead of solving the full Navier-Stokes equations, compute numerically the solution to a modified fluid mass and momentum system, using, for example, a drag term to provide particle-fluid momentum exchange. This is, in essence, the approach of Tsuji. And the juxtaposition of the two approaches suggests ways to both speed up the present computations and to provide a better basis for other numerical simulations of these problems. The real challenge for all modelers is to integrate physical and numerical experiments to provide a reliable theory for particle-fluid flows under a wide variety of industrial conditions. The numerical method proposed here may be one step down this path.

Acknowledgement

This work has been supported by the National Science Foundation under grant DMS-9802520. Computations were carried out at the Center for Computational Research at the State University of New York at Buffalo.

References

1. Clift, R.: An occamist review of fluidized bed modeling, in *AIChE Sympopsium Series 89,* 1993, p. 1.

2. Fogelson, A. and Peskin, C. S.: A fast numerical method for solving the three-dimensional Stokes equations in the presence of suspended particles, *J. Comput. Phys.* **79** (1988), 50.

3. Glasser, B. J., Kevrekidis, I. G. and Sundaresan, S.: One- and two-dimensional traveling wave solutions in gas-fluidized beds, *J. Fluid Mech.* **306** (1996), 183.

4. Glowinski, R., Pan, T. W., and Periaux, J.: A fictitious domain method for external incompressible viscous flow modeled by Navier-Stokes equations, *Comp. Meth. Appl. Mech. Eng.* **112** (1994), 113.

5. Jackson, R.: Progress towards a mechanics of dense suspensions of solid particles, in *AIChE Symposium Series 90,* 1984, p. 1.

6. Janssen, H. A.: Versuche uber Getreidedruck in Silozellen, *Zeitschrift. verein Deutscher Ingenieure* 39 (1895), 1045.

7. Johnson, A. A. and Tezduyar, T. E.: Simulation of multiple spheres falling in a liquid-filled tube, *Comp. Meth. Appl. Mech. Eng.* **134** (1996), 351.

8. Joseph, D., Hu, H., and Crochet, M.: Direct simulation of fluid-particle motions, *J. Theoretical and Computational Fluid Dynamics* **3** (1992), 285.

9. Kalthoff, W., Schwarzer, S. and Herrmann, H. J.: Algorithm for the simulation of particle syspensions with inertia effects, *Phys. Rev. E* **56** (1997), 2234.

10. Ladd, A.: Numerical simulation of particle suspensions via discretized Boltzmann equation: I & II, *J. Fluid Mech.* **271** (1995), 285.

11. Maxey, M. and Patel, B.: Force-coupled simulations of particle suspensions at zero and finite Reynolds numbers in *1997 ASME Fluids Eng. Div. Summer Meeting* (1997), p. 1.

12. Pitman, E. B. and Schaeffer, D. G.: Stability of time dependent compressible granular flow in two dimensions, *Comm. Pure Appl. Math* **40** (1987), 421.

13. Puckett, E. G., Almgren, A. S., Bell, J. B., Marcus, D. L., and Rider, W. J.: A high-order projection method for tracking fluid interfaces in variable density incompressible flows, *J. Comput. Phys.* **130** (1995), 269.

14. Rietema, K.: *The Dynamics of Fine Particles,* Elsevier, Cambridge U.K., 1991.

15. Schaeffer, D. G.: Instability in the evolution equation describing incompressible granular flow, *J. Diff. Eqns.* **66** (1987), 19.

16. Schaeffer, D. G., Shearer, M., and Pitman, E. B.: Instability in critical state theories of granular flow, *SIAM J. Appl. Math.* **50** (1990), 33.

17. Sulsky, D. and Brackbill, J. U.: A numerical method for suspension flow, *J. Comp. Phys.* **96** (1991), 339.

18. Tsinontides, S. and Jackson, R.: The mechanics of gas fluidized beds with an interval of stable fluidization, *J. Fluid Mech.* 226 (1993), 273.

19. Tsuji, Y.: Discrete particle simulations of disperse gas-solid flows, in *Powders and Grains '97*, R. P. Behringer and J. T. Jenkins (eds.), Baklema Rotterdam, 1997, p. 25.

20. Valverde, J. M., Ramos A. and Castellanos, A.: Tensile strength and void fraction as functions of consolidation stress for a set of Xerographic toners, in *Powders and Grains '97*, R. P. Behringer and J. T. Jenkins (eds.), Baklema Rotterdam, 1997, p. 163.

DYNAMICS OF A TWO SPECIES OSCILLATING PARTICLE SYSTEM

D. L. BLACKMORE and R. V. SAMULYAK
Department of Mathematical Sciences,
New Jersey Institute of Technology

R. N. DAVE and A. D. ROSATO
Department of Mechanical Engineering,
New Jersey Institute of Technology,
Newark, New Jersey 07102 - 1982

Abstract

A binary mixture of n_l identical inelastic spherical particles and n_s smaller identical inelastic spherical particles moving in a two-dimensional vibrating bed is analyzed. The vibrating bed is a rectangle, open at the top, with a horizontal floor oscillating vertically with amplitude $a > 0$ and frequency $\omega > 0$. Standard particle-particle and particle-boundary force models are used in the context of a Newtonian formulation to obtain the following differential equation in $\mathbf{R}^{4(n_l + n_s)}$ describing the motion of particles:

$$\frac{d\mathbf{x}}{dt} = X(\mathbf{x}, t),$$

where X is periodic in of period $2\pi / \omega$.

Qualitative properties of the solutions of the equation of motion are proved, including the existence of periodic and chaotic dynamics, and some properties are shown to persist as $n := (n_l + n_s) \to \infty$. The findings are compared with experimental and simulation results.

1. Introduction

Spurred by the increasing variety of important industrial applications and the diversity and complexity of exhibited phenomena -- often quite different from that observed in fluid flow -- research on granular flow dynamics has increased dramatically during the last two decades. In particular, segregation in granular flows consisting of mixtures of two or more particle types has been the subject of rather intense scrutiny from the scientific and engineering communities. Through the analysis of mathematical models, as for example in Jenkins and Mancini [4], studies of sophisticated simulation codes, including Louge [5] and Rosato at al. [8], and experimental investigations such as Vanel et al. [10], our understanding of segregation phenomena in particle flows has been greatly expanded. Yet, if we were to stand back and make an honest assessment of our current grasp of the physics and dynamics of segregation in granular flows, we would probably have to admit that there is still more that we do not know than we do know about granular

A.D. Rosato and D.L. Blackmore (eds.), IUTAM Symposium on Segregation in Granular Flows, 255–268.
© 2000 *Kluwer Academic Publishers. Printed in the Netherlands.*

segregation phenomena. Our purpose in this paper is to make a small addition to the body of knowledge on the dynamics of segregation. We do this by using the methods of modern dynamical systems theory to identify and characterize such qualitative behavior as bifurcations, chaos and mixing for fairly simple binary mixtures in a vibrating two-dimensional bed.

Direct analysis of discrete systems of differential equations and approximate continuum models of particle mixtures in vibrating beds has produced very little useful information about the nature of the dynamics associated with actual segregation phenomena. This is, of course, not surprising in view of the high level of complexity of the equations corresponding to realistic mathematical models of the process dynamics. Simulations and experimental procedures have been rather more successful in providing insights into possible separation mechanisms occurring in vibrating bed flows: they have provided strong, albeit not entirely conclusive, evidence of existence of periodic motions, bifurcation phenomena, chaotic regimes and dynamics resembling complete mixing of species corresponding to certain ranges of key parameters. By taking a qualitative dynamical systems approach to a realistic Newtonian model for granular flow in a vibrating bed, we shall obtain some additional corroborative evidence for existence of the dynamic possibilities indicated by the results of simulations and experiments.

The Newtonian dynamical system model that we develop here for a two- dimensional vibrating bed flow is actually a straightforward generalization of the one- dimensional model studied in [1]. We shall consider an arbitrarily large but finite number of particles in the vibrating bed system, thus we will be essentially dealing with pre-limiting forms of the continuum model derived in [2]. As the number of particles in granular flows of practical interest tend to be very large (usually numbering in the thousands or hundreds of thousands), we pay particular attention to those dynamical properties of vibrating granular flows of binary mixtures which can be shown to persist for arbitrarily large numbers of particles. Moreover, we shall be especially interested in identifying persisting dynamical properties of the flows that have been observed in experiments, simulations and in some approximate analytical investigations of vibrating bed flows.

This paper is organized as follows: In Section 2 we introduce some useful mathematical notation and also derive the system of ordinary equations that will serve as the model for the motion of the particles in the binary mixture. Next, in Section 3, the Newtonian equations of motion obtained in Section 2 are reformulated as a perturbation of a completely integrable Hamiltonian system which we employ as the focal point of our analysis. Our first results on the motion of the system are obtained in Section 4 for the case where inelastic and frictional forces are neglected. We prove the existence of periodic, quasiperiodic and chaotic flows for vibrations of the floor for certain amplitude and frequency ranges. In Section 5 we use the results of the preceding section to describe the short time motion of the system when inelastic and frictional forces are present. We also prove the existence of steady-state solutions entrained to the vibrating floor when the amplitude and frequency of the vibration are sufficiently small. In addition, we prove the existence of periodic and chaotic (mixing) regimes of the granular flow dynamics for certain ranges of the amplitude and frequency parameters. We also discuss the possibility of detecting a tendency for the larger particles to move to the top of the bed in this section, and compare our qualitative results with simulation and experimental observations. We conclude in Section 6 with a discussion of our results and possible directions for future research in this vein.

2. Derivation of the Model

Euclidean k-space, $k \geq 1$, will be denoted by $\mathbf{R}^k$. This space has a standard inner (dot) product $\langle .,. \rangle$ which induces the norm $\|\cdot\|$; in $\mathbf{R} = \mathbf{R}^1$, the real numbers, the norm is just the absolute value $|\cdot|$. The two-dimensional vibrating bed that we shall investigate is the time-dependent subset of the plane defined as

$$V(t) := \left\{ (x, y) \in \mathbf{R}^2 : \ |x| \leq b/2, \, a \sin \omega t \leq y \right\}, \tag{1}$$

and pictured in Figure 1. Here $b>0$ is the width of bed. We have assumed that the (vertical) walls are semi-infinite in order to avoid the possibility of particles actually leaving the bed. The walls are stationary, while the (horizontal) floor moves according to the equation $y = a \sin \omega t$ and so it vibrates with amplitude $a>0$ and frequency $\omega > 0$.

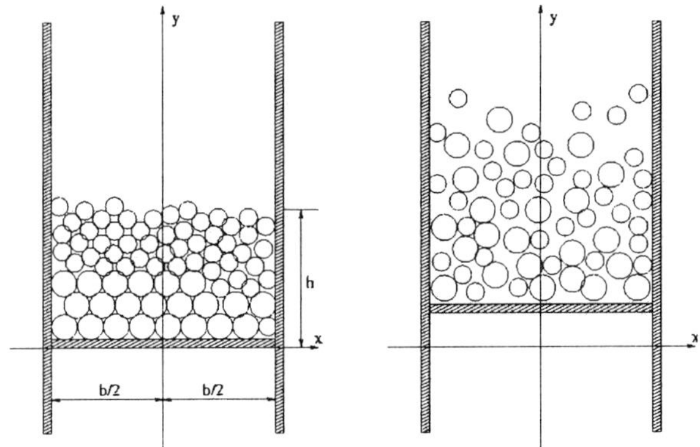

Figure 1: The vibrating bed

We assume that the system of particles in the bed is comprised of a binary mixture of n_l identical spherical balls of radius r_l and mass m_l and n_s smaller identical spherical balls of radius r_s and mass m_s, with $r_l, r_s << b$. As $r_l > r_s$ it is usually, but not always, the case that $m_l > m_s$. Define $n := n_l + n_s$. It was noted in the preceding section that the dynamical behavior of the system of particles in the vibrating bed for n large will be the focus of our investigation. It is assumed also that all of the particles are initially at rest with respect to the floor of the vibrating bed, and that the initial configuration of particles is a more or less uniform array filling out a height of $h>0$ as shown in Fig. 1. If the initial packing of the particles is quite close, then

$$h \cong \frac{\pi}{b} \left(n_l r_l^2 + n_s r_s^2 \right), \tag{2}$$

while $h > \frac{\pi}{b} \left(n_l r_l^2 + n_s r_s^2 \right)$ must hold regardless of the nature of the packing. The tendency of the larger particles to move to the top of the configuration over time as a result of the vibration of the bed has been the subject of numerous studies, for example [8] and [10]. Therefore, we shall assume that the larger particles are always at initial positions which are as close as possible to the floor.

Next we shall describe the particle-particle and particle-wall (floor) forces so that they provide a realistic model of the interaction of inelastic, frictional particles and walls (floor). We follow the developments of [1] and [2] which were based on the force models in Walton [11] that have been demonstrated to yield good agreement with experimental and simulation results. The position of the particles will be marked by their center points. Let the larger particles be positioned at $(x_i, y_i), 1 \leq i \leq n_l$, and let the centers of the smaller particles be located at the points (x_i, y_i), $n_l + 1 \leq i \leq n = n_l + n_s$. We shall find it convenient to introduce the following notation: the position vector of ith particle is denoted by

$$\mathbf{r}_i := (x_i, y_i); \tag{3}$$

the velocity of the ith particle is

$$\mathbf{v}_i := \dot{\mathbf{r}}_i = (\dot{x}_i, \dot{y}_i), \tag{4}$$

where $\cdot = d / dt$; and

$$\mathbf{r}_{ij} := \mathbf{r}_j - \mathbf{r}_i \quad \text{and} \quad \mathbf{v}_{ij} := \mathbf{v}_j - \mathbf{v}_i \tag{5}$$

denotes the relative position and velocity, respectively, of the ith and jth particles.

First we shall specify the normal interparticle forces. We decompose these normal forces into a (conservative) elastic part and a smaller (nonconservative) inelastic portion. The elastic part of the normal force of the jth particle on the jth particle is given according to the types of the interacting particles, as

$$\mathbf{N}_{ij}^{(e)} = \mathbf{N}_{ij}^{(e)}(\mathbf{r}_{ij}) := \begin{cases} -\phi_{ll}\left(\|\mathbf{r}_{ij}\|^2\right)\dfrac{\mathbf{r}_{ij}}{\|\mathbf{r}_{ij}\|}, & 1 \leq i, j \leq n_l \\[2ex] -\phi_{ls}\left(\|\mathbf{r}_{ij}\|^2\right)\dfrac{\mathbf{r}_{ij}}{\|\mathbf{r}_{ij}\|}, & 1 \leq i \leq n_l < j \leq n \\[2ex] -\phi_{sl}\left(\|\mathbf{r}_{ij}\|^2\right)\dfrac{\mathbf{r}_{ij}}{\|\mathbf{r}_{ij}\|}, & 1 \leq j \leq n_l < i \leq n \\[2ex] -\phi_{ss}\left(\|\mathbf{r}_{ij}\|^2\right)\dfrac{\mathbf{r}_{ij}}{\|\mathbf{r}_{ij}\|}, & n_l \leq i, j \leq n, \end{cases} \tag{6}$$

where $\phi_{ll}, \phi_{sl} = \phi_{ls}$ and ϕ_{ss} are smooth $(= C^\infty)$ nonnegative-valued, non-increasing functions defined on $\mathbf{R}$ such that $\phi_{ll}(u) = 0$ for $u \geq (2r_l)^2$, $\phi_{ls}(u) = 0$ for $u \geq (r_l + r_s)^2$, $\phi_{ss}(u) = 0$ for $u \geq (2r_s)^2$ and $\phi_{ll}(u), \phi_{ls}(u)$ and $\phi_{ss}(u)$ have constant positive values α_{ll}, α_{ls} and α_{ss}, respectively, when $u \leq 0$. In contrast, the inelastic contribution to the normal force of the jth particle upon the ith, which reflects a loss of energy resulting from the collision, is expressed in the form

$$\varepsilon_1 \mathbf{N}_{ij}^{(e)} = \varepsilon_1 \mathbf{N}_{ij}^{(e)}(\mathbf{r}_{ij}, \mathbf{v}_{ij}) :=$$

$$= \begin{cases} \varepsilon_1 \psi_{ll}\left(\|\mathbf{r}_{ij}\|^2\right)\langle\mathbf{r}_{ij},\mathbf{v}_{ij}\rangle\dfrac{\mathbf{r}_{ij}}{\|\mathbf{r}_{ij}\|}, & 1\le i,j\le n_l \\[2em] \varepsilon_1 \psi_{ls}\left(\|\mathbf{r}_{ij}\|^2\right)\langle\mathbf{r}_{ij},\mathbf{v}_{ij}\rangle\dfrac{\mathbf{r}_{ij}}{\|\mathbf{r}_{ij}\|}, & 1\le i\le n_l < j\le n \\[2em] \varepsilon_1 \psi_{sl}\left(\|\mathbf{r}_{ij}\|^2\right)\langle\mathbf{r}_{ij},\mathbf{v}_{ij}\rangle\dfrac{\mathbf{r}_{ij}}{\|\mathbf{r}_{ij}\|}, & 1\le j\le n_l < i\le n \\[2em] \varepsilon_1 \psi_{ss}\left(\|\mathbf{r}_{ij}\|^2\right)\langle\mathbf{r}_{ij},\mathbf{v}_{ij}\rangle\dfrac{\mathbf{r}_{ij}}{\|\mathbf{r}_{ij}\|}, & n_l\le i,j\le n, \end{cases} \tag{7}$$

where $\psi_{ll}, \psi_{sl} = \psi_{ls}$ and ψ_{ss} are smooth, nonnegative, non-increasing functions defined on $\mathbf{R}$ such that $\psi_{ll}(u) = 0$ for $u \ge (2r_l)^2$, $\psi_{ls}(u) = 0$ for $u \ge (r_l + r_s)^2$, $\psi_{ss}(u) = 0$ for $u \ge (2r_s)^2$ and $\psi_{ll}(u), \psi_{ls}(u)$ and $\psi_{ss}(u)$ have constant positive values β_{ll}, β_{ls} and β_{ss}, respectively, when $u \le 0$. Here ε_1 is a small nonnegative quantity that we shall use to control the size of the inelastic effect compared to the elastic part of the collisional restoring force.

Next we describe the interparticle frictional forces which we shall take to be relatively small since we shall ignore the rotational motion of the particles in our investigation. The frictional force exerted by the ith particle on the jth particle is tangential force of the form

$$\varepsilon_2 \mathbf{T}_{ij} = \varepsilon_2 \mathbf{T}_{ij}(\mathbf{r}_{ij}) :=$$

$$= \begin{cases} \varepsilon_2 \eta_{ll}\left(\|\mathbf{r}_{ij}\|^2\right)\chi\left(\|\mathbf{v}_{ij}^*\|^2\right)\dfrac{\mathbf{v}_{ij}^*}{\|\mathbf{v}_{ij}^*\|}, & 1\le i,j\le n_l \\[2em] \varepsilon_2 \eta_{ls}\left(\|\mathbf{r}_{ij}\|^2\right)\chi\left(\|\mathbf{v}_{ij}^*\|^2\right)\dfrac{\mathbf{v}_{ij}^*}{\|\mathbf{v}_{ij}^*\|}, & 1\le i\le n_l < j\le n \\[2em] \varepsilon_2 \eta_{sl}\left(\|\mathbf{r}_{ij}\|^2\right)\chi\left(\|\mathbf{v}_{ij}^*\|^2\right)\dfrac{\mathbf{v}_{ij}^*}{\|\mathbf{v}_{ij}^*\|}, & 1\le j\le n_l < i\le n \\[2em] \varepsilon_2 \eta_{ss}\left(\|\mathbf{r}_{ij}\|^2\right)\chi\left(\|\mathbf{v}_{ij}^*\|^2\right)\dfrac{\mathbf{v}_{ij}^*}{\|\mathbf{v}_{ij}^*\|}, & n_l\le i,j\le n, \end{cases} \tag{8}$$

where $\mathbf{v}_{ij}^*$ is the tangential component of the velocity defined as

$$\mathbf{v}_{ij}^* := \mathbf{v}_{ij} - \langle\mathbf{r}_{ij},\mathbf{v}_{ij}\rangle\dfrac{\mathbf{r}_{ij}}{\|\mathbf{r}_{ij}\|}, \tag{9}$$

where $\eta_{ll}, \eta_{sl} = \eta_{ls}$ and η_{ss} are smooth, nonnegative, non-increasing functions on $\mathbf{R}$ such that $\eta_{ll}(u) = 0$ for $u \ge (2r_l)^2$, $\eta_{ls}(u) = 0$ for $u \ge (r_l + r_s)^2$, $\eta_{ss}(u) = 0$ for $u \ge (2r_s)^2$,

$\eta_{ll}(u)$, $\eta_{ls}(u)$ and $\eta_{ss}(u)$ have constant positive values γ_{ll}, γ_{ls} and γ_{ss}, respectively, for $u \leq 0$, and χ is a smooth, nonnegative function defined on $\mathbf{R}$ such that

$$\chi(u) = \frac{d}{dx}\chi(u) = 0 \text{ for } u \leq 0, \frac{d}{dx}\chi(u) \geq 0 \text{ for } 0 < u < u_0, \frac{d}{dx}\chi(u) < 0 \text{ for}$$

$u_0 < u$ and $\chi(u) \to 1$ as $u \to \infty$. We have introduced an ε_2 analogous to ε_1 in (7) so that we can control the size of the relatively small frictional forces.

We shall also need realistic models for the normal and frictional forces imparted on the ith particle when it collides with one of the vertical walls of the vibrating bed. To describe the normal force, including both elastic and inelastic components, we take our cue from (6) for the elastic part so that we define

$$\mathbf{N}_i^{(e)} = \mathbf{N}_i^{(e)}(\mathbf{r}_i) :=$$

$$= \begin{cases} \left[\phi_{ll}\left(4|x_i + b|^2\right) - \phi_{ll}\left(4|x_i - b|^2\right)\right]\mathbf{e}_1, & 1 \leq i \leq n \\ \left[\phi_{ss}\left(4|x_i + b|^2\right) - \phi_{ss}\left(4|x_i - b|^2\right)\right]\mathbf{e}_1, & n_l \leq i \leq n, \end{cases} \tag{10}$$

where $\mathbf{e}_1$ is the unit basis vector in the direction of the x-axis. For the inelastic part of the normal force of collision with a vertical wall we look to (7) and set

$$\varepsilon_1 \mathbf{N}_i^{(i)} = \varepsilon_1 \mathbf{N}_i^{(i)}(\mathbf{r}_i, \mathbf{v}_i) :=$$

$$= \begin{cases} -\varepsilon_1 \dot{x}_i\left[\psi_{ll}\left(4|x_i + b|^2\right) - \psi_{ll}\left(4|x_i - b|^2\right)\right]\mathbf{e}_1, & 1 \leq i \leq n \\ -\varepsilon_1 \dot{x}_i\left[\psi_{ss}\left(4|x_i + b|^2\right) - \psi_{ss}\left(4|x_i - b|^2\right)\right]\mathbf{e}_1, & n_l \leq i \leq n, \end{cases} \tag{11}$$

The frictional force imparted on the ith particle will be modeled along the lines of (8), namely we take this force to be

$$\varepsilon_2 \mathbf{T}_i = \varepsilon_2 \mathbf{T}_i(\mathbf{r}_i, \mathbf{v}_i) :=$$

$$= \begin{cases} -\varepsilon_2 \dfrac{\dot{y}_i}{|\dot{y}_i|}\left[\eta_{ll}\left(4|x_i + b|^2\right) + \eta_{ll}\left(4|x_i - b|^2\right)\right]\chi\left(|\dot{y}_i|^2\right)\mathbf{e}_2, & 1 \leq i \leq n_l \\ -\varepsilon_2 \dfrac{\dot{y}_i}{|\dot{y}_i|}\left[\eta_{ss}\left(4|x_i + b|^2\right) + \eta_{ss}\left(4|x_i - b|^2\right)\right]\chi\left(|\dot{y}_i|^2\right)\mathbf{e}_2, & n_l \leq i \leq n, \end{cases} \tag{12}$$

where $\mathbf{e}_2$ is the unit basis vector in the direction of the y-axis.

Apart from the gravitational forces on the ith particle given by

$$\mathbf{G}_i = \begin{cases} -m_l g\mathbf{e}_2, & 1 \leq i \leq n_l \\ -m_s g\mathbf{e}_2, & n_l \leq i \leq n, \end{cases} \tag{13}$$

where $g = 9.8\text{m/sec}^2$ is the usual constant acceleration of gravity, it remains only to describe the normal and tangential (frictional) forces imparted on a particle by collision with the vibrating floor. The elastic part of the normal force exerted by the floor on the ith particle is modeled, in a manner similar to that of collision with the vertical walls, as

$$\tilde{\mathbf{N}}_i^{(e)} = \tilde{\mathbf{N}}_i^{(e)}(\mathbf{r}_i, t_i; a, \omega) :=$$

$$= \begin{cases} K\phi_{ll}\left(4|y_i - a \sin \omega t|^2\right)\mathbf{e}_2, & 1 \leq i \leq n_l \\ K\phi_{ss}\left(4|y_i - a \sin \omega t|^2\right)\mathbf{e}_2, & n_l \leq i \leq n, \end{cases} \tag{14}$$

where K is a positive constant introduced so that both $K\phi_{ll}(4u^2) \geq m_l g$ and $K\phi_{ss}(4u^2) \geq m_s g$ for $0 < u \leq u_1$ and $0 < u \leq u_2$, respectively, where $0 < r_l - u_1 << 1$ and $0 < r_s - u_2 << 1$. As the inelastic part of the normal force imparted on the ith particle by the oscillating floor, we take

$$\varepsilon_1 \tilde{\mathbf{N}}_i^{(e)} = \varepsilon_1 \tilde{\mathbf{N}}_i^{(e)}(\mathbf{r}_i, \mathbf{v}_i, t_i; a, \omega) :=$$

$$= \begin{cases} \varepsilon_1 K(a\omega \cos\omega t - \dot{y}_i)\psi_{ll}\left(4|y_i - a\sin\omega t|^2\right)\mathbf{e}_2, & 1 \leq i \leq n_l \\ \varepsilon_1 K(a\omega \cos\omega t - \dot{y}_i)\psi_{ss}\left(4|y_i - a\sin\omega t|^2\right)\mathbf{e}_2, & n_l \leq i \leq n, \end{cases} \tag{15}$$

To complete the description of the forces, we define the frictional portion of the force exerted on the ith particle by the floor as

$$\varepsilon_2 \tilde{\mathbf{T}}_i = \varepsilon_2 \tilde{\mathbf{T}}_i(\mathbf{r}_i, t_i; a, \omega) :=$$

$$= \begin{cases} -\varepsilon_2 \dfrac{\dot{x}_i}{|\dot{x}_i|}\eta_{ll}\left(4|y_i - a\sin\omega t|^2\right)\chi\left(|\dot{x}_i|^2\right)\mathbf{e}_1, & 1 \leq i \leq n_l \\ -\varepsilon_2 \dfrac{\dot{x}_i}{|\dot{x}_i|}\eta_{ss}\left(4|y_i - a\sin\omega t|^2\right)\chi\left(|\dot{x}_i|^2\right)\mathbf{e}_1, & n_l \leq i \leq n, \end{cases} \tag{16}$$

Using the forces (6)-(8), (10) and (11)-(16) in Newton's second law, we obtain the following equations of motion for the two species of particles in the (planar) vibrating bed:

$$\ddot{\mathbf{r}}_i = -g\mathbf{e}_2 + \frac{1}{m_i}\left[\left(\sum_{j=1}^{n'} \mathbf{N}_{ij}^{(e)} + \mathbf{N}_i^{(e)} + \tilde{\mathbf{N}}_i^{(e)}\right) + \varepsilon_1\left(\sum_{j=1}^{n'} \mathbf{N}_{ij}^{(i)} + \mathbf{N}_i^{(i)} + \tilde{\mathbf{N}}_i^{(i)}\right)\right.$$

$$\left. \varepsilon_2\left(\sum_{j=1}^{n'} \mathbf{T}_{ij} + \mathbf{T}_i + \tilde{\mathbf{T}}_i\right)\right], \qquad (1 \leq i \leq n), \tag{17}$$

where the prime on the summation sign denotes the exclusion of $j = i$. We note that these equations of motion constitute a system of $2n = 2(n_l + n_s)$, second-order, nonlinear ordinary differential equations for the coordinates (x_i, y_i), $1 \leq i \leq n$, of the centers of mass of the particles in the vibrating bed.

3. Perturbed Hamiltonian Model

Although (17) together with initial conditions for the dependent vector variables $\mathbf{r}_i, 1 \leq i \leq n$, and their derivatives suffices to uniquely determine the motion of the particles in the vibrating bed, for our purposes it is more convenient to recast (17) as a system of first-order equations that can be viewed as a perturbation of a Hamiltonian system. It is not difficult to verify that, by using standard techniques from differential equation and Hamiltonian dynamics (see [3] and [6]), we can introduce new variables represented in the vector form as

$$\mathbf{x} = (q, p) = (q_1, \ldots, q_{2n}, p_1, \ldots, p_{2n}) \in \mathbf{R}^{4n}, \tag{18}$$

such that (17) is equivalent to the (vector) differential equation

$$\dot{\mathbf{x}} = X(\mathbf{x}, t) = X_0(\mathbf{x}) + aX_1(\mathbf{x}, t) + \varepsilon_1 Y_1(\mathbf{x}, t) + \varepsilon_2 Y_2(\mathbf{x}, t) \tag{19}$$

satisfying the following properties:

(P1). The (vector) function X_0 is smooth on the $4n$-dimentional manifold

$$M := \{\mathbf{x} \in \mathbf{R}^{4n} : q_i \neq q_j \text{ for } i \neq j, 1 \leq i, j \leq 2n\}, \tag{20}$$

and X_1, Y_1 and Y_2 are smooth functions mapping $M \times \mathbf{R}$ into $\mathbf{R}^{4n}$ which are periodic in t with period $2\pi / \omega$.

(P2). The equation

$$\dot{\mathbf{x}} = (\dot{q}, \dot{p}) = X_0(\mathbf{x}) + aX_1(\mathbf{x}, t) \tag{21}$$

is Hamiltonian, i.e. there is a smooth function $H : M \times \mathbf{R} \to \mathbf{R}$ such that

$$X_0(\mathbf{x}) + aX_1(\mathbf{x}, t) = (\partial H / \partial p, -\partial H / \partial q).$$

Moreover, H is t-periodic of frequency ω.

(P3). The equation

$$\dot{\mathbf{x}} = (\dot{q}, \dot{p}) = X_0(\mathbf{x}) \tag{22}$$

is a completely integrable (in the sense of Liouville-Arnold) Hamiltonian system, i.e. there exists a smooth function $H_0 : M \to \mathbf{R}$ such that

$$X_0 = (\partial H_0 / \partial p, -\partial H_0 / \partial q),$$

and there also exists integrals of (22), $\Phi_1, \ldots, \Phi_{2n-1}$ such that $\{H_0, \Phi_1, \ldots, \Phi_{2n-1}\}$ comprises a linearly independent, involutive set; namely, the differentials $dH_0, d\Phi_1, \ldots, d\Phi_{2n-1}$ are linearly independent on M and the Poisson bracket

$$\{F, G\} := \sum_{k=1}^{2n} \left(\frac{\partial F}{\partial q_k} \frac{\partial G}{\partial p_k} - \frac{\partial F}{\partial p_k} \frac{\partial G}{\partial q_k} \right) = 0$$

whenever F and G are different integrals in this set.

4. Motion for Elastic Frictionless Case

When the system has no inelastic or frictional forces, $\varepsilon_1 = \varepsilon_2 = 0$, so (19) reduces to the Hamiltonian system (21), and we can apply the analysis of Prykarpatsky et al. [7] for small amplitudes and frequencies to obtain our first result.

Theorem 4.1 There are $\delta_1, \delta_2 > 0$, independent of n, such that there are solutions of (21) satisfying the following properties whenever $0 \leq a < \delta_1$ and $0 < \omega < \delta_2$:

(a) There are exist solutions that are periodic of period $2\pi / \omega$ and entrained to the motion of the vibrating floor.

(b) The system has quasiperiodic solutions in the following sense: There are manifolds T smoothly diffeomorphic with $S^1 \times \ldots \times S^1$ ($2n$-fold Cartesian product of circles $2n$ torus) and corresponding smooth families of $2\pi / \omega$-periodic families of smooth diffeomorphisms f_t defined for all real t, equal to the identity for $t=0$ and close to the identity for all t, such that any trajectory $\mathbf{x}(t)$ of (21) with $\mathbf{x}(0) \in T = f_0(T)$ satisfies $\mathbf{x}(t) \in f_t(T)$ for all time. Moreover, some of these tori have dense trajectories, i.e. $f_t^{-1}(\mathbf{x}(t))$ visits every neighborhood of every point on T as t is allowed to vary over the real line.

Proof Sketch: To prove (a) we first introduce $\theta (= t)$ as a dependent variable, thereby obtaining the following (autonomous) system of $4n+1$ equations:

$$\dot{\mathbf{x}} = X_0(\mathbf{x}) + aX_1(\mathbf{x},t)$$
$$\dot{\theta} = 1.$$

(23)

Owing to the periodicity of X_1 in t, we may consider (23) to be a differential equation corresponding to a smooth vector field on $M \times S^1$. Hence we can reformulate the problem of finding $2\pi/\omega$-periodic solutions of (21) to that of finding fixed points of the (stroboscopic) Poincaré map of the transversal section $M \times \{0\}$ onto itself. This map clearly has man fixed points when $a=0$ and we can adjust the initial conditions so that the Brouwer degree (index) corresponding to a particular fixed point $\mathbf{z}$ is not zero. As this degree depends continuously on the parameters of the system, it remains nonzero for sufficiently small a and ω. Hence the Poincaré map has a fixed point $\mathbf{z}(a,\omega)$ close to $\mathbf{z}$ when a and ω are small enough.

Part (b) follows directly from the main theorem of [7] regarding the preservation of invariant tori in slowly perturbed completely integarble Hamiltonian systems.

Actually (a) of Theorem 4.1 is hardly surprising since it is intuitively clear that if the particles are initially at rest in the vibrating bed, they will simply move as a rigid configuration in the same relative positions with respect to the floor when ω is small. In fact, intuition indicates that this type of motion should be independent of amplitude for sufficiently small frequencies. This can be verified via a straightforward modification of the proof of (a) of the preceding theorem; whence, we obtain:

Theorem 4.2 There exists an $\omega_0 > 0$, independent of n, such that for any $a>0$ in (21) there exists periodic solutions of period $2\pi/\omega$ whenever $0 < \omega < \omega_0$.

Statement (b) of Theorem 4.1 -- in particular the density part -- shows that, at least in some subregion of the domain of the vibrating bed, the flow is mixing, which means that the particles, both large and small, thoroughly intermingle with one another. Of course, it would be useful to be able to describe the extent of this mixing subregion, but this is a difficult problem that we shall return to in the sequel. Our next result concerning chaotic motion actually has a certain relationship to mixing phenomena.

Theorem 4.3 For each $a>0$ and $n>0$ there exists an $\omega_c(a;n) > 0$ such that (21) has a chaotic solution for some $\omega > \varepsilon_c(a;n)$. The chaotic frequency is bounded as a function of the number of particles, i.e. there exists a nonnegative functions $W(a)$ such that $\omega_c(a;n) \leq W(a)$ for all integers n greater than two.

Proof Sketch: Once again we use the stroboscopic Poincaré function $\wp$ associated to the system (23). From Theorem 4.1 and Theorem 4.2, we know that for each $a>0$ the function $\wp$ has fixed points for ω sufficiently small if we choose the initial conditions properly. A more careful selection of initial conditions enables us to locate hyperbolic fixed points having nontrivial stable and unstable manifolds. We then construct a function of Mel'nikov type (see [3], [9]) which we denote by $M_n(\omega,a)$. This function represents the distance (exterior to a small ball centered at the fixed point) between the stable and unstable manifolds of a selected hyperbolic fixed point. An analysis of $M_n(\omega,a)$ reveals that for fixed $a>0$ and $n \geq 2$ it has isolated zeros for ω greater than some (fairly large) value that we denote by $\omega_c(a;n)$. This analysis shows also that $\omega_c(a;n)$ is bounded as a function of n. This completes the proof since the "isolated" intersection of stable and unstable manifolds implies the existence of transverse homoclinic points that cause chaotic horseshoe motions.

Chaotic regimes corresponding to local Smale horseshoe behavior have dense orbits, hence Theorem 4.3 also implies the existence of mixing granular flow regions within the vibrating bed. A complete description of the region or regions of mixing flow in terms of the original physical domain of the vibrating bed, is also very difficult to obtain in this chaotic case. The main difference between the mixing behavior described by Theorem 4.1 (a) and Theorem 4.3 is that the mixing is far more vigorous and random in the latter case.

5. Motion of the Inelastic System with Friction

In this section we describe certain aspects of the motion of the complete system (19) representing the motion of the two species particle configuration in the vibrating bed. We shall, for the most part, confine our attention to cases where the inelastic and frictional effects are relatively small We start with a result which is a direct consequence of Theorem 4.1 and the continuous dependence of solutions of (21) on the system parameters.

Theorem 5.1 Given $\varepsilon > 0$ one can find a $\delta > 0$, independent of n, such that the solution of (19) satisfy the following properties whenever $0 \le a, \omega, \varepsilon_1, \varepsilon_2 < \delta$:

(a) There exists $0 < t_1 = t_1(a, \omega, \varepsilon_1, \varepsilon_2)$ such that there are solutions that are within ε of a $2\pi/\omega$-periodic function for all $0 \le t < t_*$.

(b) There exists $0 < t_2 = t_2(a, \omega, \varepsilon_1, \varepsilon_2)$ and a family f_t of imbeddings of the $2n$-torus T into M such that (19) has solutions $\mathbf{x}(t)$, with $\mathbf{x}(t) \in f_t(T)$ for all $0 \le t < t_2$, that are ε-close to functions satisfying (b) of Theorem 4.1.

In particular, there are motions that have regimes that are nearly mixing over short time intervals.

When the inelastic effects are large enough compared to the energy imparted on the system of particles by the vibrating floor, the motion of the particles tends to become entrained with that of the floor in the long run, as shown by the following theorem.

Theorem 5.2 Let $\varepsilon_1 > 0$ and suppose that the (periodic) motion of the vibrating floor is represented by $\mathbf{x} = \Phi(t)$ in terms of the coordinates employed in (19). Then if a and ω are sufficiently small, every solution $\mathbf{x}(t)$ of (19) satisfies $\left\| \dot{\mathbf{x}}(t) - \dot{\Phi}(t) \right\| \to 0$ as $t \to \infty$.

Proof sketch: The easiest way to prove this intuitively obvious result is by working with the Newtonian equations of motion (17). We first introduce a coordinate system moving with the oscillating floor by defining

$$\mathbf{R}_i := (x_i, y_i - a\sin\omega t)$$

for $1 \le i \le n$. Then (17) assumes the form

$$\ddot{\mathbf{R}}_i = \mathbf{F}_i\left(\mathbf{R}_i, \dot{\mathbf{R}}_i, t\right), \quad (1 \le i \le n) \tag{24}$$

where each $\mathbf{F}_i$ is periodic in t of period $2\pi / \omega$. Now take the inner product of (24) with $\dot{\mathbf{R}}_i$, thereby obtaining,

$$\left\langle \dot{\mathbf{R}}_i, \ddot{\mathbf{R}}_i \right\rangle = \frac{1}{2}\frac{d}{dt}\left\|\dot{\mathbf{R}}_i\right\|^2 = \left\langle \dot{\mathbf{R}}_i, \mathbf{F}_i \right\rangle. \tag{25}$$

A careful analysis of the right-hand side of (25) yields

$$\frac{1}{2}\frac{d}{dt}\left\|\dot{\mathbf{R}}_i\right\|^2 \leq -K\left\|\dot{\mathbf{R}}_i\right\|^2$$

for all t whenever t and ω are sufficiently small, where K is some positive constant. Whence an application of Gronwall's inequality yields

$$\left\|\dot{\mathbf{R}}_i\right\|^2 \leq -\left\|\dot{\mathbf{R}}_i(t)\right\|^2 e^{-2Kt}$$

for all $1 \leq i \leq n$ and $t \geq 0$. Consequently $\left\|\dot{\mathbf{R}}_i\right\|^2 \to 0$ as $t \to \infty$, and the result follows. $\square$

System (19) can also experience periodic and chaotic motion according to the following result:

Theorem 5.3 Consider (19) with ε_1 and ε_2 fixed positive numbers and n a fixed positive integer. This system has solutions with the following properties:

(a) For each $a>0$ there exist $0 < \omega_1(a;\varepsilon_1,\varepsilon_2,n) < \omega_2(a;\varepsilon_1,\varepsilon_2,n)$ such that (19) has a periodic solution of period close to $2\pi / \omega$ for each $\omega \in [\omega_1,\omega_2]$.

(b) There exists an $\omega_*(a;\varepsilon_1,\varepsilon_2,n) \in (\omega_1,\omega_2)$ such that (19) has a chaotic solution for each $\omega \in [\omega_*,\omega_2]$.

Proof Sketch: To prove (a), we rewrite (19) in the autonomous form

$$\begin{aligned} \dot{\mathbf{x}} &= X(\mathbf{x},\theta) \\ \dot{\theta} &= 1 \end{aligned} \tag{26}$$

on $M \times S^1$. Then we create a transversal of the form $\theta = G(\mathbf{x})$, where G is a smooth function such that $G(0) = G'(0) = 0$ and $\|G(\mathbf{x})\|$ is small for all $\mathbf{x} \in M$. Define $\Omega := \{(\mathbf{x}, G(\mathbf{x})) : \mathbf{x} \in M\}$ and let $H : \Omega \to \Omega$ be the Poincaré map associated to (26) and Ω. It can be shown that if G is chosen properly, H maps a closed ball in Ω into itself whenever ω is sufficiently large. Hence, it follows from Brouwer's theorem that H has a fixed point $z(\omega)$ in this ball for all $\omega \in [\omega_1,\omega_2]$ for $0 < \omega_1 < \omega_2$ sufficiently large. Obviously, the fixed point $z(\omega)$ corresponds to a periodic solution of period close to $2\pi / \omega$ for all $\omega \in [\omega_1,\omega_2]$ in virtue of our construction.

Part (b) is proved by a more careful analysis of the construction used to prove (a). In particular, one can prove that the fixed point $z(\omega)$ can be selected to be hyperbolic with nonvoid stable and unstable manifolds if ω_1,ω_2 and the transversal Ω are chosen judiciously. Then Melnikov's method can be used to prove the existence of transversal homoclinic points corresponding to $z(\omega)$ whenever $\omega \geq \omega_*$, for some $\omega_* \in (\omega_1,\omega_2)$. It follows that (19) has solutions with chaotic Smale horseshoe behavior for $\omega \in [\omega_*,\omega_2]$. Thus the proof is complete.

The transition from regular periodic motion to chaotic motion in the vibrating bed is illustrated in Figures 2 and 3 which were obtained by approximate numerical simulation of the granular flow dynamics. Notice that the regular convective behavior in Figure 2 -- which is in good qualitative agreement with simulation and experimental results such as in [5], [8] and [10] -- degenerates into chaotic convective behavior illustrated in Figure 3. This bifurcation from regular to chaotic convective motion is exactly as predicted by the approximate analysis of the continuum model of [2] carried out in [9].

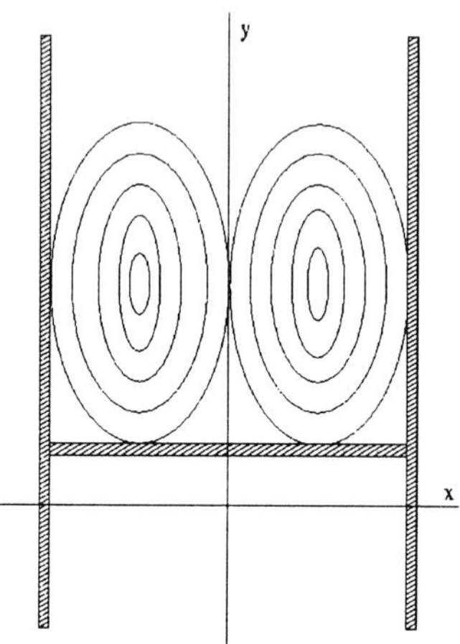

Figure 2: Regular periodic convection $\left(\omega_1 \leq \omega \leq \omega_*\right)$

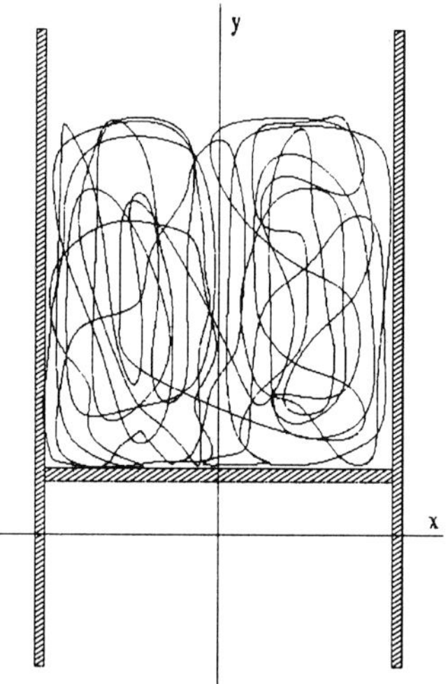

Figure 3: Chaotic convection $\left(\omega_1 \leq \omega \leq \omega_*\right)$

As a final observation, we note that our dynamical system analysis of the motion of two particle species in a vibrating bed leads to the following plausible, albeit not exactly rigorous, explanation for the observed tendency [8, 10] of the larger particles to move to the top of the bed. For simplicity, we assume that $n_l = 1$ and that n_s is so large that the small particles can completely cover the vibrating floor of the bed. If $m_l \gg m_s$ and the interaction forces in the Newtonian equations of motion (17) have an order of magnitude at least as large as the gravitational force, the acceleration of the large particle is clearly much smaller in norm than those of the small particles. Assuming that the particles are initially at rest in the bed, we infer from (17) that the speed attained by the large particle at any time greater than zero is much smaller than the speeds of the small particles. Whence we conclude from the short term mixing behavior described in Theorem 5.1 or the chaotic mixing associated with Theorem 5.3 (b) that for sufficiently large amplitudes and frequencies, the small particles tend to fill in the space between the vibrating floor and the large particle much faster than the large particle can return to a position on the floor after the onset of motion. Thus the large particle tends to move to the top of the vibrating bed when the applied oscillation is sufficiently vigorous.

6. Concluding Remarks

It has been shown that the mathematical tools of nonlinear dynamics can be marshaled to prove several qualitative properties of the motion of a binary mixture of inelastic, rough particles in a two-dimensional vibrating bed -- properties which persist regardless of the number of particles. These properties, including the existence of periodic orbits, quasiperiodic motions, chaotic regimes and bifurcations among these types of behavior for certain amplitude and frequency ranges, are in good qualitative agreement with results obtained from simulations and experimental investigations. We were even able to use our analysis to propound a plausible explanation for the observed tendency of the larger particles to move to the top of the bed.

As our results are almost exclusively qualitative, direct comparison of the vibration parameter ranges giving rise to the various types of dynamics with those found by other means such as experiments, simulations and the use of alternative mathematical approaches is not really possible. Moreover, this same lack of quantification of the features of the vibrating flow, makes useful application of our results to practical engineering problems difficult, to say the least. Consequently, we shall endeavor in our future research to attach realistic quantitative interpretations to our results so as to develop them as predictive and design tools for a variety of important applications. We shall also continue to investigate the correlations between the properties of the finite-dimensional particle systems and several of the infinite-dimensional continuum models for granular flow.

Acknowledgments. This research was supported in part by a PPRC grant from the New Jersey Commission on Science and Technology.

References

1. Blackmore, D. and Dave, R.: Chaos in one-dimensional granular flows with oscillating boundaries, in R. Behringer and J. Jenkins (eds.), *Powder & Grains 97*, Balkema, 1997, Rotterdam, pp. 409-412.
2. Blackmore, D., Samulyak, R. and Rosato, A.: New mathematical models for particle flow dynamics, *J. Nonlin. Math. Phys.*, **6** (1999), 198-221.
3. Guckenheimer, J. and Holmes, P.: *Nonlinear Oscillations, Dynamical Systems and Bifurcations of Vector Fields*, Springer-Verlag, 1997, New York.
4. Jenkins, J. and Mancini, F.: Balance laws and constitutive relations for plane flows of a dense, binary mixture of smooth, nearly inelastic discs, *J. Appl. Mech*, **109** (1987), 27-34.
5. Louge, M.: Computer simulations of rapid granular flows of spheres interacting with a flat frictional boundary, *Phys. Fluids*, **6** (1994), 2253-2269.
6. Prykarpatsky, A. and Mykytiuk, I.: *Algebraic Integrablity of Nonlinear Dynamical Systems on Manifolds*, Kluwer, 1998, Dordrecht.

7.   Prykarpatsky, Y., Samoilenko, A. and Blackmore, D. Imbeddings of integral submanifolds and associated adiabatic invariants of slowly perturbed integrable Hamiltonian systems, *Rep. Math. Phys.* (in press).

8.   Rosato, A., Strandburg, K., Prinz, F. and Swendsen, R., (1986) Monte-Carlo simulation of particulate matter segregation, *Powder Technology*, **49** (1986), 59-69.

9.   Samulyak, R. and Blackmore, D. Bifurcation from regular to chaotic motion in granular flow in a vibrating bed (to appear).

10.   Vanel, L., Rosato, A. and Dave, R.: Size-dependent segregation in a 3D vibrated bed, in R. Behringer and J. Jenkins (eds.), *Powder & Grains 97*, Balkema, 1997, Rotterdam, pp. 385-387.

11.   Walton, O.: Numerical simulation of inelastic, frictional particle-particle interactions, in M. Roco (ed.), *Particulate Two-Phase Flow*, Butterworth-Heinemann, 1992, Boston, pp. 884-911.

SHOCK WAVES AND PARTICLE SIZE SEGREGATION IN SHALLOW GRANULAR FLOWS

J. M. N. T. GRAY, Y. C. TAI, and K. HUTTER
Institut für Mechanik,
Technische Universität Darmstadt,
64289 Darmstadt, Germany

1. Introduction

Rapid shallow granular free surface flows occur when a surface layer of static granular material becomes unstable or when granular material is released by some mechanism onto an inclined surface. They are abundant all around us from salt in a salt cellar and flow at the free surface of stockpiles, to landslides, rock-falls and snow slab avalanches. Although the length scales vary dramatically the dynamics of the avalanches are similar.

Savage & Hutter [12],[13] exploited the shallowness of these flows to derive a one-dimensional depth averaged theory for the flow of an incompressible Mohr-Coulomb granular material sliding down a rigid surface. Using a slope fitted curvilinear coordinate system Oxz, with the z axis normal to the slope and the x-axis parallel to it, the depth integrated mass and momentum balances reduce to

$$\partial_t h + \partial_x(hu) \; = \; 0, \tag{1}$$

$$\partial_t(hu) + \partial_x(hu^2 + \cos\zeta K h^2/2) \; = \; hd, \tag{2}$$

where h is the avalanche depth, u is the depth averaged downslope velocity and, ∂_t and ∂_x denote differentiation with respect to time and the downslope coordinate, respectively. The net driving force is the difference between the gravitational acceleration and the Coulomb rate-independent dry friction sliding law

$$d = \sin\zeta - (u/|u|)\tan\delta(cos\zeta + \kappa u^2), \tag{3}$$

where $\zeta(x)$ is the slope inclination angle, $\kappa = -\partial_x\zeta$ is the slope curvature and δ is the basal friction angle. The system is similar to the shallow water equations of fluid dynamics, however, the earth pressure coefficient K,

269

A.D. Rosato and D.L. Blackmore (eds.), IUTAM Symposium on Segregation in Granular Flows, 269–276.
© 2000 *Kluwer Academic Publishers. Printed in the Netherlands.*

which determines the ratio of in-plane to normal pressure, is a non-linear function of $\partial_x u$ rather than a constant. The earth pressure coefficient approaches two limiting values, *active* and *passive*, dependent on whether the downslope motion is divergent or convergent. These are given by

$$K_{\mathrm{act/pas}} = 2\sec^2\phi\left(1 \mp \sqrt{1 - \cos^2\phi\sec^2\delta}\right) - 1, \tag{4}$$

where ϕ is the internal angle of friction in the Mohr-Coulomb constitutive relation. The transition between the two states can be achieved either by a jump [12] or a smooth transition [15]. When the internal and basal friction angles are equal, then $K_{\mathrm{act}} = K_{\mathrm{pas}}$, and the equations reduce to a non-strictly hyperbolic system with characteristic wavespeeds $\lambda = u \pm (\cos\zeta Kh)^{1/2}$. In this situation the *granular Froude number*, $\mathrm{F_r} = |u|/(\cos\zeta Kh)^{1/2}$, determines whether the flow is sub- or super- critical. A two-dimensional extension of the theory for modelling granular flow over complex three-dimensional topography has been derived by Gray et al [5].

2. Plane shock waves

Plane shocks waves are generated when an obstacle is placed in the flow or when there is a change in the inclination angle during super-critical flow. The shock is marked by a rapid change in the debris flow thickness and velocity and propagates in the opposite direction to the flow. Plane shocks on weakly accelerative slopes play an important role in the process of stratification pattern formation.

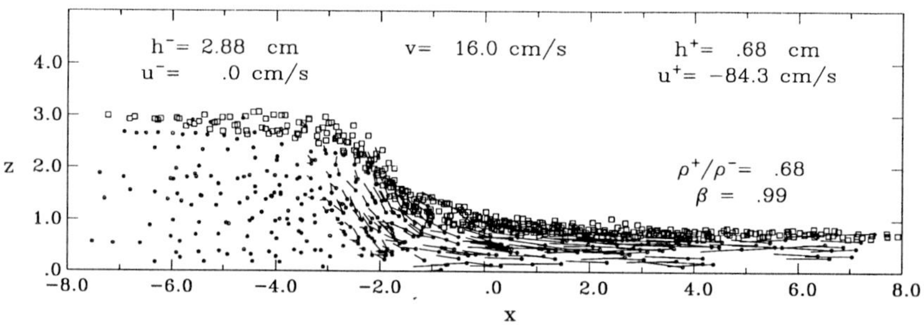

Figure 1. The flow thickness and velocity vectors determined from 16 CCD images are translated and superimposed upon one another to build up a detailed image of the travelling wave. The free surface of the granular material is indicated by a square at each data point. The initial position of each particle is indicated by a circle and its position $1/100^{\mathrm{th}}$ of a second later is indicated by a line segment. All distances are measured in cm and velocities in cm/s.

Figure 1 shows data obtained from 16 CCD images of a travelling plane shock wave. The material enters from the right with a thickness of 0.68 cm and velocity of 84.3 cm/s, and after the grains pass through the shock the layer is over 4 times thicker and $u = 0$. The wave propagates upslope at a constant speed $v = 16$ cm/s. Discontinuous solutions to the Savage-Hutter model exist, and provide a simple explanation of the jumps. Consider the case when $\phi = \delta$ and $d = 0$ in the governing equations 1–4. A simple solution of the resulting equations is that the avalanche velocity and thickness are constant, i.e. $u = u_0$ and $h = h_0$. Different constant states may be joined together by the mass and momentum jump conditions

$$
\begin{aligned}
[\![h(u - v)]\!] &= 0, & (5) \\
[\![hu(u - v)]\!] &= -[\![\cos \zeta K h^2/2]\!], & (6)
\end{aligned}
$$

where $[\![\,]\!]$ are the jump brackets and v is the normal speed of the non-material singular surface. With these conditions it is a simple exercise to construct a travelling wave solution with $u = 0$ behind the shock.

3. Shock capturing numerical methods

A moving grid (Lagrangian) finite difference scheme was developed to solve a reduced non-conservative form of the equations by Savage & Hutter [11]. This method has proved to be a particularly effective numerical integration scheme for smooth solutions [1], [6], [7], [8] and [17]. However, when there are large gradients, or non-smooth solutions, numerical oscillations develop which cause numerical instabilities.

Two one-dimensional methods have been applied to numerically capture plane shock waves. The first is a modified Total Variational Diminishing Lax Friedrichs (TVDLF) method [16], [18] and the second is a Non-Oscillating Central (NOC) scheme [9], [10]. Both of these methods use Eulerian grids and do not rely on a known solution to the Riemann problem. A comparison between the exact plane shock wave solution on non-accelerative slopes and the TVDLF and NOC schemes is illustrated in figure 2, and demonstrates that these high resolution methods are very effective at capturing the travelling wave.

4. Stratification patterns in heaps

When a granular material with grains of two differing sizes is poured into the vertical gap between thin plates from a point source a triangular heap is built up. Although the material is continuously deposited at the top of the pile, it does not flow immediately down the faces because of the difference between the static and dynamic internal friction angles. Once the static friction angle is exceeded the avalanche flows down the face of the pile and *kinetic sieving* takes place [11], [14], which sorts the granular

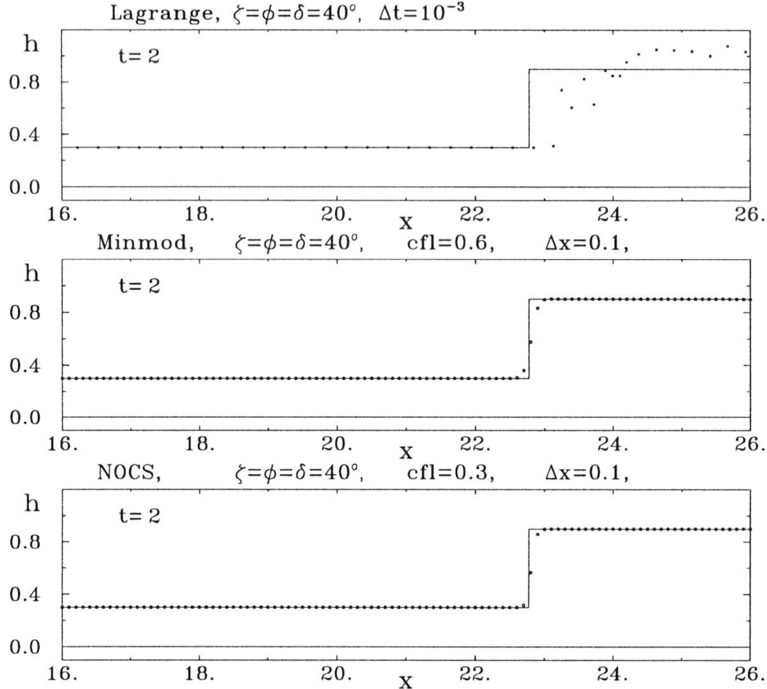

Figure 2. Comparison of analytic plane shock wave solution (solid line) for the thickness, h, with the computed thickness distribution (circles) as a function of space x. The granular material flows from left to right and increases rapidly in thickness at the shock. The results are illustrated for the original Lagrangian method (top) the modified TVDLF method (middle) and the NOC scheme (bottom).

material by grain size. Kinetic sieving is the name given to the process whereby gaps open up between the grains, as they are sheared, and the smaller particles are more likely to fall into open space beneath them, than the large particles, because they are more likely to fit into the available space. An *inverse-grading* of the particles rapidly develops with the larger particles overlying the smaller ones. In a bi-disperse granular mixture this creates a two-layer stripe close to the free surface. When the front of the avalanche reaches the run-out zone a plane shock wave is generated, which propagates upslope and *freezes* the stripe into the granular deposit as shown in figure 3. Successive releases create a large scale stratification pattern.

A simple mixture theory has been developed [4] to model stratification pattern formation. The volume fraction of small particles per unit mixture volume, $0 \leq \nu \leq 1$, is defined and it is assumed that the remaining space $1 - \nu$ is occupied by large particles. Assuming that the volume fraction of small particles has no influence on the dynamics of the bulk mixture

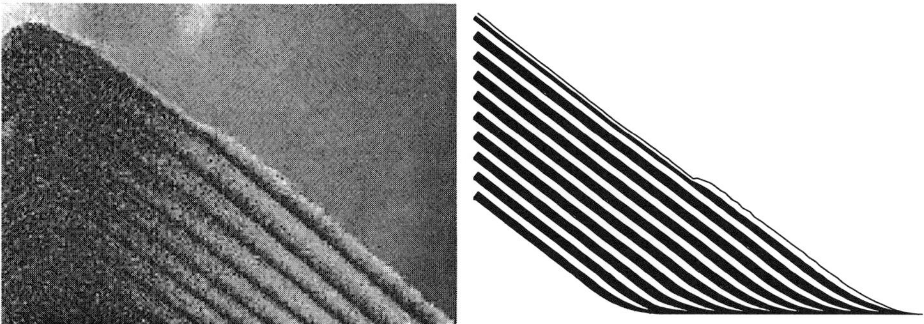

Figure 3. Stratification pattern formation in a bi-disperse grain size population of large (white) particles and small (grey) particles. The uppermost pair of dark and white layers prior to the upslope propagating shock are avalanching downslope. Behind the shock, which can be seen as a rapid increase in the avalanche thickness in the experiments (left) and the simulation (right), the material is at rest. All the material below the top two layers is also at rest.

the Savage-Hutter theory can be used to compute the avalanche thickness and the downslope velocity. The normal velocity can then be inferred from the incompressibility relation together with the basal boundary condition. The experiments suggest that the time-scale for kinetic sieving to take place is much faster than the time-scale for the avalanche motion. It is therefore assumed that the grains are initially presorted when they are input into the flow and a tracer equation is used to follow the local particle size distribution. The results of numerical simulations are shown in figure 3. The avalanche propagates downslope transporting the initial particle size distribution with it. When it reaches the run-out plane the gravity forcing ceases and a granular shock develops, which propagates upslope with approximately constant speed. Successive layers are built up by defining the old free surface of the avalanche to be the new basal topography for the next avalanche to run down. In this way large scale stratification patterns are built up.

5. Stratification and segregation patterns in thin rotating drums

In rotating drums the granular flow is divided into two distinct regions that are separated by a non-material singular surface, at which there are discontinuities (jumps) in the velocity and density fields. Close to the inclined free surface there is a thin fluid-like avalanche, and below it there is a solid layer that rotates about the axis of revolution. Strong mass transfer takes place between the fluid and solid regions. The material that is transported by the avalanche to the base of the inclined slope, is then absorbed by the solid body and rotated back up to the top of the slope, where it replenishes the avalanche.

A series of particle size segregation experiments have also been performed in thin rotating drums. To emphasise the patterns the disk was laid horizontally and gently shaken so that all the small particles fell to the bottom. Once returned to the vertical one side of the disk is filled with large (white) particles and the other side is filled with small (grey) ones. When the disk is rotated at a constant rate (110 seconds per revolution) intermittent avalanches are formed at the free surface. The intermittency again stems from the difference between static and dynamic internal friction angles. The central circular core of material remains completely undisturbed by slow rotation of the drum. Each avalanche release sorts the material, forming a stripe, which is frozen into the deposit by the shock wave and subsequently rotated and buried in the undisturbed material below the free surface. Subsequent releases create a sequence of stripes tangent to the central core, which create a *Catherine wheel* effect as shown in figure 4.

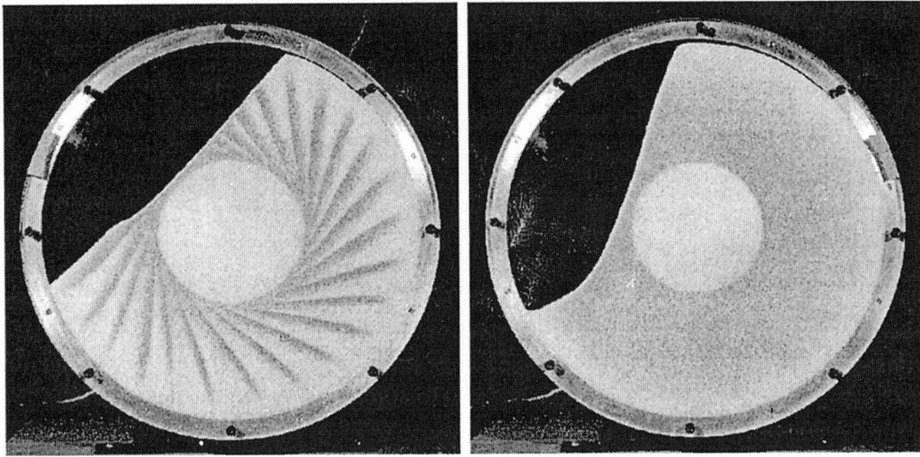

Figure 4. Pattern formation due to particle size segregation in a thin rotating disk at low (left) and fast (right) revolution rates. The large particles are white and the small particles are grey. The drum is rotated slowly counter clockwise and contains a bi-disperse mixture of grain sizes.

At faster rotation rates (< 20 seconds per revolution) the intermittency of the avalanche release, the shock waves and the stripes disappear and a steady flow regime dominates. The material is continuously released on the upper side and continuously deposited on the lower side of the concave free surface and is transported between the two positions by a quasi-steady avalanche in which kinetic sieving takes place. Since the smaller particles are concentrated at the bottom of the avalanche they are the first to get deposited on the lower half of the free surface and a new pattern develops in which the central core is undisturbed, and there is a continuous distribution of grain sizes outside the central core, starting with a high concentration of small particles near the core and ending with a high concentration of large particles near the outer wall.

Acknowledgement

This research was supported by the Deutsche Forschungsgemeinschaft through the SFB 298 project "Deformation und Versagen bei metallischen und granularen Strukturen".

References

1. Koch, T., Greve, R. and Hutter, K.: Unconfined flow of granular avalanches along a partly curved surface, Part II: Experiments and numerical computations. *Proc. R. Soc. London A* **445** (1994), 415-435.
2. Gray, J.M.N.T. and Hutter, K.: Pattern formation in granular avalanches. *Contin. Mech. Thermodyn.* **9**(6) (1997), 341-345.
3. Gray, J.M.N.T. and Hutter, K.: Physik granularer Lawinen. *Physikalische Blätter* **54**(1) (1998), 37-43.
4. Gray, J.M.N.T. and Tai, Y.C.: Particle size segregation, granular shocks and stratification patterns. NATO ASI series, H.J. Herrmann et al. (eds.), *Physics of dry granular media*, 697–702. Kluwer Academic 1998.
5. Gray, J.M.N.T., Wieland, M. and Hutter, K.: Free surface flow of cohesionless granular avalanches over complex basal topography. *Proc. Roy. Soc.* **445** (1999), 1841-1874.
6. Greve, R., Koch, T. and Hutter, K.: Unconfined flow of granular avalanches along a partly curved surface. Part I: Theory. *Proc. R. Soc. London, A* **445** (1994), 399-413.
7. Hutter, K. and Koch, T.: Motion of a granular avalanche in an exponentially curved chute: experiments and theoretical predictions. *Phil. Trans. Roy. Soc. A* **334** (1991), 93-138.
8. Hutter, K. and Greve, R.: Two-dimensional similarity solutions for finite mass granular avalanches with Coulomb and viscous-type frictional resistance. *J. Glac.* **39** (1993), 357-372.
9. Jiang, G. and Tadmor, E.: Non-oscillatory central schemes for multidimensional hyperbolic conservation laws. *SIAM J. Sc. Comp.* **19**(6) (1997), 1892–1917.
10. Nessyahu, H. and Tadmor, E.: Non-oscillatory central differencing for hyperbolic conservation laws. *J. Comput. Phys.* **87** (1990), 408–463.
11. Savage, S.B. and Lun C.K.K.: Particle size segregation in inclined chute flow of dry cohesionless granular solids. *J. Fluid Mech.* **189** (1988), 311-335.
12. Savage, S.B. and Hutter, K.: The motion of a finite mass of granular material down a rough incline. *J. Fluid Mech.* **199** (1989), 177-215.
13. Savage, S.B. and Hutter, K.: The dynamics of avalanches of granular materials from initiation to runout. Part I: Analysis. *Acta Mech.* **86** (1991), 201-223.
14. Savage, S.B.: Mechanics of granular flows. In Hutter K. (ed.) *Continuum mechanics in environmental sciences and geophysics.* CISM No. 337 Springer, Wien-New York, 1993, pp467-522.
15. Tai, Y.C. and Gray, J.M.N.T.: Limiting stress states in granular avalanches. *Annal. Glac.* **26** (1998), 272–276.
16. Tóth, G. and Odstrčil, D.: Comparison of some flux corrected transport and total variation diminishing numerical schemes for hydrodynamic and magnetohydrodynamic problems. *J. Comput. Phys.* **128**(1) (1996), 82–100.
17. Wieland, M., Gray, J.M.N.T., and Hutter, K.: Channelised free surface flow of cohesionless granular avalanches in a chute with shallow lateral curvature. *J. Fluid. Mech.*

392 (1999), 73-100.
18.   Yee, H.C.: *A class of high-resolution explicit and implicit shock-capturing methods.*
NASA TM–101088, 1989.

MIXING IN A MONO-DISPERSE GRANULAR BED SUBJECT TO VERTICAL OSCILLATIONS

CARL WASSGREN
School of Mechanical Engineering
Purdue University
West Lafayette, IN 47907-1288

Abstract

Intentional and unintentional vibrational mixing of granular materials occurs in many .industrial operations. To study the degree of mixing resulting from vibrations, two-dimensional discrete element computer simulations of a mono-disperse granular material subject to vertical, sinusoidal oscillations are performed. The degree of mixing is quantified using a mixing index and diffusion coefficient. The simulation results indicate that mixing increases with decreasing frequency for a fixed acceleration amplitude and increasing amplitude for a fixed frequency. When plotted against oscillation velocity amplitude, the mixing index and diffusion coefficient data follow S-shaped curves. The simulations show that greater mixing occurs between the upper and' lower halves of the bed than between the left and right halves. The simulations also indicate that the presence of kink convection cells results in enhanced mixing throughout the bed.

1. Introduction

Mixing of granular materials is an important operation in a number of industries including those that handle chemicals, pharmaceuticals, and foodstuffs. In order to design for better efficiency, an understanding of the mixing process is desirable. One method of mixing particulate materials involves vibrating the material. These vibrations can either be intentional or may occur unintentionally when material is transported or in storage.

Several phenomena that can induce mixing have been observed when granular beds are subject to vibration. These include side-wall convection cells, surface waves, and kink convection cells. The occurrence of these phenomena depends strongly on the dimensionless acceleration amplitude of the vibrations, $\Gamma=a\omega^2/g$ where a is the amplitude of the oscillations, ω is the radian frequency, and g is the acceleration due to gravity [1]. For $\Gamma>1$, side-wall convection cells occur adjacent to wall boundaries and are oriented such that particles move down along vertical walls and up within the bed.

A.D. Rosato and D.L. Blackmore (eds.), IUTAM Symposium on Segregation in Granular Flows, 277–286.
© 2000 *Kluwer Academic Publishers. Printed in the Netherlands.*

The direction of the convection cells can be reversed if the wall is inclined sufficiently [2]. The convective motion can induce mixing or result in particle segregation if particles are of sufficiently different size [3]. Standing surface waves occur over two ranges of Γ [4,5]. For $\Gamma<3.7$ the surface waves form at one-half the forcing frequency ($f/2$ waves) and for $\Gamma>3.7$ the waves form at one-quarter the forcing frequency ($f/4$ waves) [1,7]. Mixing can occur as a result of surface waves since particles are ejected from the wave peaks and land in different regions of the particle bed and because there is considerable internal shearing near the bed surface during wave formation. Lastly, kink convection cells occur for $\Gamma>3.7$, the Γ at which the flight time of the particle bed undergoes a period doubling bifurcation [1]. As a result of the bifurcation, "kinks" or "nodes" can form in the particle bed [1,8]. A node results when different regions of the bed oscillate out-of-phase with each other. Bracketing each node are two counter-rotating convection cells oriented such that the average motion of particles is down at the node when averaged over several oscillation cycles. The convection cells are the result of the interaction between fluidized regions of the particle bed avalanching over "solidified" regions of the bed already in contact with the container base [6]. The movement of particles in kink convection cells can also result in considerable particle mixing.

This paper describes the use of discrete element computer simulations to investigate the effect of vertical, sinusoidal oscillations on mixing of a single component granular material in a wide container. In particular, the degree of mixing, quantified by a mixing index and a diffusion coefficient, is determined for various vibration parameters. The effect of kink convection cells is also investigated.

2. Simulation Description

The simulation models a two-dimensional container with a horizontal, rigid base that is subject to vertical, sinusoidal oscillations and contains a granular material. Periodic lateral boundaries are used to simulate a very wide container and eliminate boundary effects such as side-wall convection [2,6]. The granular material is modeled as a collection of nearly mono-disperse, frictionless spheres constrained to move in two dimensions. The forces acting on the spheres include a gravity body force and normal contact forces. The model used to determine normal contact forces is a damped, linear spring [9]. The parameters used in the simulations are given in table 1.

The simulation proceeds by oscillating the container floor with a sinusoidal, vertical trajectory of amplitude, a, and cyclic frequency, f, corresponding to a radian frequency, $\omega=2\pi f$. After a specified number of oscillation cycles the container oscillations are halted and the particles are allowed to come to rest. This final state is used to determine the degree of mixing. For a particular simulation, the same initial conditions are used when investigating the degree of mixing for a different number of oscillations.

A second series of simulations were conducted to investigate the effect of kink convection cells on mixing. Kinks can be produced in experiments when $\Gamma > \Gamma_{bif}$, where Γ_{bif} is the acceleration amplitude corresponding to the first period-doubling bifurcation, by perturbing the oscillating particle bed sufficiently. Inserting a rod or blowing a jet of air into the bed can often cause kink waves to occur. To produce kink waves in the simulations, the particle bed is divided horizontally into two regions of equal width divided by a wall boundary. The initial positions of the particles in the two different regions are offset by a vertical distance such that the motion of the two separate particle beds is out-of-phase. Once the out-of-phase motion is established, the dividing wall is removed and the bed is allowed to oscillate for several additional cycles. The resulting particle states are used as initial conditions in the simulations involving kink convection cells.

3. Measurements

The degree of mixing within the bed is quantified using a mixing index, M, and a diffusion coefficient, D. The mixing index is a common metric used in industrial applications when particles have some distinguishing characteristic such as size, density, or shape (refer to Fan et $al.$ [10] for a review of more than 30 different mixing indices). The mixing index is defined here as

$$M = 1 - \frac{\sigma}{\sigma_0} \tag{1}$$

where σ is the standard deviation of the particle species area concentrations measured at different locations with respect to the mean species area concentration, and σ_0 is the standard deviation for the initial conditions. The area concentration, c, of particle species j in a particular sampling region, s, is determined by

$$c_j^s = \frac{A_j^s}{A_{\text{tot}}^s} \tag{2}$$

where A_j^s is the total area of species j in the sampling region and A_{tot}^s is the total area of all particle species in the same region. The mean area concentration of species j over all sampled regions is given by

$$\overline{c}_j = \frac{1}{S} \sum_{s=1}^{s=S} c_j^s \tag{3}$$

where S is the total number of sampling regions. The standard deviation of the particle species j area concentration, σ_j, is found using the relation

$$\sigma_j = \frac{1}{S-1} \sqrt{\sum_{s=1}^{s=S} (c_j^s - \overline{c}_j)^2} \tag{4}$$

The mixing index, M, will vary between zero and one with zero corresponding to a container of material that remains unmixed with respect to the initial state and one corresponding to a well-mixed state where the species concentration at each sampling location is equal to the average concentration. One limitation of using a mixing index is that it requires the definition of at least two different particle species. As a result, the mixing index is not, in general, independent of the initial conditions of the particle species. For example, defining all of the particles on the left side of the container as species 1 and all of the particles on the right side of the container as species 2 may give a different mixing index after a specified number of oscillations than if the upper half and lower halves of the particle bed were designated as species 1 and species 2.

To avoid the need to define different particle species a second metric, the diffusion coefficient, D, is also used to quantify the degree of mixing within the bed. The mean diffusion coefficient for the particle bed is given by

$$D = \frac{1}{S}\sum_{s=1}^{S} D_s \tag{5}$$

where D_s is the diffusion coefficient for region s and S is the total number of sampling regions. The diffusion coefficient for region s is found using

$$D_s = \frac{1}{6(t_2 - t_1)}\sum_{n=1}^{N_s^{t_1}} \frac{A_{n,s}^{t_1}}{A_{tot,s}^{t_1}} \left| (\Delta \mathbf{x}')_{n,s}^{t_1 \to t_2} \right|^2 \tag{6}$$

where t_1 and t_2 are the initial and final times between which the diffusion coefficient is determined, $N_s^{t_1}$ is the total number of particles in region s at time t_1, $A_{n,s}^{t_1}$ is the area of particle n in region s at time t_1, $A_{tot,s}^{t_1}$ is the total area of particles in region s at time t_1, and $(\Delta \mathbf{x}')_{n,s}^{t_1 \to t_2}$ is the deviation of particle n's displacement, $(\Delta \mathbf{x})_{n,s}^{t_1 \to t_2} = \mathbf{x}_{n,s}^{t_2} - \mathbf{x}_{n,s}^{t_1}$, from the mean displacement of particles, $(\overline{\Delta \mathbf{x}})_s^{t_1 \to t_2}$, in region s between times t_1 and t_2:

$$(\Delta \dot{\mathbf{x}}')_{n,s}^{t_1 \to t_2} = (\Delta \mathbf{x})_{n,s}^{t_1 \to t_2} - (\overline{\Delta \mathbf{x}})_s^{t_1 \to t_2} \tag{7}$$

The mean displacement of particles in region S is found using an area-weighted average of all particle displacements in region s:

$$(\overline{\Delta \mathbf{x}})_s^{t_1 \to t_2} = \sum_{n=1}^{N_{si}^{t_1}} \frac{A_{n,s}^{t_1}}{A_{tot,s}^{t_1}}(\Delta \mathbf{x})_{n,s}^{t_1 \to t_2} \tag{8}$$

In these simulations, the diffusion coefficient is calculated over each oscillation cycle, *i.e.* $t_2 - t_1 = 1/f$. The sampling regions in the simulations were determined by dividing the container into a grid of equal sized squares with a non-dimensional length of L/d on a side where d is the mean particle diameter. Note that both the mixing index and diffusion coefficient depend on the number of samples made and the area of the sampling regions. Using different sampling L/d values gives an indication of the degree of mixing at different length scales.

When using the mixing index, M, two methods of defining two distinct particle species are used. In the first method all of the particles in the left half of the container at the initial conditions are tagged as being species 1 while all of the particles in the right half of the container are tagged as species 2. The second method defines all of the particles in the upper half of the particle bed as species 1 and the particles in the lower half as species 2. The top plots in figures 1 and 2 show the two definitions. Other than being labeled as different species, all of the particles within the container are identical.

4. Results

The mixing index is measured as a function of the number of oscillation cycles for frequencies between 5 Hz and 30 Hz at various dimensionless acceleration amplitudes, $\Gamma = a\omega^2/g$. Figures 1 and 2 show two frames each from the same simulation where $\Gamma = 6.0$ (0 nodes), $f = 20$ Hz, and the container width, W, to particle diameter, d, ratio is $W/d = 150$. The only difference between simulations is how the two particle species are defined. In each figure, the top plot shows the initial conditions of the granular bed and lower plot shows the bed after 59 oscillation cycles. The grid in the figures represents the sampling regions used in the mixing index calculations with $L/d = 5.0$ where L is the length of the sampling square side. A value of $L/d = 5.0$ was chosen to give a sampling region large enough to include many particles over which to average, yet small enough to give a measure of the degree of mixing at a scale on the order of a particle diameter.

It is apparent from the two figures that the degree of mixing depends on how the particle species are initially defined. The degree of mixing between the upper and lower halves of the beds is greater than the degree of mixing between the left and right halves. The mixing index data also reflects this trend. Figure 3 plots the mixing index using the left/right half particle species definitions against oscillation cycles for $W/d = 150$, 20 Hz, and various acceleration amplitudes. The data indicates that the degree of mixing increases as the number of oscillation cycles increases and that greater mixing occurs as Γ increases. In addition, the presence of a node in the particle bed enhances mixing compared to when no nodes appear. Similar trends are also observed using the upper/lower half species definition; however, the mixing index for this species definition is generally greater than the mixing index for the left/right species definition. To compare the degree of mixing for various vibration parameters, the mean diffusion coefficient for the bed, D, averaged over all oscillation cycles is plotted against dimensionless oscillation amplitude, a/d, in figure 4. The trends shown for diffusion coefficient are similar to those observed for the mixing index. The data indicate that for a given frequency, diffusion increases as oscillation amplitude increases. In addition, for a given dimensionless acceleration amplitude, Γ, diffusion increases as frequency decreases. As shown in figure 5, the data for frequencies greater than 10 Hz and $W/d = 50$ follow an S-shaped curve when plotted against dimensionless velocity amplitude, $a\omega/(gd)^{1/2}$. Although exhibiting similar trends, the data for frequencies of 5 and 7 Hz fall outside the otherwise tight grouping of data.

The data for $W/d = 150$ (0 and 2 node) also deviates from the trend at larger velocity amplitudes. Observations of the simulations show that for a container width of $W/d = 50$ (0-nodes) only a single $f/4$ wave occurs that is not well-defined due to the narrow container width. For $W/d = 150$ (0-nodes), however, two distinct $f/4$ waves occur. The amplitude and the internal shearing of the well-formed waves is greater than that for the indistinct wave and as a result, the well-formed waves result in greater mixing.

The occurrence of surface waves also explains the apparent discrepancy between the mixing index data in figure 3 and diffusion coefficient data in figure 5 for 0-node and 2-node simulations. The data in figure 3 indicates that the presence of a node increases mixing as defined by the mixing index while the data in figure 5 indicates nodes decrease mixing as defined by a diffusion coefficient. A closer inspection of the diffusion coefficient for each sampling region explains this result. When no nodes are present, the diffusion coefficient is large near the free surface of the bed due to particles being ejected from the wave peaks and large relative movement of particles resulting from the shearing action of the waves. Thus, the diffusion coefficients for regions near the free surface are much larger than the diffusion coefficients for the remainder of the bed and skew the overall average diffusion coefficient. When a node is present, the surface waves are weaker and fewer particles are ejected from the free surface. Although the overall average diffusion coefficient is smaller when nodes are present, the bed is more uniformly mixed throughout. Figure 6 shows the initial state (top plot) and state after 59 oscillation cycles (bottom plot) for a simulation with $\Gamma=6.0$, $f=20$ Hz and 2 node. The concentration of species is more uniform than for a simulation using the same vibration parameters but having 0 nodes as shown in figure 1.

The dependence of the diffusion coefficient and mixing index on the size of the sampling region was also examined. The data indicate that, in general, the degree of mixing, measured using both the mixing index and diffusion coefficient, increases as L/d increases until a value of approximately 10. For large values of L/d the degree of mixing remains essentially constant.

5. Conclusions

Results from two-dimensional discrete element computer simulations of a mono-disperse granular bed subject to vertical, sinusoidal oscillations indicate that the degree of mixing increases with decreasing frequency for a fixed acceleration amplitude and increasing amplitude for a fixed frequency. When plotted against velocity amplitude the mixing index and diffusion coefficient data for frequencies greater than 10 Hz are closely grouped and follow an S-shaped curve. For frequencies less than 10 Hz, the data also follow S-shaped curves but fall outside the otherwise tight grouping of data. The simulations also indicate that greater mixing occurs between the upper and lower halves of the bed than between the left and right halves. An average diffusion coefficient was used in addition to a mixing index to quantify the degree of mixing in the particle bed since the diffusion coefficient does not require the definition of different particle species as is required by the mixing index. However, a diffusion coefficient averaged over for the entire bed can be skewed by sampling regions having significantly larger diffusion coefficients. For example, surface waves can have much larger diffusion coefficients in regions near the free surface of a bed than in the remainder of the bed. Although the diffusion coefficient averaged over all of the regions is large, only the free surface mixes well while the remainder of the bed remains poorly mixed. To produce greater mixing within the bed for acceleration amplitudes greater than the period doubling bifurcation acceleration amplitude, kinks or nodes should be introduced into the bed. Although not investigated here, it is expected that as the number of nodes in the bed increases, so will the degree of mixing.

6. Tables and Figures

TABLE 1. Parameters used in the discrete element computer simulations.

container width/mean particle diameter	50, 150
number of particles	1000, 3000
approximate bed depth/ mean particle diameter	20, 20
range of particle diameters	0.9 – 1.1 mm (uniform distribution)
particle density	2500 kg/m^3
particle/particle coefficient of restitution	0.8
particle/particle contact stiffness	5.289*10^3 N/m
particle/particle contact damping coefficient	8.337*10^{-3} N/(m/s)
particle/wall coefficient of restitution	0.8
particle/wall contact stiffness	1.058*10^4 N/m
particle/wall contact damping coefficient	1.667*10^{-2} N/(m/s)
simulation time step	5.707*10^{-6} s

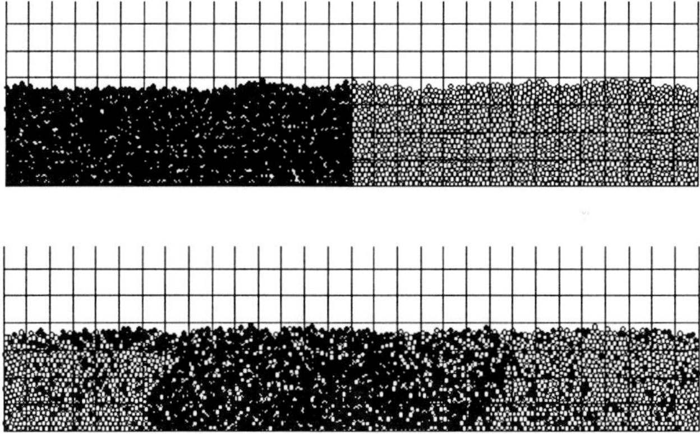

Figure 1. Initial particle states (top) and particle states after 59 oscillation cycles (bottom) for a simulation with Γ=6.0, f=20 Hz (0 nodes), and periodic boundaries. The particles on left side of the container in the top figure are defined as species 1 and the particles on the right side are species 2. The grid represents the sampling areas (squares of length L/d=5 on a side) used in determining the mixing index and diffusion coefficient.

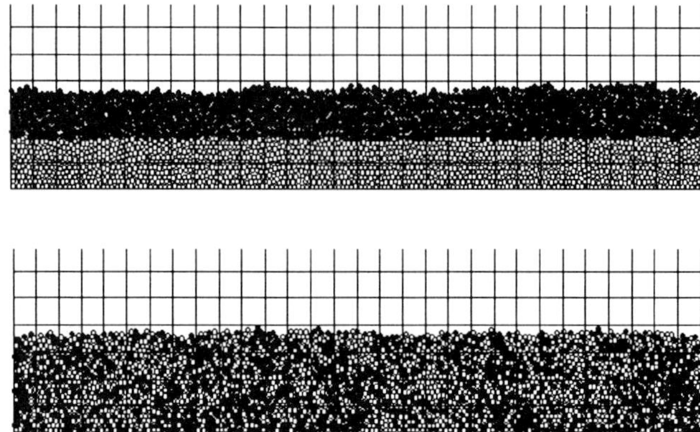

Figure 2. Initial particle states (top) and particle states after 59 oscillation cycles (bottom) for a simulation with $\Gamma=6.0$, $f=20$ Hz (0 nodes), and periodic boundaries. The particles in the upper half of the granular bed in the top figure are defined as species 1 and the particles in the lower half are species 2. The grid represents the sampling areas (squares of length $L/d=5$ on a side) used in determining the mixing index and diffusion coefficient.

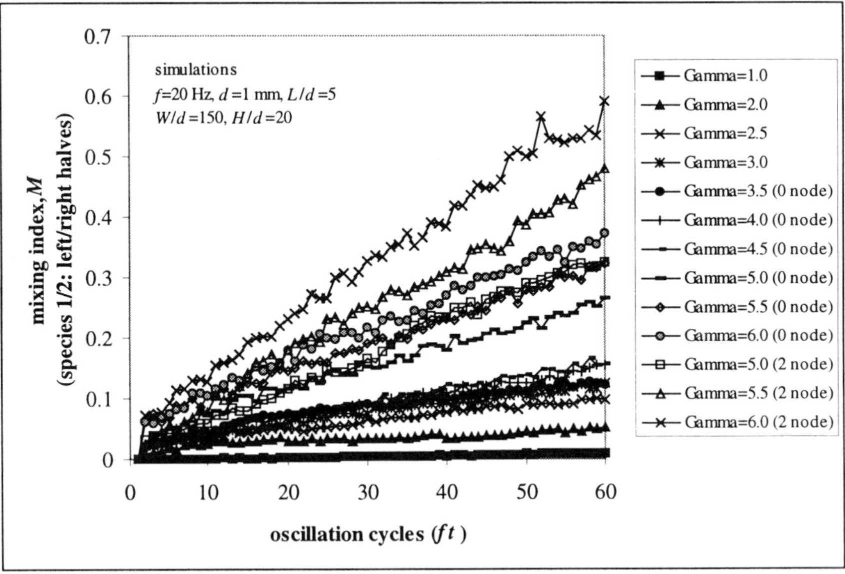

Figure 3. Mixing index, M, using the left/right definition of particle species plotted against oscillation cycle for a frequency of $f=20$ Hz and varying dimensionless oscillation acceleration amplitudes, $\Gamma=a\omega^2/g$.

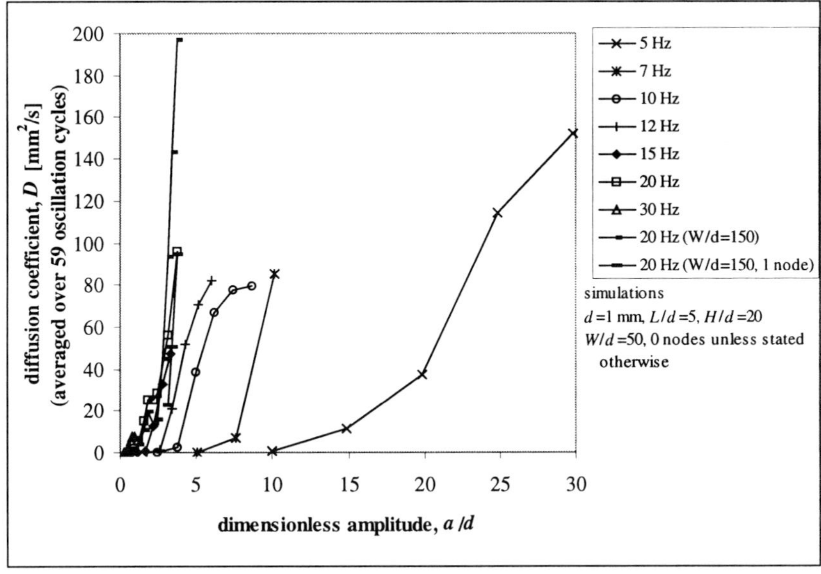

Figure 4. The mean diffusion coefficient, *D*, averaged over 59 oscillation cycles plotted against dimensionless oscillation amplitude, *a/d*, for various oscillation parameters.

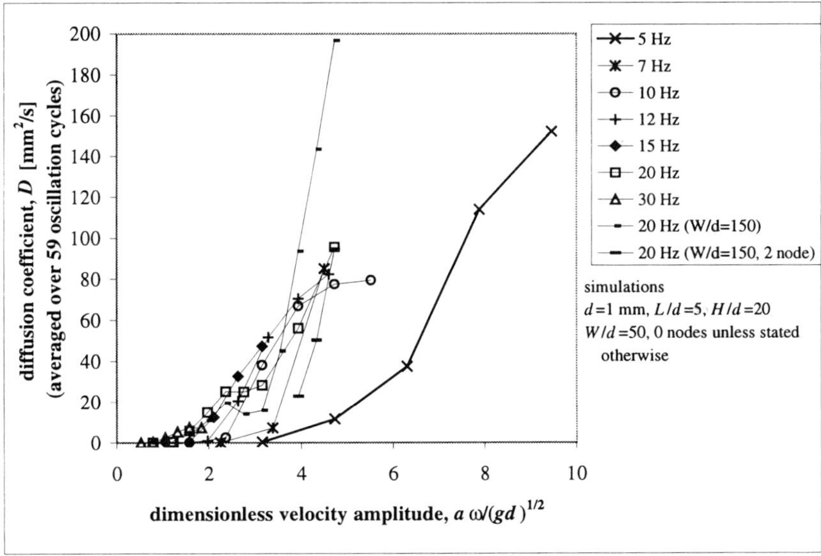

Figure 5. The mean diffusion coefficient, *D*, averaged over 59 oscillation cycles plotted against dimensionless velocity amplitude, $a\omega/(gd)^{1/2}$, for various oscillation parameters.

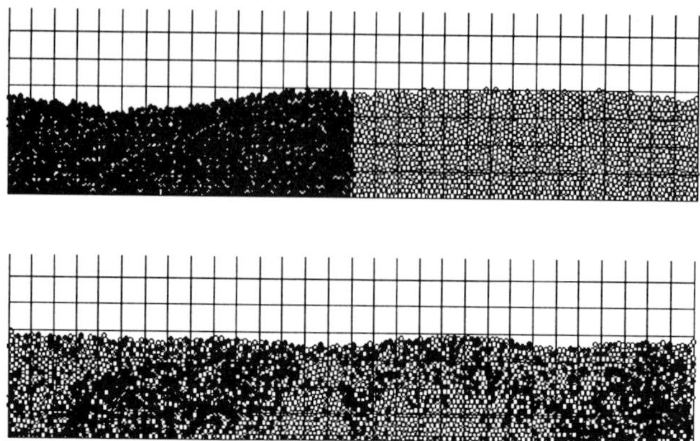

Figure 6. Initial particle states (top) and particle states after 59 oscillation cycles (bottom) for a simulation with Γ=6.0, *f*=20 Hz (2 node), and periodic boundaries. The particles on the left half of the granular bed in the top figure are defined as species 1 and the particles on the right half are species 2. The grid represents the sampling areas (squares of length *L/d*=5 on a side) used in determining the mixing index and diffusion coefficient.

7. References

1. Wassgren, C.R., Brennen, C.E., and Hunt, M.L.: Vertical vibration of a deep bed of granular material in a container, *Journal of Applied Mechanics* 63[3] (1996), 712-719.
2. Knight, J.B.: External boundaries and internal shear bands in granular convection, *Physical Review E* **55** [5] (1997), 6016-6023.
3. Knight, J.B., Jaeger, H.M., and Nagel, S.R.: Vibration-induced size separation in granular media: The convection connection, *Physical Review Letters* **70** (1993), 3728-3731.
4. Melo, F., Umbanhower, P., and Swinney, H.L.:Transition to parametric wave patterns in a vertically oscillated granular layer, *Phys. Rev. Lett.*72[1] (1994), 172-175.
5. The exact range of Γ values for surface waves also appears to depend on the friction between particles and the depth of the particle bed [6,7].
6. Wassgren, C.R.: *Vibration of Granular Materials*, Ph.D. Thesis, California Institute of Technology, Pasadena, CA. (1997).
7. Wassgren, C.R., Hunt, M.L., and Brennen, C.E.: Investigation of *f*/2 and *f*/4 waves in granular beds subject to vertical, sinusoidal oscillation, in Behringer & Jenkins (eds), *Powders & Grains 97*, Balkema, 1997, Rotterdam, pp. 433-436.
8. Douady, S., Fauve, S., and Laroche, C.: Subharmonic instabilities and defects in a granular layer under vertical vibrations," *Europhys. Lett.* **8**[7], (1989), 621-627.
9. Cundall, P.A. and Strack, O.D.L.: A discrete numerical model for granular assemblies, *Geotechnique* **29**[1], (1989), 47-65.
10. Fan, L.T., Chen, S.J., and Watson, C.A.: Solids mixing, *Ind. Eng. Chem.*6[7], (1970), 53-69.

IMPACT-REGULATED MIXING AND SEGREGATION IN POURED GRANULAR HEAPS

U.TÜZÜN[†], L.SMITH[†], J.BAXTER[†] and D.M.HEYES[‡]
[†]*Department of Chemical and Process Engineering, School of Engineering in the Environment;*
[‡]*Department of Chemistry, School of Physical Sciences;*
University of Surrey, Guildford, GU2 5XH, U.K.

1. Introduction

A large body of work, for example [1-4] has concentrated on the observation that mixtures of granular material with dispersity in size and/or shape tend to separate in response to external perturbations, which are inevitable in almost all solids storage and handling operations. The focus of much of this work has been to determine the sensitivity of segregation phenomena to mixture properties. An underlying principle of much of this work seems to be that segregation is an innate property of granular mixtures; that it is in some sense unavoidable. This is certainly true if the mixture properties allow spontaneous percolation of fines through coarse; microstructural analyses [4] enable such mixtures to be identified. However, segregation problems in industry are often evident for mixtures where the dispersity in size or shape is sufficiently modest to preclude percolation. In such cases, we maintain that segregation is strongly influenced by process conditions *as well as* mixture properties, and so can (at least in principle) be controlled to some extent by careful manipulation of the process conditions.

Recently [2] we demonstrated experimentally that the stratification of poured mixtures into layers could be controlled by regulating the impact (the incident kinetic energy) of the feed stream on the growing heap, in this case via the feed rate. For low impact (slow feeding) conditions, geometric effects of particle size and shape are primarily responsible for segregation. However, high impact feeding can restrict the degree of segregation in two ways; by causing incident particles to embed themselves within the growing heap (thus limiting their freedom to move and restricting the geometric effects) or by causing bulk mixing within the poured assembly. In this paper the effects of impact are studied using granular dynamics (GD) computer simulations. The methodology for these simulations has been described in detail elsewhere [5]. The most important feature of GD is its explicit force-basis; unlike for cellular automata which have been used elsewhere in the study of segregation [1], the effects of impact can be directly taken into account.

Perhaps the most useful feature of GD is the wide range of parameters which can be extracted routinely from simulation data (stresses and velocities of individual particles, local and bulk voidage measurements). It is arguably the case that published work on GD often fails to take advantage of the full range of data available, instead relying upon snapshots of the assembly and/or the temporal evolution of bulk properties such as flow rates. This is not to deny the usefulness of such work, merely to suggest that much more can be achieved by careful statistical analysis of the vast amount of data available from GD simulations.

A.D. Rosato and D.L. Blackmore (eds.), IUTAM Symposium on Segregation in Granular Flows, 287–296.

Using snapshots of GD simulations as an overall perspective, allied to detailed analysis of the internal workings of heaps as they evolve, this paper attempts to investigate in greater detail the experimental observation that impact has a strong influence on segregation propensity. Several methods of regulating impact in simulations are summarised, with illustrative examples of how impact regulates segregation. Two cases are then studied in greater statistical detail, using a range of analysis tools developed for the purpose.

2.   Regulating Impact During Pouring of Heaps

This section gives visual illustration of how segregation in the simulations was affected by various factors. The basic simulation technique was to place a flat-bottomed hopper some distance above a free surface, fill the hopper with a mixture of particles of different sizes, and allow the hopper to discharge and the material to form a heap. The effect of impact on the growing heap is regulated in a number of ways:

- by pouring onto an inclined plane rather than a flat plate; rather than regulating the impact itself, this regulates the *response* of the assembly to impact
- by varying the feed rate (*e.g.*, the orifice size), or removing the hopper altogether and allowing particles to 'trickle' one at a time onto the heap from a short distance above
- by controlling the density of particles (especially coarser particles), hence their mass and kinetic energy on impact
- by varying the free-fall height (placing the hopper at different heights above the surface, or by minimising the free-fall height (initially placing the hopper just above the heap and moving it as the heap grows)

Figures 1 and 2 show the same material (a binary mixture of size ratio 3:1) poured at the same feed rate onto a flat plane and an inclined surface. There is clearly a much greater degree of segregation in the latter case; the inclined plane acts as a buttress for the outer layers of the structure, and the resultant rigidity appears to strongly promote segregation. Pouring onto an inclined plane has been previously used in experiments [3] as an analogue of pouring onto a heap, which certainly seems reasonable as segregation is almost certainly limited to the outermost few layers only. For these simulations, it is arguably the case that finite-size effects and two-dimensionality may have overly inhibited the propensity for segregation in free pouring.

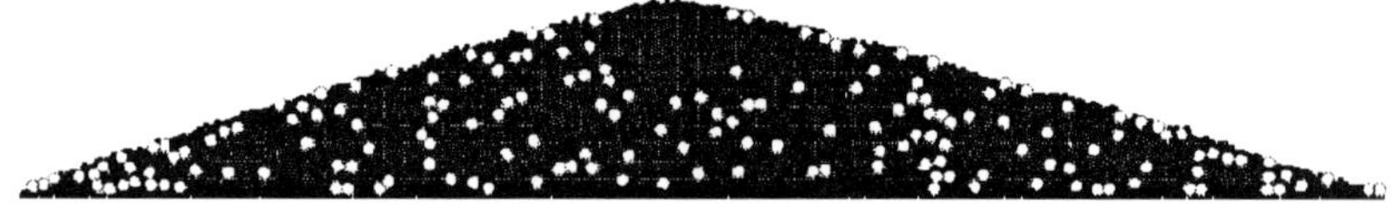

Figure 1: Binary mixture, size ratio 3:1, poured onto flat surface

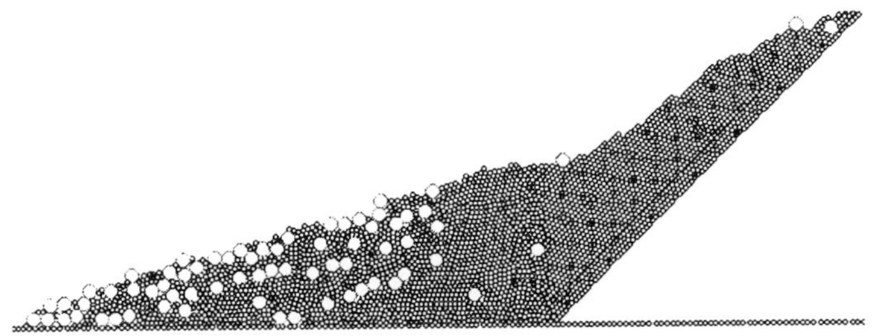

Figure 2: Binary mixture, size ratio 3:1, poured onto inclined plane

The effects of feed rate are illustrated using a mixture with less "natural" tendency to segregate (a 2:1 size ratio binary mixture). Figures 3 and 4 show the same mixture poured onto an inclined plane at different rates; the feed rate for the latter was about three times that for the former. This shows a marked effect of feed rate; there was a considerable degree of segregation for the slower feed case, whereas almost the entire assembly for the fast feed case appears to be well mixed.

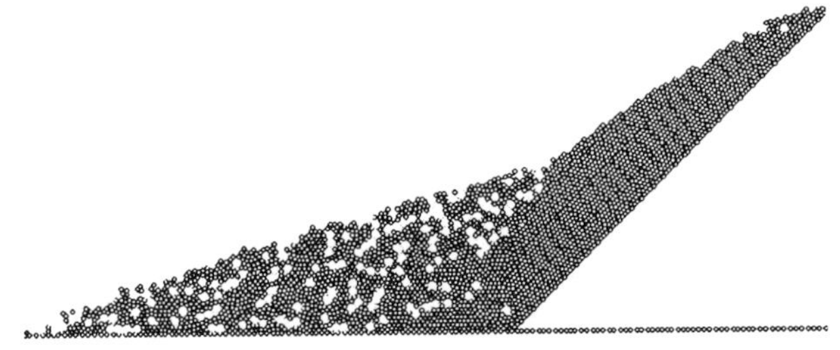

Figure 3: 2:1 size ratio mixture poured slowly onto inclined plane

Figure 4: 2:1 size ratio mixture poured fast onto inclined plane

In the simulation shown in Figure 6, all particles were assigned equal masses, irrespective of their size. Effectively this meant a much-reduced solid density for the coarser particles. Very striking effects on segregation lead us to conclude that impact effects, particularly due to the more massive particles, have a very strong influence on segregation. Figures 5 and 6 show freely poured heaps of 5:1 size ratio binary mixtures. Note that the latter heap is considerably larger overall than the former; the size ratio is equal for each heap although each particle looks smaller in the latter heap.

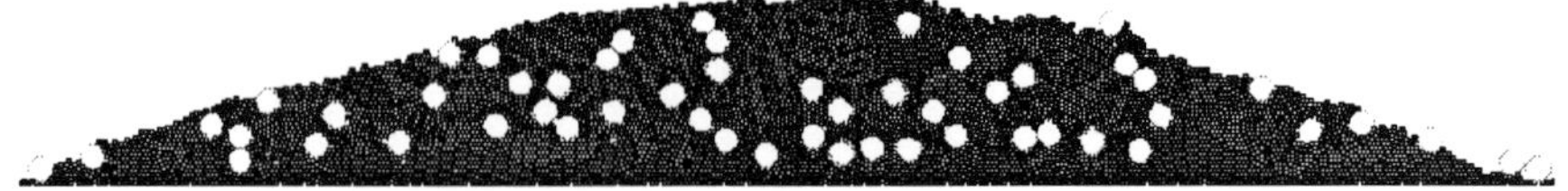

Figure 5: Freely poured 5:1 size ratio binary mixture, uniform solid density

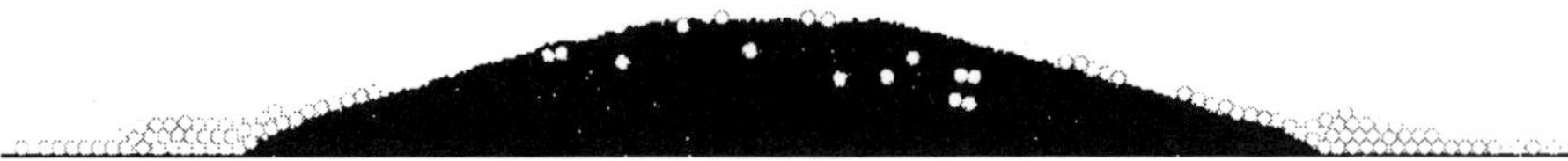

Figure 6: Freely poured 5:1 size ratio binary mixture, less dense coarse particles

This series of snapshots of heaps is useful in illustrating the important effects of impact on segregation. However, the information is presented in a highly qualitative fashion. As mentioned in the Introduction, one significant feature of GD is the range of data which can be extracted from the simulations. We now turn to analysis of the effects of drop height on segregation, using a range of tools which have been developed for detailed quantitative analysis of simulation data.

3. Detailed analysis of heaps for high and low impact conditions

3.1. HIGH AND LOW IMPACT CONDITIONS

Simulations described henceforth in this paper as "high impact case" refer to conditions whereby the mixture (of size ratio 3:1) was fed from a hopper, fixed in position about 100 small particle diameters above the fixed surface. Simulations were also performed with fixed hoppers at lower heights, but to lower the impact sufficiently to distinguish between high and low impact, it was necessary to dispense with the hopper altogether. The low impact simulations were performed by "trickling" particles, several at a time, onto the top of the heap, by placing them a small distance (6 small particle diameters) above the current top of the heap and allowing them to fall under gravity.

3.2. THE POST-PROCESSING TOOLSET

Periodic output from the simulation procedure consisted of time series of positions and velocities (x,y and rotational) for all particles. These raw results were investigated by a set of visual tools developed for the purpose. The toolset contains two main sets of routines:
- routines for time series analysis to investigate significant or periodic phenomena
- routines for characterisation and colour-coded display of particles according to various criteria (for example, components of instantaneous or time-averaged velocity or acceleration vectors)

Often, time series analysis was used in the first instance to identify interesting points in time, in terms of significant periodic phenomena, and then snapshots were taken at or around these points in time.

3.3. TIME SERIES ANALYSIS FOR HIGH IMPACT CASE

Figures 7 to 9 demonstrate the application of spectral analysis in a time series to identify periodicity of significant activity within the heap. Figure 7 shows a time series for the mean lateral velocity of particles over a time period following the initial transient phase of heap formation.

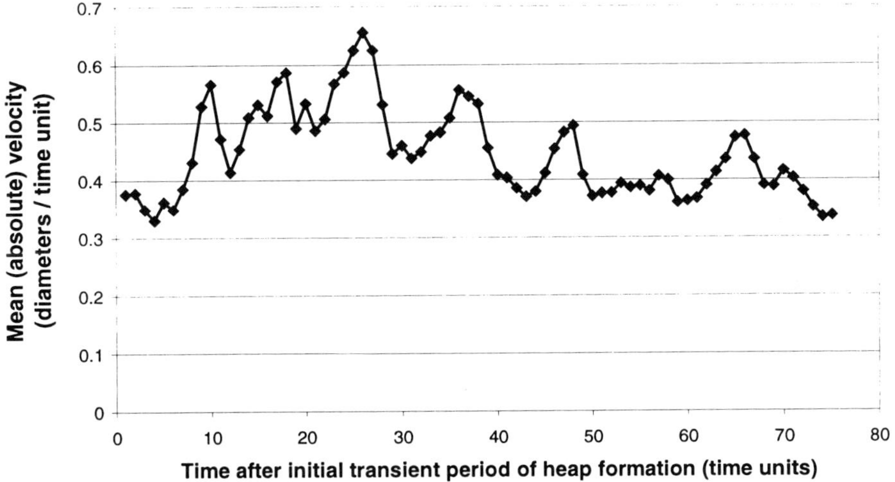

Figure 7: Time series of mean lateral particle velocity after initial transient phase of heap formation

As is often the case in time series analysis, the next stage was to compute the corresponding autocorrelation function (Figure 8), which measures the degree to which the value of a time measured variable is correlated with its value after a given lag. The autocorrelation function is often instructive in its own right. In this case, the periodicity in the autocorrelation function itself might indicate periodicity in the original time series. However, for our purposes its calculation is more important as an intermediate step of the analysis.

In order to quantify periodic behaviour in the original signal, the Fourier transform of the autocorrelation function was then computed. This gives a spectral density function (or "power spectrum") for the original signal, shown in Figure 9.

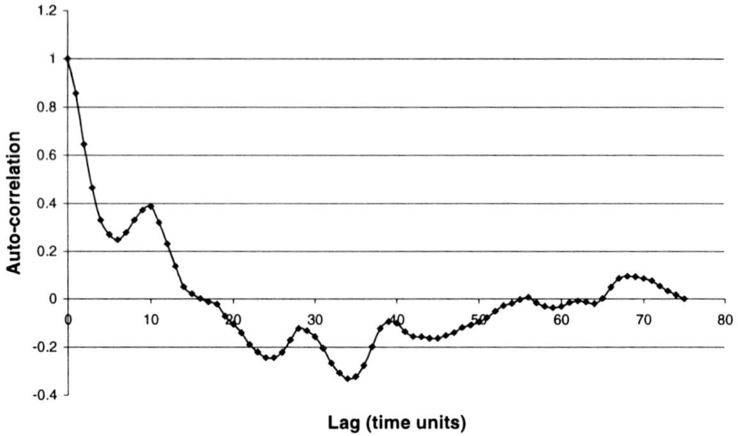

Figure 8: Autocorrelation function of the signal in Figure 7

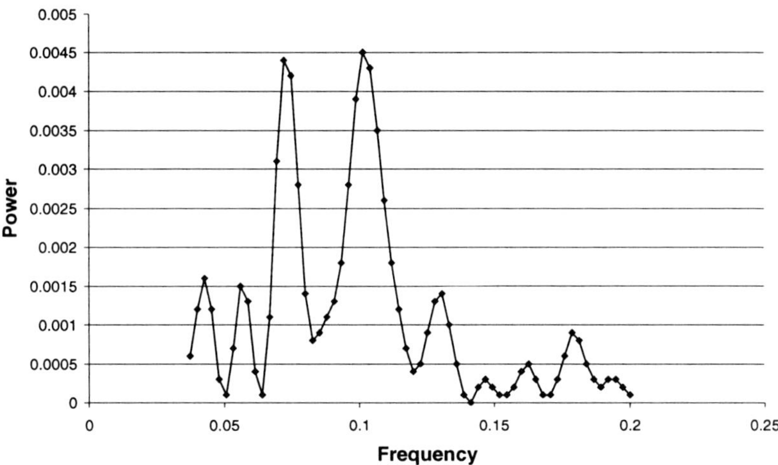

Figure 9: Spectral density function relating to the signal in Figure 7

The peaks in the spectral density function correspond to the frequencies of the most significantly periodic behaviour in the original signal (the inverse of these frequencies being the period). There is a strong peak in the spectrum at a frequency just above 0.1; the original time series appears to show a number of peaks approximately 10 time units apart. So, at least in terms of the variable in question, there is evidence of periodic behaviour within the heap, with a "time constant" of approximately 10 units. Determining the time constants for periodic events is perhaps the most potentially useful application of this type of analysis in the study of granular mechanics. The implications of these results are discussed more fully below.

3.4. COLOUR-CODED SNAPSHOTS OF HIGH AND LOW IMPACT CASES

Applying filters to simulation data allow "regions of stability" to be identified for the high and low impact cases. These are coloured darkly in Figures 10(a) and 10(b) and consist of particles which do not move more than one particle diameter from the time of the snapshot to the end of the simulation.

(a)   (b)

Figure 10: "Regions of stability" for (a) low impact; (b) high impact simulations

For the low impact case there is a well-defined region of stability extending outwards from the centre, broadly resembling the heap as a whole in shape. In contrast, very little of the high impact heap is stable except for a very small region near the bottom centre; thus there is bulk motion over much of the assembly.

A second set of filters were used to identify and locate the most recent entrant particles within the heap at a given point in time. For the high impact case, new entrants embed themselves on entry, displacing existing particles and give rise to a central 'bowling' (Figure 11a). By way of comparison, for the low impact case, new entrants to the heap are more spread out and do not become initially trapped in the central region to anywhere near the same degree as for the high impact case (11b).

(a)   (b)

Figure 11: (a) "bowling" for new entrants in high impact case;
(b) dispersion of new entrants along surfaces in low impact case

The high impact case is characterised by periodic pluming, where new entrants are eventually ejected from the bowl region towards the free surfaces. It emerged that this is the probable source of the periodicity in the time series analysis described in the previous section. Figure 12 highlights those particles which have large vertical accelerations <u>and</u> are moving upwards at the time, apparently illustrating particles moving upwards out of the bowl region towards the free surface. This activity is strongly time-dependent; such activity was completely absent as suggested by a snapshot only two time units later (Figure 12b). Furthermore, the snapshot in Figure 12(a) shortly precedes one of the peaks in activity in the time series analysed above. Thus, the increased activity in the time series can be understood as extra activity

near the free surfaces, presaged by the ejection of particles from the central bowl
region.

(a)   (b)

Figure 12: Upward-moving particles with highest accelerations (high impact case):
(a) immediately prior to peak in activity in time series; (b) a short time later

The high impact case is further characterised by a bulk flow in the horizontal
direction and displays a reasonably well defined horizontal velocity gradient, with
increasing velocity towards the free surface. In Figure 13, darker shading indicates
greater lateral absolute velocity. By way of comparison, Figure 14 for the low
impact case shows localised regions of higher horizontal velocity component and a
significantly lower degree of symmetry indicating more localised surface activity.

Figure 13: Lateral velocity map (darker shading = higher velocity), high impact case

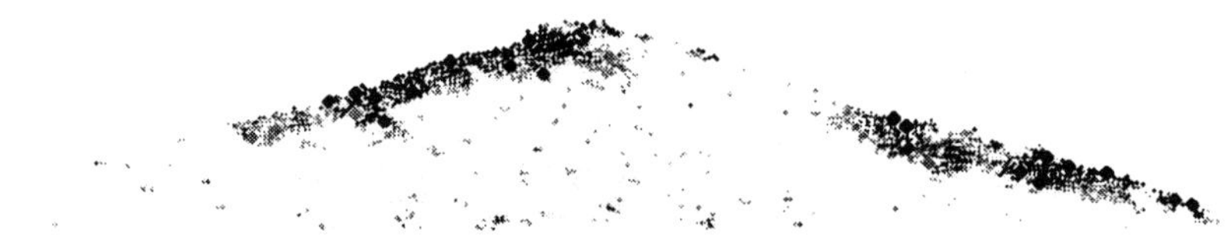

Figure 14: Lateral velocity map (darker shading = higher velocity), low impact case

Figure 14 clearly shows that avalanching down the free surfaces of the heap
proceeds in a discrete fashion for the low impact case, whereas continuous
avalanching is evident for the high impact case. In the high impact case, the
assembly appears to flow in an almost fluid-type fashion; this gives rise to a high
degree of bulk convection and mixing, preventing segregation as discussed below.

3.5. IMPLICATIONS OF IMPACT EFFECTS FOR SEGREGATION

Figures 15 and 16 show snapshots of the heaps at an advanced stage of low and high
impact simulations respectively. At first glance, it might appear that both are quite

well-mixed. However, closer inspection shows that for the high impact case there are more large particles in the middle of the heap. Perhaps more importantly, for the low impact case, there are many more large particles towards the base. These particles have effectively "segregated" since they were at the edges when the heap was smaller. From the above evidence, they could only have reached this position by rolling down the free surfaces, not by penetrating through the bulk. For the high impact case, there were comparatively very few coarse near the base; it seems that bulk mixing / convection in the freely flowing boundary layers prevented any preferential movement of coarse particles down the free surfaces towards the edges.

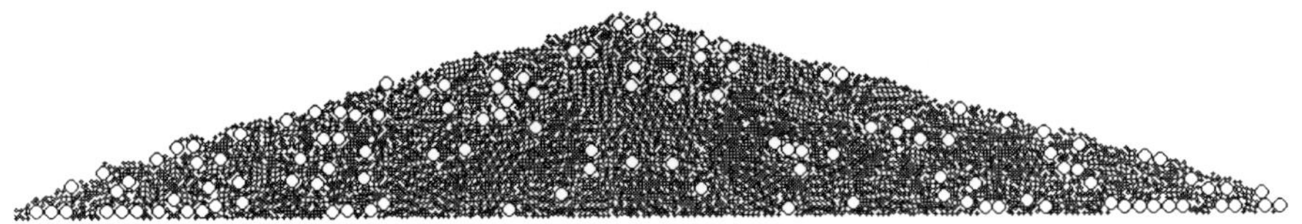

Figure 15: Snapshot at the end of low impact feed simulation

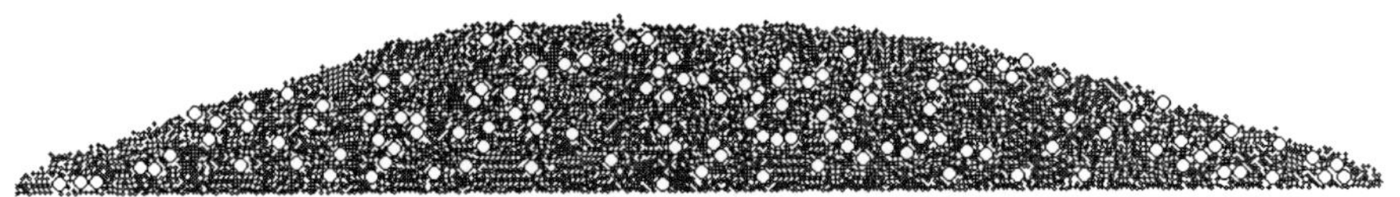

Figure 16: Snapshot at the end of high impact feed simulation

Figure 17 shows the distributions of coarse particles for the high and low impact cases as functions of distance from the base. For this purpose the heaps are subdivided into horizontal strata, 3 small particle diameters in height. In the high impact simulation, the concentration of coarse directly next to the base is very low; elsewhere the concentration is quite uniform. For the low impact case, the proportion of coarse in the lowermost stratum (approx. 50%) is much higher than for the high impact case; otherwise, as for the high impact case, the heap is overall relatively "well-mixed". Nevertheless, as discussed above, the markedly different coarse fractions in the lowermost regions of the heap are indicative of markedly different segregation behaviour for the two simulations.

4. Conclusions

In this paper we have demonstrated the importance of impact for segregation of discrete granular mixtures in pouring to form heaps, in a series of case studies involving granular dynamics simulations. We have outlined a set of numerical tools which we have developed for the visual and time-series analysis of simulation data, and have given an example of how these tools are useful in identifying significant periodic phenomena for segregation in heaps. High impact conditions result in new entrant particles initially embedding themselves within the heap rather than flowing down the free surface. Furthermore, when the particles eventually reach the free surface regions, they flow very easily, rather like a conventional fluid, and this appears to give rise to mixing and convection in the flowing regions. In these ways the propensity for size segregation is greatly reduced.

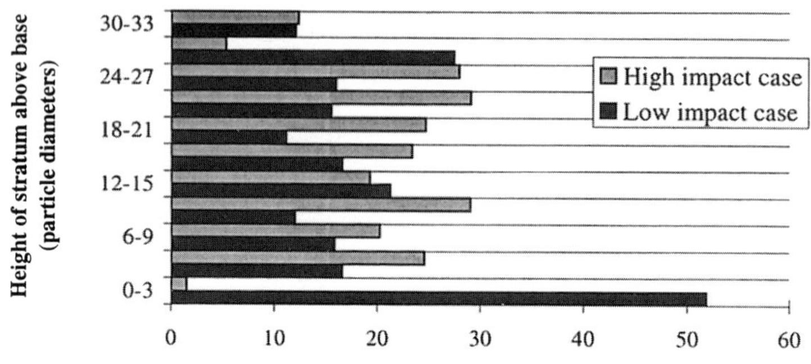

Figure 17: Distributions of coarse particles across horizontal strata for low and high impact simulations

The toolset for analysis is in an embryonic form; we continue to develop methods of processing the simulation data in a useful way. Statistical mechanical treatments of simulation data are of particular interest, particularly for granular flows with a higher degree of bulk activity than the "quasi-static" nature of heap formation. Segregation may be described as a self-diffusive process (for example, [6]); the tools we are continuing to develop allow us to calculate fluctuating velocities and ultimately self-diffusion coefficients for different species in a mixture. Allied to continuing development of the granular dynamics simulation technique itself, this development promises many further useful insights into the segregation of granular mixtures.

5. References

1. Makse, H.A., Havlin, S., King, P.R. and Stanley, H.E.: Spontaneous stratification in granular mixtures, *Nature* **386** (1997), 379-
2. Baxter, J., Tüzün, U., Heyes, D.M., Hayati, I. and Fredlund, P.: Stratification in poured granular heaps, *Nature* **391** (1998), 136-
3. Drahun, J.A., and Bridgwater, J.: The mechanisms of free surface segregation, *Powder Technol.* **36** (1983), 39-53.
4. Tüzün, U. and Arteaga, P.A.: A microstructural model of flowing ternary mixtures of equal-density granules in hoppers, *Chem. Eng. Sci.* **47** (1992), 1619-1634.
5. Baxter, J., Tüzün, U., Burnell, J. and Heyes, D.M.: Granular dynamics simulations of two dimensional heap formation, *Phys. Rev. E* **55** (1997), 3546 – 3554.
6. Nikitidis, M.S., Tüzün, U. and Spyrou, N.M.: Measurement of size segregation by self-diffusion in slow-shearing binary mixture flows using dual photon gamma-ray tomography, *Chem. Eng. Sci.* **53** (1998), 2335-2351.

SEGREGATION OF POLYDISPERSE GRANULAR MEDIA IN THE PRESENCE OF A TEMPERATURE GRADIENT

S. LUDING, O. STRAUSS
Institute for Computer Applications 1
Pfaffenwaldring 27, 70569 Stuttgart, Germany

S. MCNAMARA
Levich Institute, Steinman Hall T-1M
140th St. and Convent Avenue
New York, NY 10031, USA

Abstract

Granular media are examined with the focus on polydisperse mixtures in the presence of two localized heat-baths. If the two driving energies are similar, the large particles prefer to stay in the 'cold' regions of the system – as far away from the energy source as possible. If one of the temperatures is larger than the other, the cold region is shifted towards the colder reservoir; if the temperature of one source is much higher, a strong, almost constant temperature gradient builds up between the two reservoirs and the large particles are found close to the cold reservoir. Furthermore, clustering is observed between the heat reservoirs, if dissipation is strong enough.

1. Introduction

The segregation of granular materials is an effect of eminent importance for industrial operations and has been a subject of research for decades. The behavior of powders within the industrial environment, e.g., silos, hoppers, conveyor belts or chutes, displays interesting effects – one of them is size segregation. Segregation or the mixing properties of granular media are not yet completely understood and thus cannot be controlled under all circumstances. For a review of experimental techniques, theoretical approaches, and numerical simulations, see Refs. [1, 2] and the references therein.

A lot of effort has been invested in the understanding of size-segregation (see this proceedings for a recent overview of the state of the art). It turns out that segregation can be driven by geometric effects, shear, percolation and also by a convective motion of the small particles in the system [3]. Segregation due to convection, in rather dilute, more dynamic systems appears to be orders of magnitude faster than segregation due to purely geometrical effects in dense, quasi-static situations [4, 5]. However, there are still many open questions that are subject to current research on model granular media [1, 6-9].

297

A.D. Rosato and D.L. Blackmore (eds.), IUTAM Symposium on Segregation in Granular Flows, 297–303.
© 2000 *Kluwer Academic Publishers. Printed in the Netherlands.*

Most of the segregation phenomena are obtained in the presence of gradients of density or temperature. Here we isolate the latter case, i.e., we examine the segregation of two species of grains in the presence of local heat reservoirs, but in the absence of external forces like gravity.

2.   The Inelastic Hard Sphere Model

In this study, we use the standard interaction model for the instantaneous collisions of particles with radii a_k, and mass m_k, with the subscripts $k = S$ or L, for small and large particles, respectively. This model accounts for dissipation, using the restitution coefficient r, and is introduced and discussed in more detail in Refs. [9-13]. The post-collisional velocities $\mathbf{v}'$ are given in terms of the pre-collisional velocities $\mathbf{v}$ by

$$\mathbf{v}'_{1,2} \;=\; \mathbf{v}_{1,2} \mp \frac{(1+r)m_{red}}{m_{1,2}}\,\mathbf{v}_n, \tag{1}$$

with $\mathbf{v}_n \equiv \left[(\mathbf{v}_1 - \mathbf{v}_2)\cdot\hat{\mathbf{n}}\right]\hat{\mathbf{n}}$, the normal component of $\mathbf{v}_1 - \mathbf{v}_2$, parallel to $\hat{\mathbf{n}}$, the unit vector pointing along the line connecting the centers of the colliding particles. The reduced mass is here $m_{red} = m_1 m_2/(m_1 + m_2)$. If two particles collide, their velocities are changed according to Eq. (1).

If a particle i crosses a line of fixed temperature T_j, its velocity is changed in magnitude, but not in direction, according to the rule

$$\mathbf{v}'_i = \pm v^{T_j}\,\frac{\mathbf{v}_i}{|\mathbf{v}_i|} \tag{2}$$

with the random thermal velocity $\pm v^{T_j}$ drawn from a Maxwellian velocity distribution, and with random sign. (If the '+' sign is used in Eq. (2), a large net mass flux occurs when a cluster of particles crosses a line of fixed temperature. If the '-' sign is used, the two subsystems have conserved particle numbers, but are still coupled via collisions across the boundaries). In 2D, one has $v^{T_j} = \sqrt{v_x^2 + v_y^2}$, where v_x and v_y are the components of the thermal velocity vector, each distributed according to a Gaussian distribution. Eq. (2) is also applied to particles after a collision, if the particle's center of mass is closer than a_L to one of the heat reservoirs j.

In order to obtain the two random velocities v_x and v_y, two random numbers r_1 and r_2, homogeneously distributed in the inverval [0, 1] are used. With the desired typical thermal velocity $\bar{v} = \sqrt{2T_j/m_i}$, one has

$$v_x = \sqrt{-\bar{v}^2 \ln r_2}\;\cos(2\pi r_1),\;\text{ and} \tag{3}$$

$$v_y = \sqrt{-\bar{v}^2 \ln r_2}\;\sin(2\pi r_1), \tag{4}$$

using the method of Box and Muller, as described in Ref. [14].

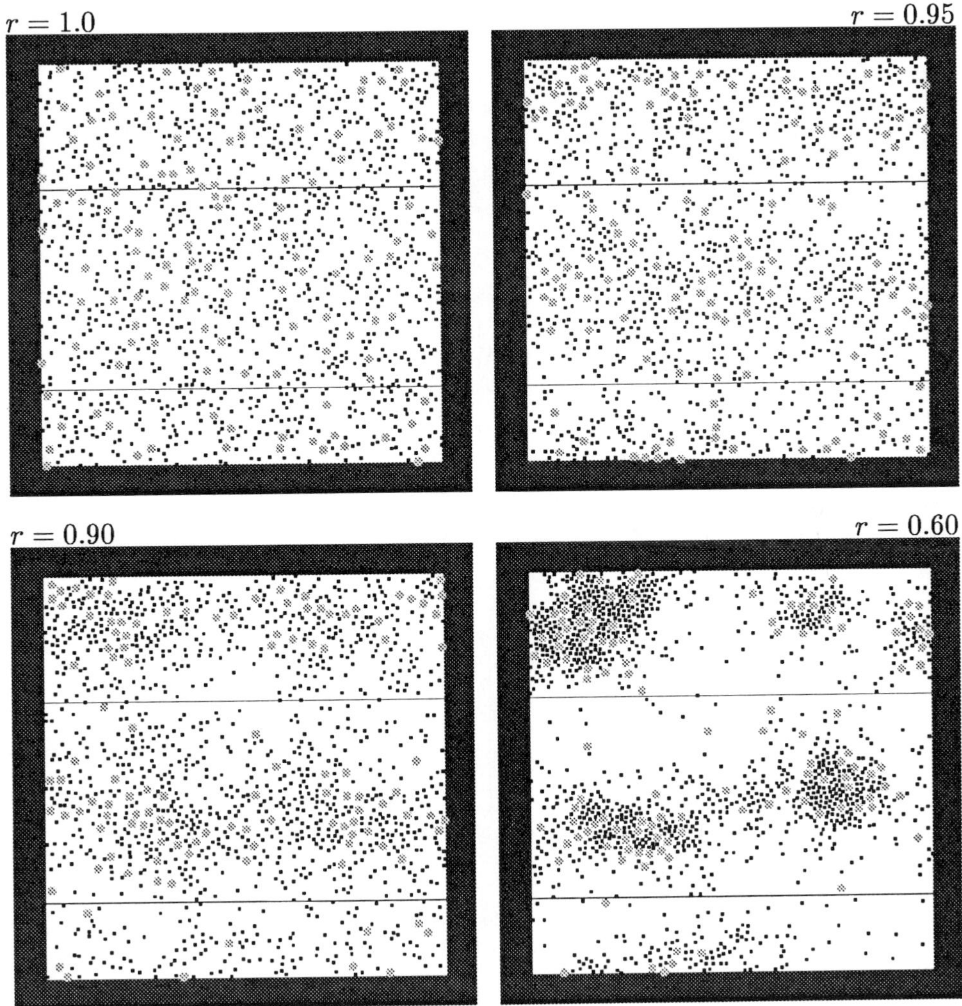

Figure 1. Snapshots from different simulations with $r = 1.0$, 0.95, and 0.60. The heat reservoirs are indicated by horizontal lines; the large particles are displayed in light gray, the small in black.

If the velocity of the particle would be set to $\mathbf{v}' = (v_x, v_y)$, artificial peaks in the density at the positions Z_1 and Z_2 would be observed. This artifact was the reason to choose the above describe way of thermal coupling. Note, that our coupling to a reservoir does not guarantee a certain temperature in a small volume around the reservoir [15, 16]. It rather adjusts the velocity of all particles that came along the reservoir and touched it with their centers of mass. Those particles that approach the reservoir usually have a lower temperature and thus reduce the mean. As r decreases, the reduced increases. Our choice of thermal coupling is arbitrary, however, the discussion of thermostating is far from the scope of the present paper – therefore, we restrict ourselves to the method introduced here.

Our method of imposing a temperature gradient does not have any simple physical analog, but it does allow us to isolate the effects of the temperature gradient. More physically feasible energy sources, such as vibrating walls perturb the motion more than the method used here.

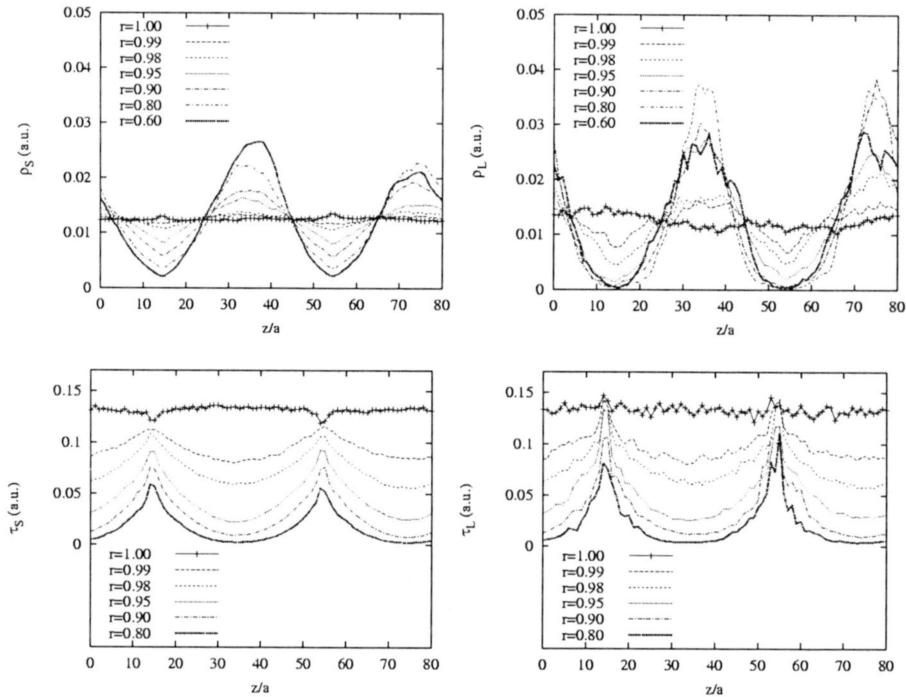

Figure 2. Density ρ (top) and temperature τ (bottom), plotted as a function of the vertical position in the system $Z = z/a$. Data for small (Left) and large (Right) particles are presented for $T_2/T_1 = 1$ and different restitution coefficients r as given in the inset. The data at $z/a = 0$ and $z/a = 80$ are identical due to periodic boundaries.

3.　The Event-Driven Simulation Method

For the simulation of hard spheres, we use the event-driven algorithm, originally introduced by Lubachevsky [17], and applied to the simulation of granular media [13]. In these simulations, the particles follow an undisturbed translational motion until an event occurs. Either an event is the collision of two particles, the crossing of one particle with the boundary of a cell (in the linked-cell structure which is used for algorithmic optimization only), or the crossing of one of the lines with fixed temperature. A particle-particle collision is treated as described in the previous section, a cell-boundary crossing has no effect on the particle motion, and the crossing of a fixed temperature location leads to a change of velocity according to Eq. (2). Only the particle(s) involved in the last event (are) treated and their next event is computed. In the next step (which does not imply a fixed step in time), the next event out of all possible events is treated.

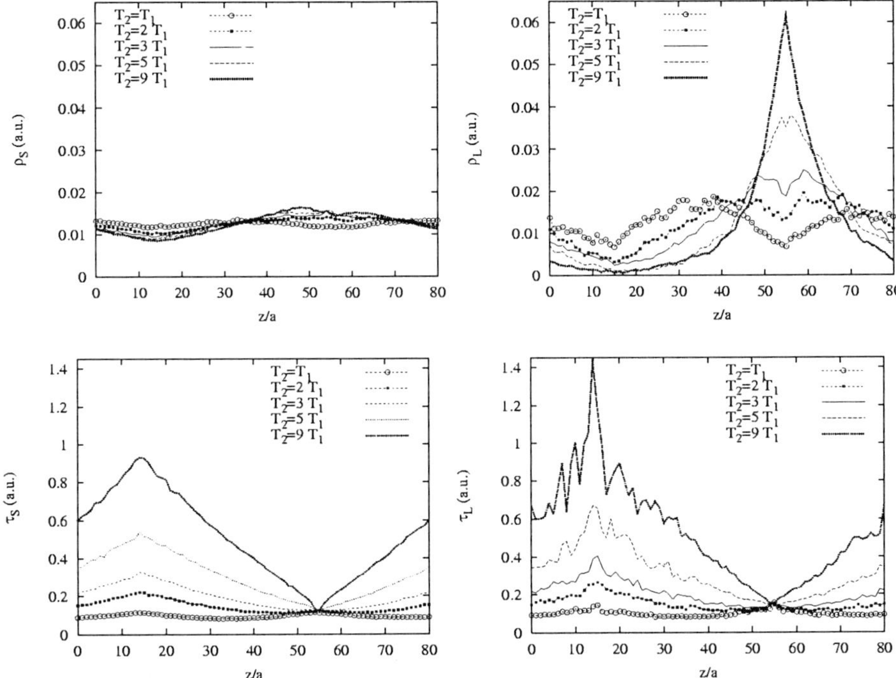

Figure 3. Density ρ (top) and temperature τ (bottom), plotted as a function of the vertical position in the system $Z = z/a$. Data for small (Left) and large (Right) particles are presented for $r = 0.99$ and different temperature ratios T_2 / T_1.

4. Boundary Conditions and System Parameters

The simulations presented here (typical situations are displayed in Fig. 1) were performed with $N = 1296$ particles in a two-dimensions (2D) box of size $l = l_x = l_y = 0.04$ m. The box has periodic boundaries, i.e., a particle that leaves the volume at the box (left), immediately enters it at the top (right) and vica-versa. Two particle types (S, L) are used with $a_S = a = 2.5 \times 10^{-4}$ m and $a_L = 2a = 5 \times 10^{-4}$ m. In the system with dimensionless size $L = l/a = 80$, $N_S = 1176$ particles of the small species and $N_L = 120$ particles of the large species coexist. Note that we used a rather arbitrary choice for the calculation of the particle masses (see above) so that $m_L = 8m_S$ for the size ratio used here. The fraction of the area that is occupied by the particles, i.e., the total volume fraction, is $\nu = \pi\left(N_S a_S^2 + N_L a_L^2\right)/l^2 \approx 0.2$.

The hot and cold temperature reservoirs are situated at the vertical positions $Z_2 = z_2/a = 15$ and $Z_1 = z_1/a = 55$, respectively, and reach over the whole width of the system. The temperatures of the reservoirs are T_1 and T_2 – but since not external body force like gravity is involved, the behavior of the system does not depend on the absolute value of T_1 or T_2; only the ratio T_2/T_1 is important, and it is varied in the range from $T_2/T_1 = 1$ to 9 in the following. In Fig., snapshots of simulations with $T_2/T_1 = 1$ and different values of r are plotted. For small r, clustering is observed and the averaging over horizontal slices (data are presented in Fig. 2) becomes questionable.

5. Results

A quantitative measure of segregation is the particle density of the different species. We present $\rho_k = n_k/N_k$, the particle number density $n_k(z)$ weighted by the number of particles N_k of species k in Fig. 2. The strength of segregation increases with decreasing r, only for small $r = 0.60$, segregation is rather weak due to clustering. Fluctuations in density are correlated to the heat reservoirs: close to a "hot" region in the vicinity of an energy source, the density is lower than in the "cold" regions in between. Note that the large particles segregate, while the small particles make up a "background fluid" with a comparatively small density variation, if dissipation is weak. Thus, all particles prefer regions of low temperature, but the large ones are attracted by the cold regions more strongly.

Now, the restitution coefficient is fixed to $r = 0.99$ and the temperature ratio T_2/T_1 is modified. In Fig. 3 the situation is presented for different ratios $T_2/T_1 = 1$, 2, 3, 5, and 9. The large particles segregate from the "background fluid" made up by the small particles and the quality of segregation increases with the magnitude of the temperature gradient. For $T_2 > T_1$, the large (and heavier) particles can be found close to the colder heat reservoir, as different to the situation discussed above. When heavier particles have the same temperature as the light ones, their mean velocity is reduced, so that they cannot diffuse away from the cold heat reservoir.

6. Summary and Conclusions

In summary, we presented simulations of different size particles in the presence of a temperature gradient. If the temperature gradient is due to the dissipative nature of the material, the large particles move towards the cold regions, as far away from an energy source as possible. If the gradient in temperature is externally imposed, most of the large particles prefer to move towards the colder heat reservoir. When dissipation is strong enough, clustering is observed in the locally driven system.

Further possible studies involve the prediction of the density and temperature profiles with kinetic energy – first for the monodisperse and later also for the polydisperse case.

7. Acknowledgements

We gratefully acknowledge the support of IUTAM, the National Science Foundation, the Department of Energy, the Office of Basic Energy Sciences, the Geosciences Research Program, the Deutsche Forschungsgemeinschaft (DFG), and the Alexander-von-Humboldt Foundation.

8. References

1. Herrmann, H.J., Hovi, J.-P., and Luding, S., (eds.): *Physics of dry granular media*, NATO ASI Series E 350, Kluwer Academic Publishers, Dordrecht, 1998.
2. Chowhan, Z. T.: Segregation of particulate solids Part I., *Pharm. Technol.* **19** (1995), 56.
3. Knight, J. B., Jaeger, H. M., and Nagel, S. R.: Vibration-induced size separation in granular media: The convection connection, *Phys. Rev. Lett.* **70** [24] (1993), 3728-3731.
4. Duran, J., Mazozi, T., Clement, E., and Rajchenbach, J.: Size segregation in a two-dimensional sandpile: convection and arching effects, *Phys. Rev. E* **50** [6] (1994), 5138-5141.
5. Dippel, S. and Luding, S.: Simulations on size segregation: Geometrical effects in the absence of convection, *J. Phys. I France* **5** (1995), 1527-1537.
6. Rosato, A.D. Strandburg, K.J., Prinz, F., and Swendsen, R. H.: Why the brazil nuts are on top: Size segregation of particulate matter by shaking, *Phys. Rev. Lett.* **58**[10] (1987), 1038-1041.
7. Clement, E., Rajchenbach, J., and Duran, J.: Mixing of a granular material in a bidimensional rotating drum, *Europhys. Lett.* **30** [1] (1995), 7-12.
8. Cantelaube, F., Duparcmeur, Y. L., Bideau, D., and Ristow, G. H.: Geometrical analysis of avalanches in a 2d drum, *J. Phys. I France* **5** (1995), 581-596.
9. Arnarson, B. and Willits, J. T.: Thermal diffusion in binary mixtures of smooth, nearly elastic spheres with and without gravity, *Phys. Fluids* **10** (1998), 1324.
10. Jenkins, J. T. and Richman, M. W.: Kinetic theory for plane flows of a dense gas of identical, rough, inelastic, circular disks, *Phys Fluids* **28** (1985), 3485-3494.
11. Lun, C. K. K.: Kinetic theory for granular flow of dense, slightly inelastic, slightly rough spheres, *J. Fluid Mech.* **233** (1991), 539-559.
12. Goldshtein, A. and Shapiro, M.: Mechanics of collisional motion of granular materials. Part 1. General hydrodynamic equations, *J. Fluid Mech.* **282** (1995), 75-114.
13. Luding, S., Huthmann, M., McNamara, S., and Zippelius, A.: Homogeneous cooling of rough dissipative particles: theory and simulations, *Phys. Rev. E* **58** (1998), 3416-3425.
14. Press, W. H., Teukolsky, S. A., Vetterling, W. T., and Flannery, B. P.: *Numerical recipes*, Cambridge University Press, Cambridge, 1992.
15. Cause, P. J. and Mareschal, M.: Heat transfer in a gas between parallel plates: moment method and molecular dynamics, *Phys. Rev. A* **38** (1988), 4241.
16. Mareschal, M., Monsour, M., Sonnino, G., and Kestamont, E.: Dynamics structure factor in a nonequilibrium fluid: a molecular-dynamics approach, *Phys. Rev. A* **45** (1992), 7180.
17. Lubachevsky, B. D.: How to simulate billards and similar systems, *J. Comp. Phys.* **94** [2] (1991), 255.

A SIMPLE METHOD TO MIX GRANULAR MATERIALS

S. MCNAMARA[+] and S. LUDING[+#]
+Levich Institute, Steinmann Hall T-1M
140th St. and Convent Avenue
New York, NY 10031, USA

#Institute for Computer Applications 1
Pfaffenwaldring 27
70569 Stuttgart, Germany

Abstract

We show that a mixture of two species of granular particles with equal sizes but differing densities can be either segregated or mixed by adjusting the granular temperature gradient and the magnitude of the gravitational force. In the absence of gravity, the dense, heavy particles move to the colder regions. If the temperature gradient is put into a gravitational field with the colder regions above the hotter, a uniform mixture of light and heavy particles can be attained. This situation can be realized in a container of finite height with a vibrating bottom, placed in a gravitational field. We present a relation between the height of the container, the particle properties, and the strength of gravity required to minimize segregation.

1. Introduction

Segregation and mixing of granular materials is of eminent importance for industrial operations and it has been subject to research since decades. However, both effects are not yet completely understood and thus cannot be controlled under all circumstances. Traditional experimental methods and theoretical approaches are nicely complemented by numerical simulations which in the last few years have developed tremendously [1]. For a review that covers a broad practical experience of segregation, see Ref. [2] and references therein.

Segregation can be driven by geometric effects, shear, percolation and also by a convective motion of the small particles in the system [3]. In vibrated systems, the segregation due to convection appears to be orders of magnitude faster than segregation due to purely geometrical effects [4, 5]. In rotating drums, another archetype of many industrial devices, segeral segregation processes acting in parallel are reported [6]; in three-dimensional devices, axial and longitudinal segregation are observed [7, 8, 9] simultaneously. For axial segregation, particle percolation is reported to be responsible

305

A.D. Rosato and D.L. Blackmore (eds.), IUTAM Symposium on Segregation in Granular Flows, 305–310.

[8], while longitudinal segregation is related to different surface flow properties in the cylinder [10, 11].

In this paper, we investigate a model segregation problem which suggests a simple way to obtain a uniform mixture of two species. We show a sketch of the system in Fig. 1.

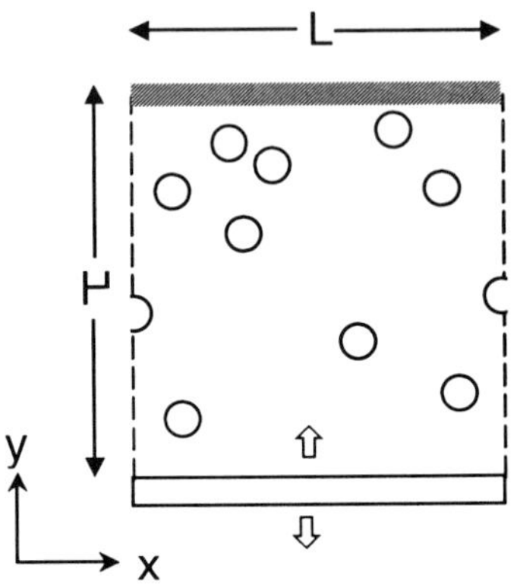

Figure 1. A sketch of the studied system

N particles are placed in a container of height H in the presence of gravity. The side walls have been replaced with periodic boundary conditions. Energy is supplied to the container by vibrating the bottom using a symmetric sawtooth wave with velocity V. The top wall is stationary. N_A of the particles have mass m_A, and the rest have mass m_B. We will take $m_A > m_B$. Though the particles have different mass, they all have the same radius a. We model the loss of energy during collisions with a restitution coefficient $r < 1$. Our detailed study of a similar system with identical particles is in Ref. [12].

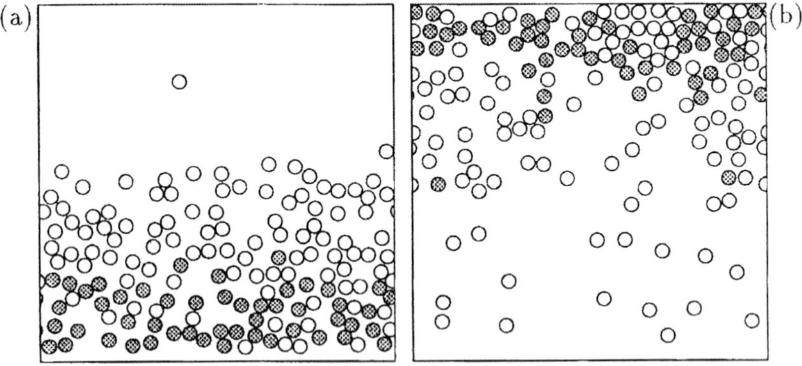

Figure 2. (a) A simulation with $N = 150$, $N_A = 50$, $L = H = 50$, $m_A / m_B = 10$, $g = 1$, $V = 1$, $r = 0.9$. The dense particles (mass m_A) are shaded. (b) same as (a), except $g = 0$.

2. The Mixing Mechanism

We find that it is possible to obtain uniform mixtures of the two species by pitting two segregation mechanisms against each other. When the particles rarely touch the top of the container, all the dense particles are found near the bottom of the plate (see Fig. 2a). A similar effect occurs in the upper atmosphere, where different molecular species are sorted by weight [13].

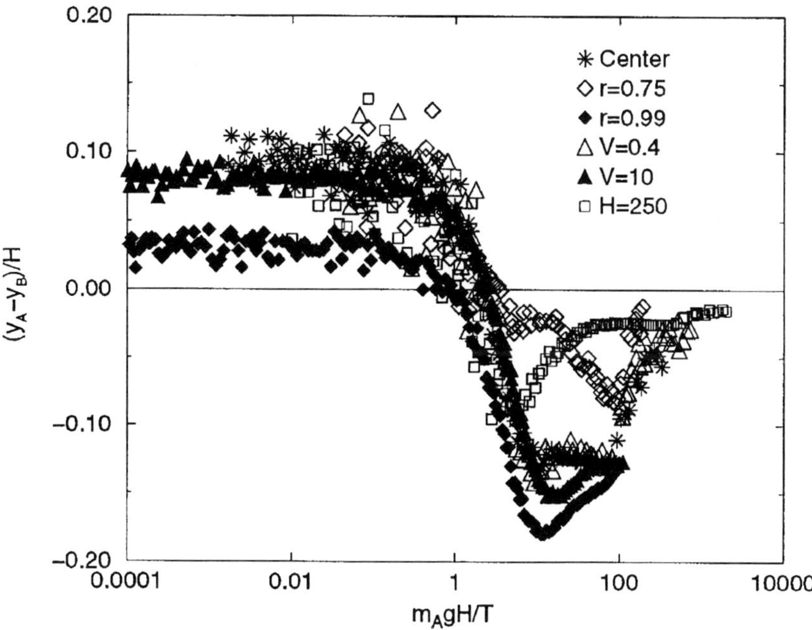

Figure 3. Segregation of the two species measured by the difference between y_A, the center of mass of species A and y_B, the center of mass of species B. This difference is plotted as a function of $m_A gH / T$, where T is the granular temperature, here defined as the average kinetic energy per particle. The points marked "Center" have $V = 2$, $m_A / m_B = 2$, $N_A = 50$, $N_B = 100$, $H = L = 50a$, and $r = 0.95$, with g swept over six orders of magnitude. The other points are the same, except for the parameter values marked on the graph. In all cases, the time unit is fixed by the wall vibration period, and the mixture is equilibrated for 300 time units, and then data is averaged for 1500 time units.

On the other hand, when gravity is turned off, the particles are pushed against the top plate and the dense particles are found close to the upper plate (Fig. 2b). By smoothly varying between these two situations, it is possible to obtain a situation where the two species are uniformly mixed. In Fig. 3, we plot the difference between y_A, the center of mass of the heavy particles, and y_B, the center of mass of the light particles, normalized by the height of the container. Mixing is optimal when the difference between the species' centers of mass vanishes.

We see that the state of maximum mixing is obtained hear $m_A gH / T \sim 2$ for all values of the parameters, except for almost elastic particles ($r = 0.99$). Here, T is the granular temperature, defined as the average kinetic energy per particle:

$T \equiv (1/2N)\sum m_i v_i^2$. When the ratio of both energies is near unity, it means that the kinetic and potential energies of the particles are comparable. The tendency of $(y_A - y_B)/H$ to approach 0 for large gravities $m_A gH/T > 10$ is due to the initial conditions. Initially, all particles are arranged in a lattice just above the vibrating floor. Due to the large gravity, it is very difficult for particles to change places, and the mixture keeps its original configuration for a very long time. For $m_A gH/T < 10$, the particles change places often, and $(y_A - y_B)/H$ is independent of initial conditions.

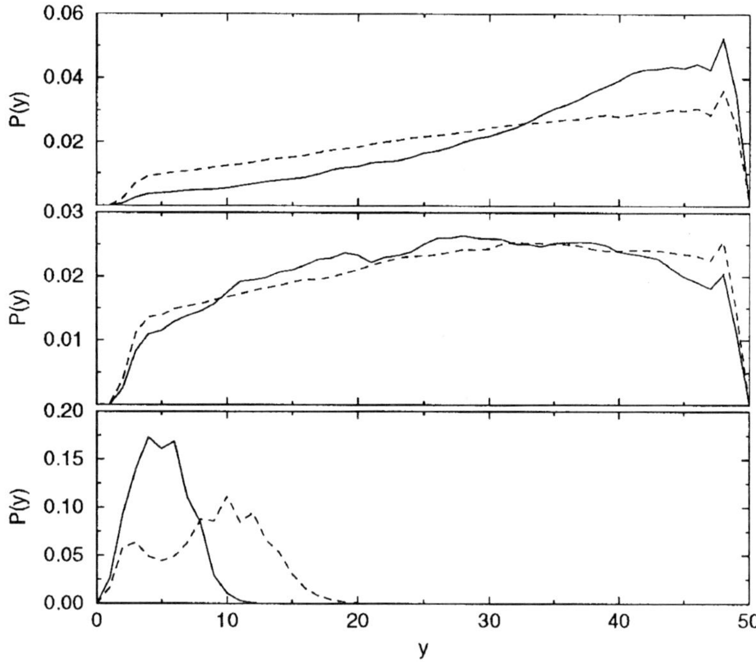

Figure 4. Profiles of the density of each species as a function of y. The solid curve represents the heavy species with mass m_A, and the dashed curve is the light species with mass m_B. All simulations have $m_A/m_B = 2$, $V = 2$, $N_A = 50$, $N_B = 100$. The top figure has $g = 10^{-4}$, the middle has $g = 0.3169$, and the bottom has $g = 10^2$. The densities are given in terms of the probability of a single particle of each species to have a given height; the area under all curves integrates to 1.

To show more closely what happens with the densities of the different species, we show in Fig. 4 the concentration of each as a function of height for three different simulations; one at small g, one at large g, and one where the particles are nicely mixed. In the situations with extremal g values, we obtain rather strong density gradients, while in the case of optimal mixing, the density gradients are small, i.e., the density is almost constant.

3. Discussion and Conclusion

Each of the two segregation mechanisms can be observed also with perfectly elastic (dissipationless) particles. In Fig. 5(a), we show a binary gas of elastic particle under gravity in the absence of forcing. The heavy particles accumulate at the bottom. In Fig. 5(b), we show a binary gas in the absence of gravity, subjected to a thermal gradient. Now the particles accumulate against the upper, cold wall. Therefore, neither segregation mechanism relies on the dissipation of energy during collisions. This dissipation serves only to set up the necessary gradients which drive the segregation of the particles (see also the paper by Luding, Strauss and McNamara in this proceedings).

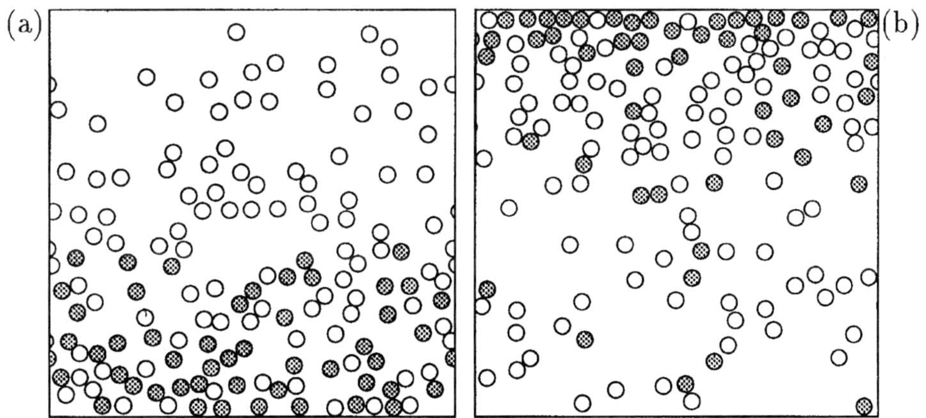

Figure 5. (a) A simulation with $N = 150$, $N_A = 50$, $L = H = 50a$, $m_A / m_B = 2$, $g = 1$, $V = 1$, $r = 1.0$. All the walls are stationary, so no energy is added or subtracted. Note that the dense particles sink to the bottom. (b) same as (a) except that $g = 0$ and when a particle touches the upper or lower wall, a new velocity is selected from a Maxwellian velocity distribution with a certain temperature. The temperature of the lower wall is 160 times as large as the temperature of the upper wall. Small temperature gradients also sort the particles by density, but it is much less visually striking.

To see this method to mix granular materials, the particles could be put into a chamber like the one shown in Fig. 1. To obtain the proper value of $m_A gH/T$, it is perhaps most convenient to adjust the height of the container H. It is also possible to control T by changing the vibration velocity V [12]. One possible advantage is that only a small amount *of* material can be mixed at one time. It also may be difficult in practice to adjust H or T correctly. Replacing the periodic boundaries with side walls may also introduce new effects.

4. Acknowledgements

Inspiring discussions with H. J. Herrmann are appreciated, and we gratefully acknowledge the support of IUTAM, the National Science Foundation and the Department of Energy. S. L. also thanks the Deutsche Forschungsgemeinschaft, and the S. M. Alexander-von-Humboldt Foundation, and the Geosciences Research Program, Office of Basic Energy Sciences, US Department of Energy.

5.   References

1. *Physics of Dry Granular Media – NATO ASI Series E 350*, edited by H. J. Herrmann, J.-P. Hovi, and S. Luding, Kluwer Academic Publishers, Dordrecht, 1998.
2. Chowhan, Z. T.: Segregation of particulate solids part I, *Pharm. Technol.* **19** (1995), 56.
3. Knight, J. B., Jaeger, H. M., and Nagel, S.R.: Vibration-induced size separation in granular media: the convection connection, *Phys. Rev. Lett.* **70** (1993), 3728.
4. Duran, J., Rajehcnbach, J., and Clément, E.: Arching effect model for particle size segregation, *Phys. Rev. Lett.* **70** (1993), 2431.
5. Duran, J., Mazozi, T., Clément, E., and Rajchenbach, J.: Size segregation in a two-dimensional sandpile: convection and arching effects, *Phys. Rev. E* **50** (1994),·5138.
6. Cooke, M. H., Stephens, D. J., and Bridgwater, J.: Powder mixing – a literature survey, *Powder Technol.* **15** (1976), 1.
7. Oyama, Y.: Horizontal rotating cylinder, *Bull. Inst. Phys. Chem. Res. (Tokyo) Rep.* **18** (1939), 600.
8. Gupta, S. D., Kharkar, D. V., and Bhatia, S. K.: Axial transport of granular solids in horizontal rotating cylinders, *Chem. Eng. Science* **46** (1991), 1531.
9. Nakagawa, M.: Axial segregation in a horizontal rotating cylinder, *Chem. Eng. Science* **49** (1994), 2544.
10. Zik, O. et al.: Rotationally induced segregation of granular material in a horizontal rotating cylinder, *Phys. Rev. Lett.* **73** (1994), 644.
11. Hill, K. M. and Kakalios, J.: Reversible axial segregation of binary mixtures of granular materials, *Phys. Rev. E* **49** (1994), R3610.
12. McNamara, S. and Luding, S.: Energy flows in vibrated granular media, *Phys. Rev. E* **58** (1998), 813.
13. Jeans, J.: *The dynamical theory of gases*, Dover Publications, 1925, paragraph 448.

MIXING AND SEGREGATION IN ROTATING DRUMS

A Numerical Study

GERALD H. RISTOW

Universität des Saarlandes
Fachrichtung Theoretische Physik
D–66041 Saarbrücken, Germany

Abstract.
In this contribution, the radial and axial segregation dynamics in rotating drums are discussed using discrete element simulations. It is shown that radial segregation takes place for arbitrarily small size differences and how mixing and segregation processes interact with each other. The stability of axial bands is investigated by using the calculated diffusion coefficient for each particle component.

1. Introduction

In industrial material handling, a common device for mixing materials with different properties is a rotating drum. However, it is also well-known that especially for materials that differ largely in *size* or *density*, a strong tendency to segregate, i.e. demix [1,2], is observed. This demixing process depends on the flowing regime of the material in the drum which is controlled by the external rotation speed. For slow rotation speeds, individual avalanches are easily detectable and their temporal separation decreases with increasing rotation speed until they can no longer be separated, leading to the transition to the *continuous flow regime*. In the beginning of this regime, the free-particle surface is flat but exhibits a S-shape which becomes more pronounced with increasing rotation rate [3]. For even higher rotation speeds, the particles centrifuge to the circumference where they form a ring.

In the avalanching and the beginning of the continuous flow regime, the smaller and denser particles have a strong tendency to segregate in a region close to the rotation axis. This is called *radial* segregation. In a long drum, this radial segregation takes place everywhere along the rotation axis, giving rise to a core of smaller or denser particles. Depending on the system parameters, this core might become unstable, leading to the formation of bands which is called *axial* segregation.

In this article, we will briefly review the numerical results on the radial segregation mechanism and illustrate the interplay of mixing and segregation. Concerning the axial segregation process, we show that our algorithm captures the end-longitudinal segregation

311

A.D. Rosato and D.L. Blackmore (eds.), IUTAM Symposium on Segregation in Granular Flows, 311–320.

and we investigate the stability of the bands by using the calculated diffusion coefficient for each particle component.

2. Numerical Model

A very successful and flexible numerical technique for modelling granular materials is the Discrete Element Method (DEM); for a recent review see e.g. [4]. This method uses an explicit, constant time step to integrate Newton's second law of motion for each particle and has the ability to model in a simple way the change of geometrical contacts during collisions.

Each particle i is approximated by a sphere with radius R_i. Only contact forces during collisions are considered and the particles have the ability to rotate. Our coordinate system during a particle-particle collision between particles i and j is sketched in Fig. 1, where $\hat{n}$ is the unit vector joining both centres of mass and $\hat{s}$ a unit vector perpendicular to it.

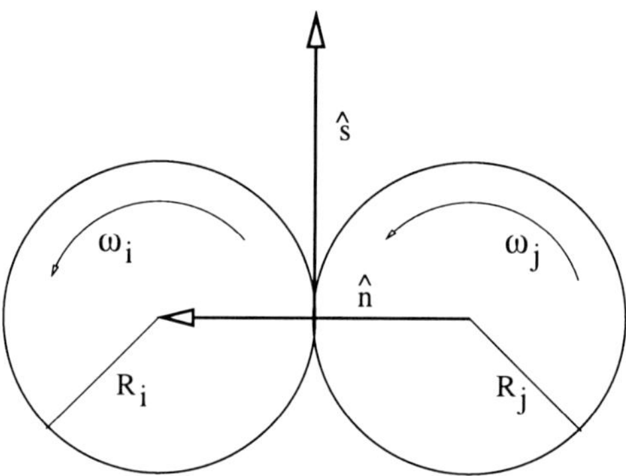

Figure 1.   Sketch of the coordinate system during a particle-particle collision.

The force acting in the normal direction ($\hat{n} = \frac{\vec{r}_i - \vec{r}_j}{|\vec{r}_i - \vec{r}_j|}$) during a collision is given in our simulations as

$$F_{ij}^n = -K_n \delta - \gamma_n (\vec{v}_i - \vec{v}_j)\, \hat{n} \, ,$$

where $\delta = R_i + R_j - (\vec{r}_i - \vec{r}_j)\, \hat{n}$ which gives the virtual overlap of the two particles and $\vec{v}_i$ stands for the velocity of particle i. The model parameter K_n controls the stiffness of the material and is related to the Young modulus of the material, whereas γ_n stands for the dynamic damping coefficient which is related to the commonly used coefficient of restitution, e_n, via

$$e_n = \exp\left(-\frac{\pi \gamma_n}{\sqrt{4 K_n m - \gamma_n^2}}\right) .$$

Even though a more complex contact behaviour, e.g. à la Hertz and Midlin, can be implemented in a straight-forward fashion, a linear relation is often found for small deformations and has the advantage of a constant collision time of

$$t_c = \frac{2m\pi}{\sqrt{4K_n m - \gamma_n^2}} \, .$$

For the contact forces in the tangential direction, $\hat{s}$, we use viscous damping for particle-particle interactions:

$$F_s^{pp} = -\text{sign}(\hat{s}) \, \min(\gamma_s |\vec{v}_{ij}\hat{s}|, \mu|F_n|)$$

and static friction for particle-wall interactions:

$$F_s^{pw} = -\text{sign}(\hat{s}) \, \min(k_s |\int \vec{v}_{ij}\hat{s}dt|, \mu|F_n|) \, .$$

The latter leads to a static angle of repose. Values of γ_s and k_s are chosen such that the dynamic angle of repose agrees with experiment over a wide range of rotation speeds [3]. In the above equations, μ denotes the (static) Coulomb friction coefficient.

3. Results and Discussion

Radial segregation is illustrated in Fig. 2 for a half-filled, two-dimensional drum of spherical particles having equal density and a size ratio of the small to the large particle sizes of 2:3. As can be seen, after only a few rotations the smaller particles are mostly found close to the rotation axis, directly below the fluidized layer. This demonstrates that radial segregation is a rapidly occurring process.

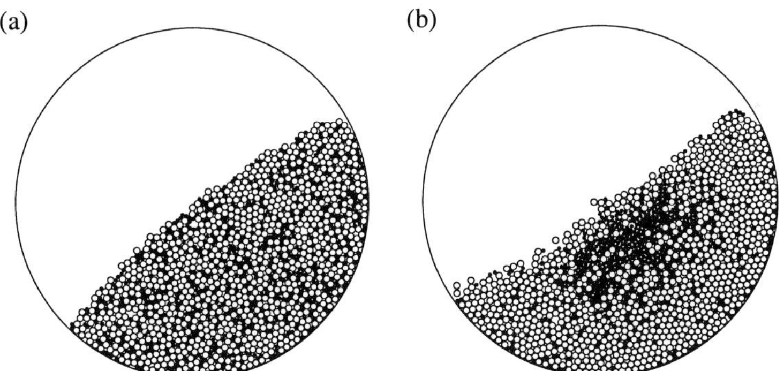

Figure 2. Radial segregation: (a) immediately before first avalanche, (b) after 9 rotations. Small (large) particles are drawn as filled (open) circles.

Close to the centrifugal regime, the particles start to remix again, and in the centrifugal regime the smaller particles will actually segregate to the circumference, i.e. they are more likely found close to the wall [5].

The numerical results presented in this article were obtained by using a rotation speed which corresponds to the beginning of the continuous flow regime where the surface is still rather flat.

3.1. RADIAL SIZE SEGREGATION

Probably the most pronounced segregation is found when particles differ in size, see Fig. 2. In order to capture the dynamics of this process and to quantify the demixing properties of different set-ups, a suitable parameter is needed. It can be defined via the radial distribution of small particles. For this, a two-dimensional cross-section of the drum is divided into concentric rings and the relative concentration of small particles is determined for each ring. All values above the theoretical value for a homogeneous mixture are integrated in order to give a quantitative parameter that chracterizes the amount of segregation. This parameter gives a value of zero for an ideally mixed configuration and a value of one for a completely segregated configuration [6,7].

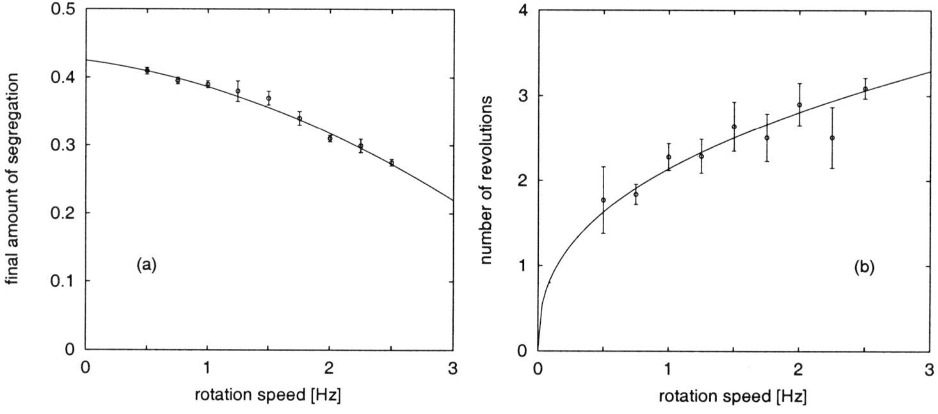

Figure 3.   Dependence of the radial size segregation on the rotation speed: (a) final amount of segregation and (b) characteristic number of revolutions.

Our segregation parameter usually shows a behaviour that is exponentially saturating from which we can extract the final amount of segregation for this particular set-up and the characteristic number of revolutions that are needed before significant segregation is visible. Here, *significant* could mean that the amount of segregation has reached half of its final value. These two quantities are shown in Fig. 3 as a function of the external rotation speed for a size ratio of 0.59 and a 45% volume fraction of small particles. The solid lines were drawn merely to guide the eye. The final amount of segregation *decreases* with rotation rate due to the increase of the width of the fluidized layer which leads to an increasing overlap with the region where the smaller particles want to segregate into [6]. To the right, we show the characteristic number of revolutions which *increases* with rotation speed. It should be noted that only a very few rotations are needed to achieve a significant and very visible segregation, which is in qualitative agreement with experiments.

It is well-known that the segregation process is faster and more pronounced if the particle size ratio, Φ, becomes smaller [2]. However, it does not seem to be clear if a threshold value exists for segregation to occur in general. The results from a two-dimensional rotating drum model indicate that segregation is observed for an arbitrarily small size ratio [8], whereas the data from vertical shaking experiments suggest a cut-off ratio around $\Phi = 0.5$ [9]. In order to address this question for the rotating drum, we show in Fig. 4 the final amount of segregation as a function of the particle size ratio for a volume filling fraction of 50% of small particles. Even though obtaining accurate data for values of Φ close to one is rather difficult due to the long segregation time, the data shown in Fig. 4 support the hypothesis from Ref. [8] that segregation will be present for any finite size difference. This was determined by using a linear fit of the form

$$q_\infty(\Phi) = c(\Phi - \Phi_0)$$

where $c = 1.6 \pm 0.1$ and $\Phi_0 = 1.00 \pm 0.02$ when all seven data points are used for the fit. The linear fit is shown as a solid line in Fig. 4.

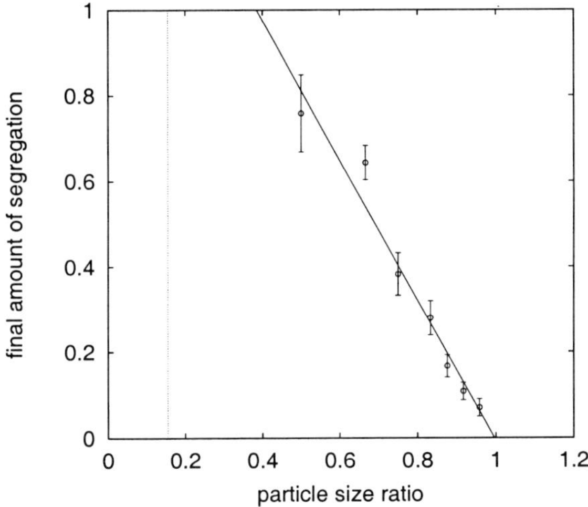

Figure 4. Final amount of segregation as function of the particle size ratio.

Since the maximum value for the final amount of segregation is one, the behaviour must deviate from the linear dependence for small size ratios. For values of $\Phi \leq \frac{2}{\sqrt{3}} - 1$ (vertical dashed line in Fig. 4), we expect a completely different behaviour since the small particles are then sufficiently small to propagate through the voids of a three-dimensional hexagonal packing.

3.2. RADIAL DENSITY SEGREGATION

Segregation is also possible for a mono disperse particle size distribution when the particles have different friction coefficients, which normally leads to a different dynamic angle of repose, or if the particles differ in density. The latter effect seems to be more pronounced and was first investigated in detail for particles in a rotating drum in Ref. [10]. It is also possible to counterbalance the tendency of the smaller particles to segregate into the middle region by making them lighter, which was demonstrated experimentally in Ref. [11].

In order to study the density segregation mechanism numerically in detail, 12 tracer particles were monitored in an ensemble of 600 particles. Their normalized distance from the drum's centre was recorded after each revolution and it was averaged over all 12 particles and two independent runs. For heavier tracer particles, the average distance decreases over time and the dynamics can be well approximated by an exponential decay [10]. From this decay, one can extract the normalized radius of the segregated cluster of denser particles and an average initial segregation velocity, which is shown in Fig. 5 as function of the density ratio defined as the ratio of the density of the tracer particles to the density of the rest of the ensemble. The normalized radius decreases strongly with increasing density ratio and seems to saturate for values greater than eight.

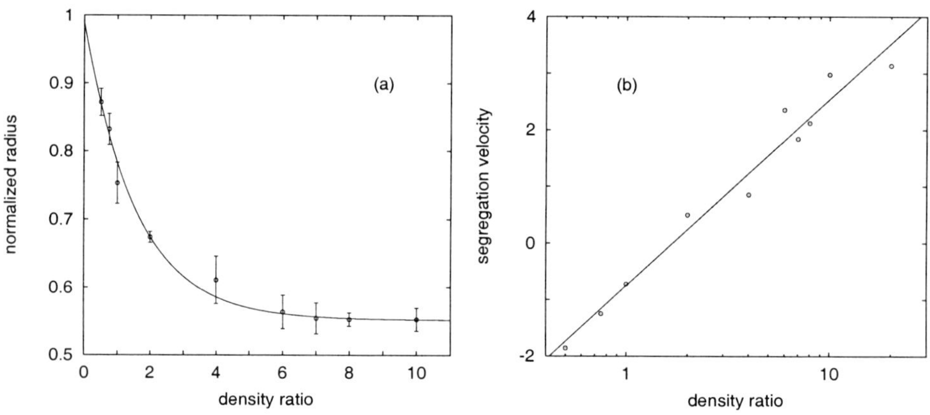

Figure 5. Dependence of the segregation on the density ratio: (a) normalized radius of the segregated cluster and (b) average initial segregation velocity.

When presented in a semi-log plot, the values for the average initial segregation velocity all lie close to a straight line and we calculate its slope from a least-squares fit. The agreement of the numerical data with a straight line strongly suggests that the initial segregation velocity is proportional to the logarithm of the density ratio.

When the gravitational constant is decreased, the normalized radius of the segregated cluster of denser particles and the initial segregation velocity will both also decrease [10].

3.3. INTERPLAY OF MIXING AND SEGREGATION

When working in the avalanching regime, it was found in experiments [12] that *no* particle mixing occurs for an exactly half-filled drum, which could be explained by the mixing of wedges caused by the avalanches. In order to illustrate the interplay of mixing and segregation, we investigate the proposed set-up for a half-filled drum numerically and use particles of different sizes having a size ratio of 2:3 [7]. The initial configuration is shown in Fig. 6(a) where the larger particles are shown in black. After turning the drum clockwise for 1.6 s at 15 rpm, which would simply interchange the regions occupied by large and small particles if no mixing were present, the interface is still well-defined and nearly a straight line (Fig. 6(b)). After turning for 2.8 s, the interface between the large and small beads is still quite sharp, though it is not a straight line anymore (Fig. 6(c)).

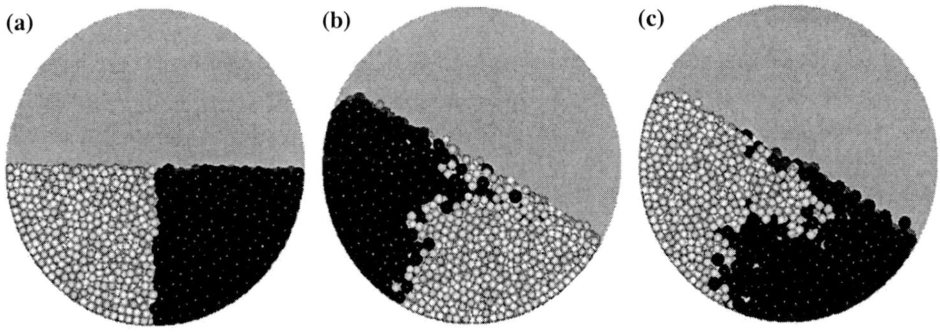

Figure 6. Competition of mixing and segregation: initial configuration (a) after interchanging the black and white region once (b) and twice (c).

The tongue of small particles into the region mostly occupied by large particles close to the centre of the cylinder is the starting point of the formation of a core of small particles. Due to the very special initial condition, it takes many rotations to fully form the segregated core of small particles. After fourteen rotations, hardly any large particles are found in the segregated core of small particles, whereas smaller particles are still found close to the wall of the drum [7]. We expect the latter effect to disappear when the drum is rotated for long enough times.

3.4. END-LONGITUDINAL SIZE SEGREGATION

The numerical results presented so far were either obtained using two-dimensional drums or short cylinders with periodic boundary conditions along the rotation axis. If cylinder end-caps are used in the simulation to represent the experimental conditions more closely, care has to be taken that the range of the boundary effects is shorter than the cylinder length. It was found that the boundary range is only a few particle diameters for sufficiently large particles and converges to a constant value when the particle size decreases [3].

When a binary particle mixture is put in a half-filled extended rotating cylinder, the rapid radial size segregation will take place in each plane perpendicular to the rotation axis. This will lead to a radially-segregated core throughout the length of the cylinder, shown in Fig. 7 in the middle. For this particular simulation, the drum had a diameter and a length of 7 cm and was filled with a binary particle mixture having diameters of 2 and 3 mm, respectively. Thus the length of the drum was 35 small particle diameters which makes boundary effects negligible in the central part of the drum. The initial volume fraction of the smaller particles shown in white was 45% and the cylinder rotated at 30 rpm. When the drum is rotated longer, the additional friction at the end-caps favours smaller particles which leads to a higher concentration of smaller particles close to both end-caps, shown in Fig. 7 to the right. Since our cylinder only contained a total number of 13300 particles, the statistics are not good enough in order to give more quantitative results which will also depend on the material properties of the particles and the drum.

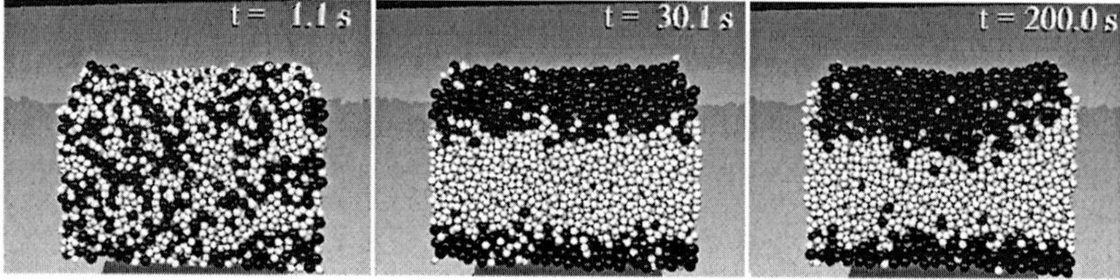

Figure 7. Cross-sectional view of an initially well-mixed cylinder after 1/2 (left), 15 (middle) and 100 rotations (right).

3.5. AXIAL SEGREGATION

The *axial* segregation process is more complex due to the three-dimensional particle motion. Small changes in the initial mixture seem to have a large effect on the band formation process. The final number of bands, their positions and widths varied from experiment to experiment. For example Nakagawa [15] found that a three-band configuration was the most stable set-up after an extended number of rotations. These bands are normally not pure, and a radially segregated core of smaller or/and denser particles might still be present [16]. Chicharro et al. [17] rotated two sizes of Ottawa sand for two weeks at 45 revolutions per minute (rpm) and found a final state of two *pure* bands each filling approximately half of the cylinder, i.e. *no* radial core was found, which is sketched in Fig. 8.

Since pure bands seem to be a stable configuration in some experiments, it is very instructive to study their stability for different material properties. In order to do so, we start with the system depicted in Fig. 8, where the left half is filled with small and the right half with large particles. The large particles have a diameter of 3 mm and a density of $\rho_l = 1.3 \frac{g}{cm^3}$. The material properties of the large particles were chosen to correspond

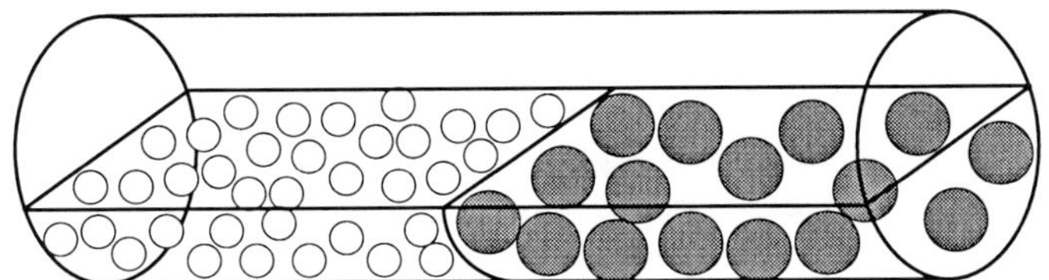

Figure 8. Axial segregation into two pure bands.

to the measured values of mustard seeds, giving $\mu_l = 0.2$. For the details of the numerical analysis on how to determine the diffusion coefficient, we refer the reader to Ref. [14].

The dependence of the diffusion coefficient, D, on the friction coefficient of the small particles is rather small and is shown in Fig. 9(a). The tendency of lower D for higher μ persists for quite large friction coefficients, where the small particles have a much higher angle of repose than the large particles.

The dependence of D on the density ratio is shown in Fig. 9(b), showing a minimum value for $\rho/\rho_l = 1$ and a large increase for lower and higher values. It should be noted that radial segregation will be present towards both sides of the graph.

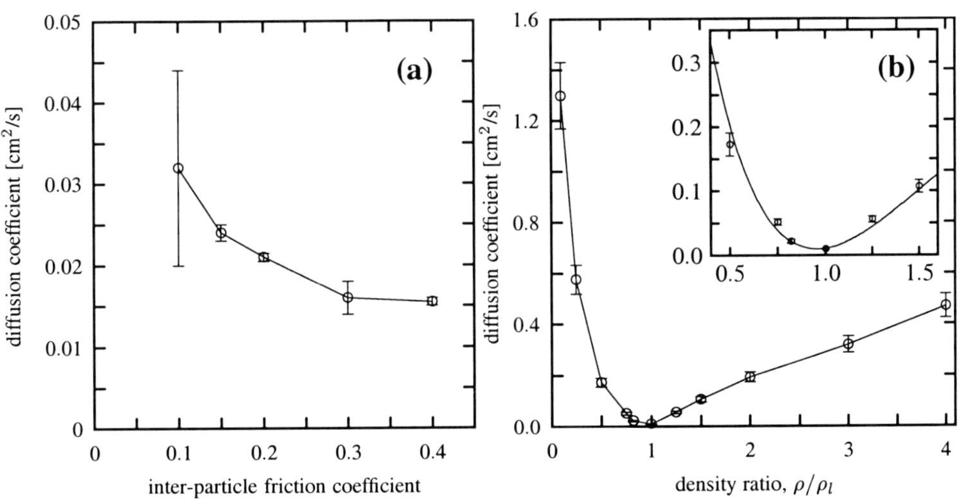

Figure 9. Diffusion coefficient as function of inter-particle friction (a) and density ratio (b).

4. Conclusions

We have shown that the radial segregation process in rotating drums is well captured by DEM simulations and can give results in quantitative agreement with experiments, see e.g. the dependence of the angle of repose on the rotation speed of the drum given in Ref. [3].

In addition, a detailed study revealed that size segregation takes place for arbitrarily small size differences, and we also showed the dynamics of the interplay between mixing and segregation.

Unfortunately, our numerical simulations using DEM currently are not able to reproduce the experimentally observed bands, besides the *end-longitudinal* segregation mentioned in the preceding section. Amongst the possible reasons for this are: (i) the time scales are still too small, up to 50 revolutions at 30 rpm using 13 300 particles were used [14]; (ii) the aspect ratio of the cylinder is too small and the motion is dominated by boundary effects, e.g. an aspect ratio of 1:1 was used in Ref. [14]; (iii) axial segregation does not occur for *perfectly spherical* particles, which were the only ones used in simulations so far. Investigations in all three directions are underway and some of the mysteries will hopefully be resolved in the near future.

References

1. Bridgewater, J.: Fundamental Powder Mixing Mechanisms, *Powder Technol.* **15** (1976), 215.
2. Williams, J. C.: The Segregation of Particulate Materials. A Review, *Powder Technol.* **15** (1976), 245.
3. Dury, C. M., Ristow, G. H., Moss, J. L., and Nakagawa, M.: Boundary Effects on the Angle of Repose in Rotating Cylinders, *Phys. Rev. E* **57** (1998), 4491.
4. Ristow, G. H.: Granular Dynamics: a Review about recent Molecular Dynamics Simulations of Granular Materials, in D. Stauffer (ed.), *Annual Reviews of Computational Physics I*, World Scientific, Singapore, 1994, p. 275.
5. Nityanand, N., Manley, B., and Henein, H.: An Analysis of Radial Segregation for Different Sized Spherical Solids in Rotary Cylinders, *Metall. Trans. B* **17** (1986), 247.
6. Dury, C. M., and Ristow, G. H.: Radial Segregation in a Two-Dimensional Rotating Drum, *J. Phys. I France* **7**, (1997) 737.
7. Dury, C. M., and Ristow, G. H.,: Competition of Mixing and Segregation in Rotating Cylinders, *Phys. Fluids* **11** (1999), 1387.
8. Baumann, G., Janosi, I., and Wolf, D. E.: Particle Trajectories and Segregation in a Two-Dimensional Rotating Drum, *Europhys. Lett.* **27** (1994), 203.
9. Vanel, L., Rosato, A. D., and Dave, R.: Rise-Times Regimes of a Large Sphere in Vibrated Bulk Solids, *Phys. Rev. Lett.* **78** (1997), 1255.
10. Ristow, G. H.: Particle Mass Segregation in a Two-Dimensional Rotating Drum, *Europhys. Lett.* **28** (1994), 97.
11. Metcalfe, G., and Shattuck, M.: Pattern Formation during Mixing and Segregation of Flowing Granular Materials, *Physica A* **233** (1996), 709.
12. Metcalfe, G., Shinbrot, T., McCarthy, J. J., and Ottino, J. M.: Avalanche Mixing of Granular Solids, *Nature* **374** (1995), 39.
13. Ristow, G. H., and Nakagawa, M.: Shape Dynamics of Interfacial Front in Rotating Cylinders, *Phys. Rev. E* **59** (1999), 2044.
14. Dury, C. M., and Ristow, G. H.: Axial Particle Diffusion in Rotating Cylinders, *Granular Matter* **1** (1999), 151.
15. Nakagawa, M.: Axial Segregation of Granular Flows in a Horizontal Rotating Cylinder, *Chem. Engineering Science-Shorter Communications* **49** (1994), 2544.
16. Hill, K. M., Caprihan, A., and Kakalios, J.: Bulk Segregation in Rotated Granular Material Measured by Magnetic Resonance Imaging, *Phys. Rev. Lett.* **78** (1997), 50.
17. Chicharro, R., Peralta-Fabi, R., and Velasco, R. M.: Segregation in Dry Granular Systems, in R. P. Behringer and J. T. Jenkins (eds.), *Powders & Grains 97*, Balkema, Rotterdam, 1997, p.479.

DEM STUDY OF SEGREGATION IN A ROTATING CYLINDER

KENJI YAMANE
Quality Control Department
Taiho Pharmaceutical Co., Ltd.
Tokusima 770-0194
Japan

Abstract

In the pharmaceutical industries, mixing is a critical process. Pharmaceutical products consist of various components, including active ingredient and excipient. After mixing, we should get a homogeneous mixture to ensure the quality of the final products such as tablets or capsules. For that reason, segregation is unwanted phenomena in the mixing process. In this study, the DEM (Discrete Element Method) was applied to investigate the mechanism of segregation in a rotating cylinder, the basic equipment of powder mixing. The DEM simulation showed the segregation in a rotating cylinder three-dimensionally. In particular, radial segregation, which was quickly observed, was quantified by computing the granular temperature of the system.

1. Introduction

Engineers have been interested in complicated granular flows, and many have studied granular as a phenomenon since mixing is an important process in industries where powders and grains are handled, especially in pharmaceutical manufacturing Here, it is important to ensure the blend uniformity of mixtures. Segregation should be avoided to maintain the quality of drug products. The phenomenon of segregation in a rotating drum has captured the attention physicists for some time. In 1939, Oyama [1] first reported that rotating cylinders partially filled with a mixture of granular media may serve to separate the individual species into bands along the rotational axis. This phenomenon is termed axial segregation. Since Oyama's experiments, a number of studies have shown that mixtures of granular media exhibit a wide range of axial segregation (see, for example [2]-[7]). It has been found from recent studies that the reversible axial segregation of different-sized granular media in a drum mixer is related to variations of the dynamic angle of repose [8], but it is also clear that subsurface effects are critical, because axial bands that do not extend to the surface have been observed within the bulk. In order to study the dynamics beneath the surface involved in the segregation process,

A.D. Rosato and D.L. Blackmore (eds.), IUTAM Symposium on Segregation in Granular Flows, 321–326.

DEM simulation was used in the present study to demonstrate granular flow patterns within a drum mixer as a function of granular temperature. In particular, surface shape, and diffusion of the rotating mixture were simulated. In this paper, the study of segregation mechanism was conducted by DEM simulation. DEM was also used to calculate granular temperature.

2. DEM Simulation

The DEM simulation has been explained in many articles (see [9,10]). We give only a brief explanation. In the absence of interstitial fluids, the equations for the translational and rotational motion of a spherical particle are

$$\ddot{\vec{r}} = \vec{f}/m + g \tag{1}$$

and

$$\dot{\vec{\omega}} = \vec{T}/I \,, \tag{2}$$

where $\vec{r}$ is the position vector of the particle center, m is the particle mass, $\vec{f}$ is the sum of all contact forces, $\vec{g}$ is the gravitational acceleration, $\vec{\omega}$ is the angular velocity, $\vec{T}$ is the net torque due to contact forces, I is the moment of inertia, and a dot indicates the time derivative.

Contact forces between spherical particles are modeled by springs, dash-pots, and a friction slider, as originally considered by Cundall and Strack [11]. The springs account for elastic repulsion, dash-pots express the damping effect, and friction sliders express the tangential friction force in the presence of a normal force. The effects of these mechanical elements on particle motion appear through stiffness k, damping coefficient η, and friction coefficient μ_f. The normal component of the contact force $\vec{f}_n$ by particle 2 on particle 1 is given by Hertzian contact theory as,

$$\vec{f}_n = \left(-k_n \delta^{3/2} - \eta_n \vec{V}_r \cdot \hat{n} \right) \hat{n} \,, \tag{3}$$

where k_n is the spring constant, δ is the normal deformation defined as the overlap distance between particle centers, η_n is the normal damping coefficient, $\vec{V}_r$ is the relative velocity of two particles and $\hat{\eta}$ is the unit vector directed from the center of particle 1 to particle 2.

The tangential component of the contact force $\vec{f}_t$ is given by

$$\vec{f}_t = -k_t \vec{\delta} - \eta_t \vec{V}_s \tag{4}$$

where k_t is the tangential spring constant, $\vec{\delta}$ is the tangential deformation vector, η_t is the tangential damping coefficient, and $\vec{V}_s$ is the slip velocity of the contact point. When the sliding condition,

$$\left| \vec{f}_t \right| > \mu_f \left| \vec{f}_n \right| \vec{V}_s \tag{5}$$

is satisfied, the Coulomb-type friction law applies and the tangential contact force of Eq. (4) is replaced by

$$\left|\vec{f}_t\right| > -\mu_f \left|\vec{f}_n\right| \hat{\vec{V}}_s \tag{6}$$

where $\hat{\vec{V}}_s$ is the unit vector defined by $\hat{\vec{V}}_s = \vec{V}_s / \left|\vec{V}_s\right|$.

The parameters specific to the present study are as follows. Each time step is 10^{-4} s, while Young's modulus and Poisson ratio are 1.0×10^5 N/m^2 and 0.3, respectively. The diameter of cylinder is 69 mm and periodic boundary conditions in the longitudinal direction are used to alleviate the severe computing demands of 3D simulations. Therefore, the simulation takes place in a 15mm long, half-filled cylinder with two periodic end boundaries. About 1000 particles of 4 mm and about 2000 particles of 2 mm were used for the calculation.

3. Results and Discussions

The initial conditions of the simulation are as follows. The large and small particles are separated as shown in Figure 1(a). Then, the drum is made to rotate with high rotation speed $\Omega = 170$ rpm in order to cause the particle to mix, but not segregate (Fig. 1b). After this, the rotation speed $\Omega = 17$ rpm was chosen in order to maintain the top surface at a nearly flat constant angle of inclination. After some

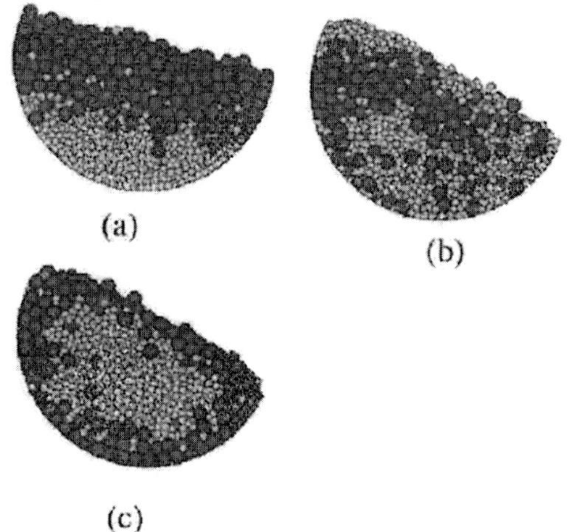

(a)

(b)

(c)

revolutions a clearly segregated core composed of small particles is observed (Fig. 1c).

Figure 1. (a) initial condition, (b) mixture condition, (c) radial segregation

In order to quantify the degree of segregation, an index based on the density profile of small particles is computed by the following procedure. We note that this index was also used to evaluate the degree of mixing for DEM simulation of a rotary vessel mixer [12]. The particle bed segment was divided into cells. The standard deviation σ for the proportion of small particles in the cell is given by

$$\sigma = \sqrt{\dfrac{\sum\limits_{i=1}^{N}(x_i - \bar{x}_c)^2}{N-1}} \qquad (7)$$

where x_i is the proportion of small particles in the i-th cell, $\bar{x}_c$ is the overall proportion of small particles, and N is the number of cells. The segregation index is defined by,

$$\sigma/\sigma_o \qquad (8)$$

where the subscript "o" indicates the completely unmixed state, and is given by

$$\sigma_o = \sqrt{\bar{x}_c(1-x_c)} \qquad (9)$$

Table 1 shows the segregation index on mixture condition and radial segregation condition shown in Fig. 1.

Table 1. Segregation index

Status	Mixture (at 170 rpm)	Radial segregation (at 17 rpm)
σ/σ_o	0.33	0.55

In order to investigate the segregation condition, the author proposes a further parameter that can be computed from the simulation: the fluctuation velocity defined as

$$\langle v \rangle = \sqrt{\dfrac{1}{m}\sum\limits_{i=1}^{m}(v_i - \bar{v})^2} \qquad (10)$$

Here m is the average number of particles in the voxel, v_i is the velocity of i-th particle, and $\bar{v}$ is the average velocity of the particles in the voxel. $\langle v \rangle$ is a parameter that is proportional to the granular temperature used in kinetic theory treatments of such problems [10, 13]. The computed granular temperature for the situation of Figure 1 (c), except for the rotation speed, is presented in Figure 2 (b).

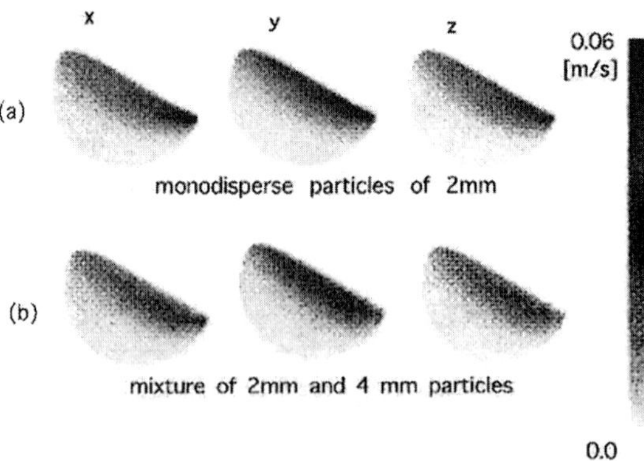

Figure 2. Fluctuation of velocity components

The rotation speed in Figure 2 is 30 rpm. In addition, the calculated granular temperature of monodisperse particles of 2 mm is shown in Fig 2 (a) to compare the segregation condition. The image picture of monodisperse particle of 2 mm is shown in Fugure 3, where 7000 particles of 2 mm were used for DEM simulation. The difference of granular temperature between monodisperse particles condition and segregation condition on two different size particles is observed. Total granular temperature was given by

$$< v_p >^2 =< v_x >^2 + < v_y >^2 + < v_z >^2 \tag{11}$$

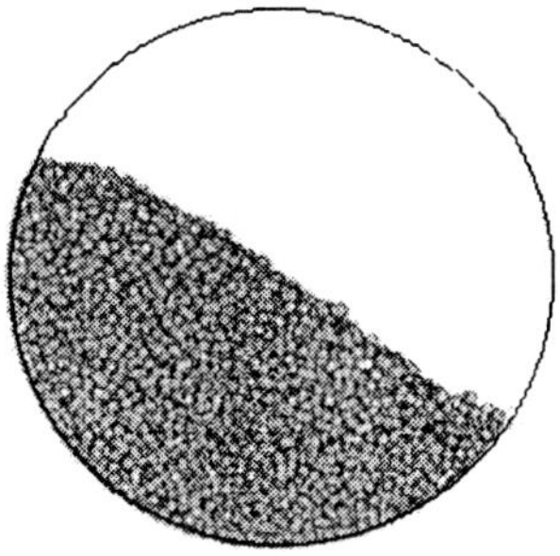

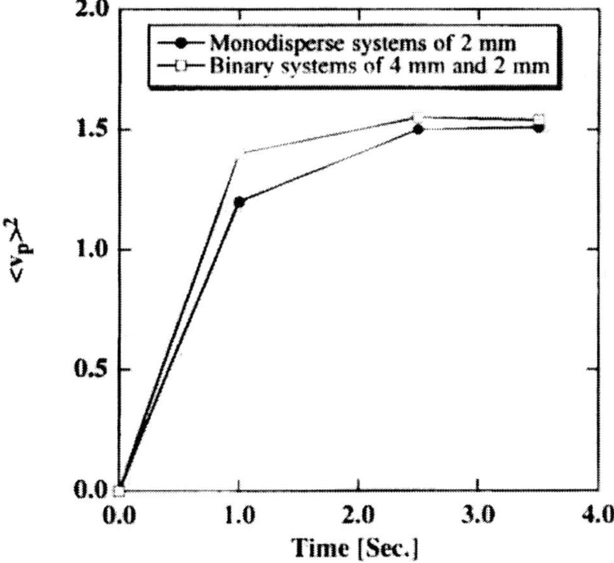

Figure 3. Image of monodisperse particles of 2 mm at rotation speed of 30 rpm

The transition of total granular temperature at rotation speed of 30 rpm is shown in Figure 4.

Figure 4. Transition of total granular temperature

4. Conclusions

Radial size segregation in a rotating cylinder was simulated using the discrete element method (DEM). The system consisted of large and small spheres such that the diameter ratio (small to large) was 1:2. After only a few revolutions of the cylinder, the initial well-mixed configuration of large and small particles, evolved to a radial segregated state in which the small particles occupied a central core surrounded by larger particles. Quantification of the segregation was done by comparing the evolution of the granular temperature for this system to that of a monodisperse system.

5. Acknowledgements

The author would like to acknowledge Tsuji's group in Osaka University for their technical advice.

6. References

1. Oyama, Y.: *Bull.Inst.Phys.Chem.Rep.* (Tokyo), **18** (1939), 600 (in Japanese).

2. Nakagawa, M.: Axial segregation of granular flows in a horizontal rotating cylinder, *Chem. Eng. Sci.* **49** (1994), 2540-2544

3. Hill, K.M., Caprihan, A., and Kakalios, J.: Bulk segregation in rotated granular material measured by magnetic resonance imaging, *Phys. Rev. Lett.*, **78** (1997), 50-53.

4. Donald, M.B., and Roseman, B.: Mechanisms in a horizontal drum mixer, *British Chem. Eng.* **7** (1962), 749-753; Effects of varying the operating conditions of a horizontal drum mixer. *Brit. Chem. Eng.* **7** (1962), 823-827.

5. Savage, S.B.: Banding or pattern formation in horizontal drum mixers, in D. Bideau and A. Hansen (eds.) *Disorder and Granular Media*, Amsterdam: North-Holland, 1993, pp. 255-285.

6. Bridgwater, J., Sharpe, N.W., and Stocker, D.C.: Particle mixing by percolation. *Trans. Inst. Chem. Eng.* **47** (1969), T114-T119.

7. Zik, O., Levine, D., Lipson, S.G., Shtrikman S., and Stavans, J.: Rotationally induced segregation of granular materials. *Phys. Rev. Lett.* **73** (1994), 644-647.

8. Hill, K.M. and Kakalios, J.: Reversible axial segregation of rotating granular media, *Phys. Rev. E* **52** (1995), 4393-4400.

9. Tsuji, Y., Tanaka, T., and Ishida, T.: Lagrangian numerical simulation of plug flow of cohesionless particles in a horizontal pipe. *Powder Technol.* **71** (1992), 239-250.

10. Yamane, K., Nakagawa, M., Altobelli, S.A., Tanaka, T., and Tsuji, Y.: Steady particulate flows in a horizontal rotating cylinder. *Phys. Fluids* **10** (1998), 1419-1427.

11. Cundall, P.A. and Strack, O.D.: A discrete numerical model for granular assemblies. *Geotechnique* **12** (1976), 47-65.

12. Muguruma, Y., Tanaka, T., Kawatake, S., and Tsuji, Y.: Discrete particle simulation of a rotary vessel mixer with baffles. *Powder Technol.* **93** (1997), 261-266.

13. Jenkins, J.T. and Hanes, D.M.: The balance of momentum and energy at an interface between colliding and freely flying grains in a rapid granular flow, *Phys. Fluids A* **5** [3] (1993), 781-783.

AIR ENTRAINMENT AND SEGREGATION IN POWDER FLOWS

R. MUDRYY, P. SINGH and A. D. ROSATO
Department of Mechanical Engineering
New Jersey Institute of Technology
University Heights
Newark, New Jersey 07102

1. Abstract

A finite element code based on the distributed Lagrange multiplier/method (DLM) is used to study the motion of flowing powders in the regime where the aerodynamic forces are important. In our implementation of the DLM method the air-particle system is treated implicitly by using a combined weak formulation in which the forces and moments between the particles and fluid cancel, as they are internal to the combined system. The governing Navier-Stokes equations are solved everywhere in the domain, including inside the particles. The flow inside the particles is forced to be a rigid body motion using a distributed Lagrange method. The direct numerical simulations show that the aerodynamic forces are important in determining the flow properties of powders with particles of size of the order of 1000 μm or smaller. Specifically, we find that when a granular system slides downwards inside a closed channel inclined at 60° to the horizontal the particles on the surface are lifted away by the aerodynamic forces which causes a partial blockage of the channel. The back flow of the air is hindered and an adverse pressure gradient develops in the channel that decelerates the downward motion of the powder. The adverse pressure gradient, in fact, is so strong that the magnitude of downward velocity of the center of mass decreases during the time interval for which the channel is partially blocked. We also find that when all particles are of the same density the aerodynamic forces enhance mixing. In a mixture containing particles of two different densities, on the other hand, the aerodynamic forces cause segregation.

1. Introduction

The aerodynamic forces acting on the particles of flowing powders may be comparable to the inter-particle collision forces, especially when the particle size is smaller than ~1000 μm. There are several experimental studies that show that the aerodynamic forces can change the flow behavior of fine powders [1,17]. In this paper, direct numerical simulations are used to study the role of entrained air in segregation and mixing behavior of flowing powders.

It is well known that when a particle of diameter d is dropped from a state of rest it accelerates downwards until it reaches its maximum free fall velocity, called the

A.D. Rosato and D.L. Blackmore (eds.), IUTAM Symposium on Segregation in Granular Flows, 327–336.
© 2000 *Kluwer Academic Publishers. Printed in the Netherlands.*

terminal velocity U_t. After reaching the terminal velocity, the particle descends with an approximately constant velocity. Let the time taken to reach the terminal velocity be T_t, and the distance traveled during this time be X_t. The velocity U_t, the time T_t and the dimensionless distance X_t/d are important velocity, time and length scales that may be used for determining the role of aerodynamic forces in powder flows. Specifically, the terminal velocity can be used to define the Reynolds number Re for the particle, which is a measure of the relative importance of the inertial and viscous forces. When Re $<<$ 1 the inertial forces are small and the drag coefficient is given by the Stokes formula. On the other hand, when Re$>>$1 inertial forces are dominant and it can be shown that the drag coefficient in this case is of O(1). Also, when T_t is smaller than the time for inter-particle collisions, the aerodynamic drag may cause significant deceleration of the particles in-between the collisions.

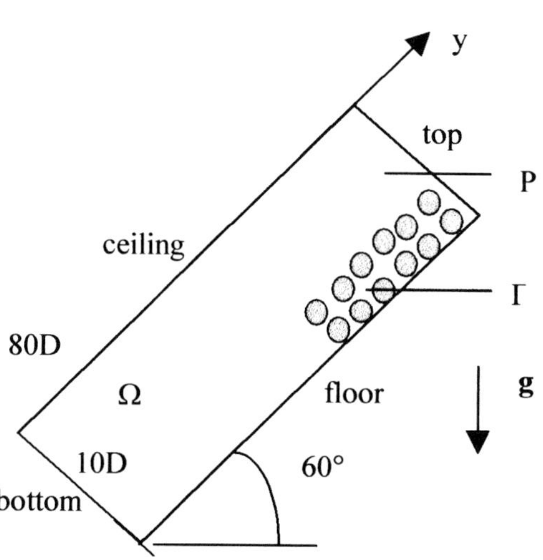

Figure 1. A typical rectangular domain used in the simulations.

There have been several numerical studies where the role of entrained air in powder flows was investigated using ad-hoc models for the coupling between the air and solid phases [6,7]. The models are necessary for approximating the forces acting on the particles when the solids fraction is large, as is the case for powder flows. This is the first direct numerical simulation study where the aerodynamic forces acting on the particles are calculated by solving the Navier-Stokes equations. The no-slip boundary condition is applied at the computational domain boundaries, and on the particles surface the air velocity is constrained to be equal to the particle velocity. That is, a numerical solution of the exact coupled initial value problem is obtained, i.e., up to the numerical truncation error.

The motion of particles through air is numerically simulated using a distributed Lagrange multiplier/fictitious domain method (DLM) [3,4,16,17]. One of the key features of our numerical method is that the fluid-particle system is treated implicitly by using a combined weak formulation where the forces and moments between the particles and fluid cancel, as they are internal to the combined system [5]. The other key feature is that the Marchuk-Yanenko operator splitting technique is used to decouple the difficulties associated with the incompressibility constraint, the nonlinear convection term and the rigid body motion constraint.

In order to study the role of aerodynamic forces in powder flows, the transient motion of 102 non-cohesive particles sliding in a closed rectangular channel is investigated (see Figure 1). The channel floor is inclined at 60° to the horizontal. Clearly, in the absence of aerodynamic forces the powder particles are expected to slide downwards with only a small relative motion that may arise due to the inter-particle collisions. This however is not the case when the aerodynamic forces are important. The aerodynamic forces acting on the particles are different, and this causes them to move relative to each other. For example, when the powder slides in a closed channel the air moving in the opposite direction lifts the particles on the surface. Furthermore, the aerodynamic forces may cause a strong interaction between the particles moving relative to each other, which, as we will discuss later, may lead to mixing when they are of the same density and segregation when they have different densities.

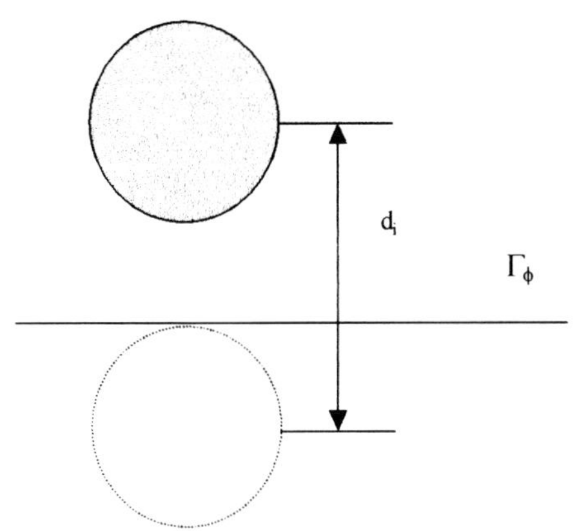

Figure 2. The imaginary particle used for computing the repulsive force acting between a particle and a wall.

2. Problem Statement and the Numerical Method

The airflow is modeled using the Navier-Stokes equation. In this paper we will present results for the two-dimensional flows. Let us denote the domain containing the air and N particles by Ω, and the interior of the ith particle by $P_i(t)$. For simplicity we will assume that the domain is rectangular, with sides Γ (see Figure 1). The governing equations for the air-particle system are:

$$\rho_L \left[\frac{\partial \mathbf{u}}{\partial t} + \mathbf{u}.\nabla\mathbf{u} \right] = \rho_L \mathbf{g} - \nabla p + \nabla.(2\eta\mathbf{D}) \qquad \text{in } \Omega \backslash P(t) \tag{1}$$

$$\nabla \cdot \mathbf{u} = 0 \qquad \text{in } \Omega \backslash P(t) \tag{2}$$

$$\mathbf{u} = \mathbf{u}_L \qquad \text{on } \Gamma \tag{3}$$

$$\mathbf{u} = \mathbf{U}_i + \omega_i \times \mathbf{r}_i \qquad \text{on } \partial P_i(t), \; i=1,\ldots,N, \tag{4}$$

Here $\mathbf{u}$ is the velocity, p is the pressure, η is the viscosity, ρ_L is the density, and $\mathbf{D}$ is the symmetric part of the velocity gradient tensor. The above equations are solved with the following initial condition:

$$\mathbf{u}\,|_{t=0} = \mathbf{u}_0 \tag{5}$$

where $\mathbf{u}_0$ is the known initial value of the velocity. The particle velocity $\mathbf{U}_i$ and angular velocity ω_i are governed by the equations,

$$M_i \frac{dU_i}{dt} = M_i g + F_i \tag{6}$$

$$I_i \frac{d\omega_i}{dt} = T_i \tag{7}$$

$$U_i \big|_{t=0} = U_{i,0} \tag{8}$$

$$\omega_i \big|_{t=0} = \omega_{i,0} \tag{9}$$

where M_i and I_i are the mass and moment of inertia of the ith particle, respectively. In this investigation we will assume that the particles are circular, and therefore we do not need to keep track of the particle orientation. The particle positions are obtained from

$$\frac{dX_i}{dt} = U_i \tag{10}$$

$$X_i \big|_{t=0} = X_{i,0} \tag{11}$$

where $X_{i,0}$ is the position of the ith particle at time $t = 0$. The drag coefficient for a non-accelerating particle falling under gravity is defined to be, $C_d = \dfrac{\frac{1}{4}\pi D^2(\rho_p - \rho_L)g}{\frac{1}{2}\rho_L U^2 D}$,

where U is the particle velocity, ρ_p is the particle density and D is the diameter. The drag acting on a particle depends on the Reynolds number $Re = \dfrac{UD\rho_L}{\eta}$ where ρ_L is the fluid density.

2.1 COLLISION STRATEGY

The collisions among the particles, and the particles and the domain walls, are modeled by applying an elastic body force that acts when the distance between two particles, or between a particle and a wall is smaller than one and half times the velocity mesh size. This additional body force, which is repulsive in nature, is added to equation (6). The particle-particle repulsive force is given by

$$F_{i,j}{}^P = \begin{cases} 0, & \text{for } d_{i,j} > R_i + R_j + \rho \\ \dfrac{1}{\varepsilon_p}(X_i - X_j)(R_i + R_j + \rho - d_{i,j})^2, & \text{for } d_{i,j} < R_i + R_j + \rho \end{cases} \tag{12}$$

where $d_{i,j}$ is the distance between the centers of the ith and j^{th} particles, R_i is radius of the ith particle and ρ is the force range, and ε_p is a small positive stiffness parameter. In our simulations, ρ is assumed to be equal to the size of an element. The repulsive force keeps the particles apart and does not allow their boundaries to be closer than ρ. The repulsive force between a particle and the wall is given by,

$$\mathbf{F}_{i,j}{}^{W} = \begin{cases} \mathbf{0}, & \text{for } d_i > 2R_i + \rho \\ \dfrac{1}{\varepsilon_w}(\mathbf{X}_i - \mathbf{X}_j)(2R_i + \rho - d_i)^2, & \text{for } d_i < 2R_i + \rho \end{cases} \tag{13}$$

where $d_{i,j}$ is the distance between the centers of the ith particle and the imaginary particle on the other side of the wall, and ε_w is another small positive stiffness parameter (see Figure 2). The above particle-particle repulsive forces and the particle-wall repulsive forces are added to equation (6) to obtain,

$$M_i \frac{d\mathbf{U}_i}{dt} = M_i\mathbf{g} + \mathbf{F}_i + \mathbf{F}_i' \tag{14}$$

where $\mathbf{F}_i' = \sum\limits_{\substack{j=1 \\ j \neq i}}^{N} \mathbf{F}_{i,j}^{P} + \sum\limits_{j=1}^{4} \mathbf{F}_{i,j}^{W}$ .is the repulsive force exerted on the ith particle by the other particles and the walls.

2.2 NUMERICAL METHOD

One of the key features of the DLM method is that the air-particle system is treated implicitly by using a combined weak formulation where the forces and moments between the particles and fluid cancel, as they are internal to the combined system. These internal aerodynamic forces are not needed for determining the motion of particles. In our combined weak formulation we solve fluid flow equations everywhere in the domain, including inside the particles. The flow inside the particles is forced to be a rigid body motion using the distributed Lagrange multiplier method. This multiplier represents the additional body force per unit volume needed to maintain rigid-body motion inside the particle boundary, and is analogous to the pressure in incompressible fluid flow, whose gradient is the force needed to maintain the constraint of incompressibility.

The initial value problem (2-11) is solved by using the Marchuk-Yanenko operator splitting scheme which allows us to decouple the following three primary problems:

1. The incompressibility condition, and the related unknown pressure p_h,
2. The nonlinear advection term,
3. The constraint of rigid-body motion in $P_h(t)$, and the related distributed Lagrange multiplier.

The resulting three sub-problems are:

1. The first step gives rise to a Stokes-like problem for the velocity and pressure distributions, which is solved by using a conjugate gradient method [2].
2. The second step is a nonlinear problem for the velocity which is solved by using a least square conjugate gradient algorithm [2].

3.   The third step is used to obtain the distributed Lagrange multiplier that enforces rigid body motion inside the particles. This problem is solved by using a conjugate gradient method described in [4,5].

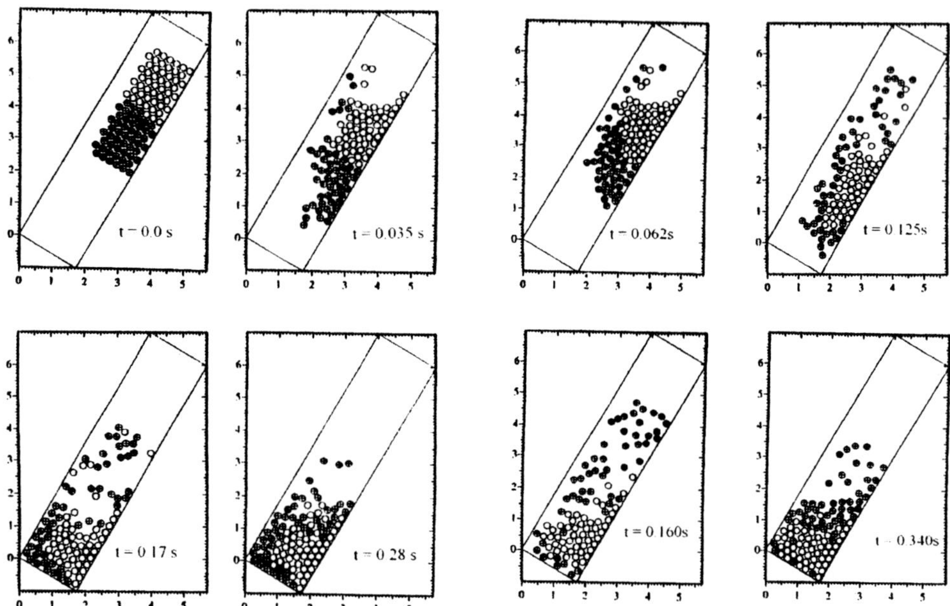

Figure 3. The position of particles is shown at four different times. The bottom 51 particles are marked by +. Notice that the aerodynamic forces cause mixing of particles. (a) All particles are of the same density. The particle positions are shown for t=0, 0.035,0.17 and 0.28. (b) The density of unmarked particles is twice of that of the marked particles. The particle positions are shown for t=0.062, 0.125, 0.16 and 0.34.

3. Results and Conclusions

In order to study the role of entrained air in powder flows, the DLM method is used to simulate the motion of 102 particles in a closed channel of length 80D and height 10D as shown in Figure 1, where the diameter D is equal to 0.2 cm. The ratio of the particle and fluid densities is assumed to be of the order of 1000. The air is assumed to be Newtonian with kinematic viscosity 0.003 cm^2/s and density 1.0 g/cm^3.

The simulations are started at t = 0, by placing the particles in a periodic lattice near the top of the channel, as shown in Figure 3. The initial linear and angular velocities of the particles are zero. The particles slide downward under gravity on a surface inclined at 60° to the horizontal. The collective motion of particles is studied in terms of the velocity of the system center of mass. The time evolution of the x- and y-components of mass center velocity is shown in Figure 4 for the case where the density of all particles is 1000 g/cm^3. The analogous plot is shown in Figure 5 for the case where the density of the bottom 51 particles (marked by a "+") is 500 g/cm^3, and that of the top 51 particles is 1000 g/cm^3.

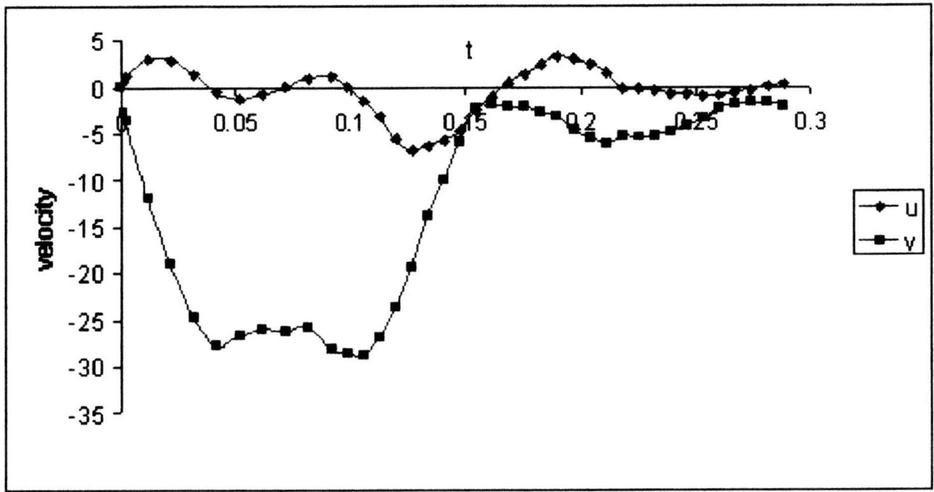

Figure 4. The time evolution of the x- and y- components of the center of mass velocity, u and v, respectively, are shown. All particles have the same density. Note that |v| does not increase monotonically with time due to an adverse pressure gradient created by a partial blockage of the channel.

The spatial positions of particles at four different times are shown in Figures 3a and 3b. In the former both marked and unmarked particles are of the same density, and in the latter the density of the marked particles is one half of that of the unmarked particles. From these figures we note that for t < 0.04 the mixing of marked and unmarked particles is negligible, as the aerodynamic forces are relatively smaller than the gravitational forces. For t > 0.06, the mixing of particles begins, especially in the top layers where the aerodynamic forces are important. In the case where all particles are of the same density, we note that there is a core of unmarked particles, which remains essentially unmixed with the marked particles. Also note that a large fraction of the marked particles—that were initially at the bottom—are on the top of the pile after the motion stops. This is because of the aerodynamic forces that chip away the particles that were initially in the front and on the surface. On the other hand, when the density of particles is different almost all marked (lighter) particles move to the top, as they fall relatively slowly. In this case the marked and unmarked particles are almost completely segregated from each other. These results imply that the aerodynamic forces cause mixing when particles are identical, as the particles move and interact with each other through the air. But, when particles have different densities, the heavier particles settle to the bottom and the lighter particles rise to the top, and thus the aerodynamic forces cause segregation. This occurs because the aerodynamic force acting on a particle depends on its size and velocity, but is independent of its weight.

From Figures 4 and 5 we note that for t < ~ 0.02s, the magnitude of y-component of velocity, |v|, increases approximately linearly with time. Since the particle velocities are small during this time interval, the aerodynamic drag is also small and therefore the motion is determined by the acceleration due to gravity. For t > ~ 0.02, the rate of increase of |v| decreases with time, as the aerodynamic drag force which acts in the direction opposite to the velocity reduces the net downward acceleration. In Figure 4, |v| reaches a locally maximum value of 28 mm/s at t ~0.045, as the net sum of the gravitational and aerodynamic forces is zero. During ~0.04 < t < ~0.11, |v| decreases and then after reaching a locally minimum value it starts to increase again. The decrease in |v| during this time interval is due the adverse pressure gradient which develops as the particles lifted from the top layers block the back flow of the air (see Figure 3a). For t > 0.11, |v| decreases again as particles start to approach the bottom, and becomes zero when all particles have reached the bottom.

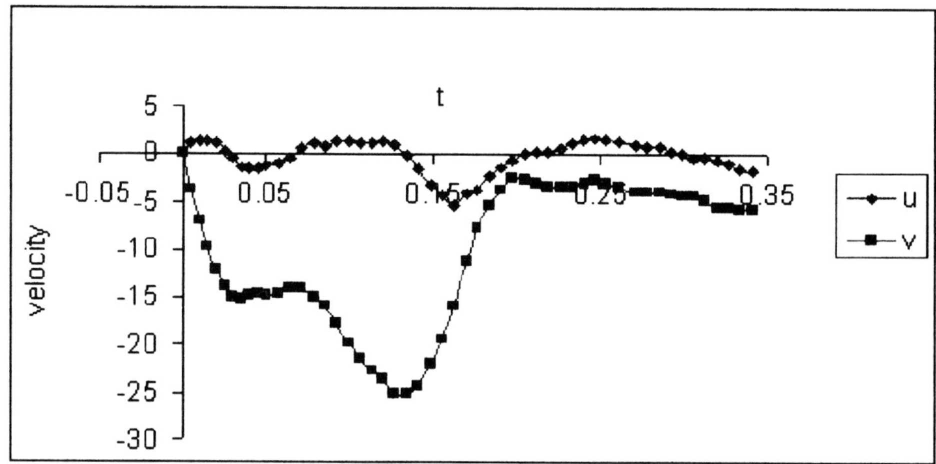

Figure 5. The time evolution of u and v for the center of mass are shown. The density of particles marked by "+" is one-half of that of the unmarked particles (See Figure 4). Note that |v| does not increase monotonically with time, and reaches a local maximum value at an earlier time than that in Figure 2.

In Figure 5, on the other hand, |v| reaches a locally maximum value of 15 mm/s at t ~ 0.025. During ~ 0.025 < t< ~ 0.06, the nearly constant value of |v| is due to the balance between the downward gravitational force and the adverse pressure gradient developed as particles lifted from top layers block the back flow of the air (see Figure 3b). The velocity |v| starts to increase again after the partial blockage in the channel is cleared. It begins to decrease at t = 0.15 as particles start to approach the bottom, and it goes to zero when all particles have reached the bottom. These time dependent changes in |v| can be explained by examining the variation of the x-component of velocity and the distribution of particles in the channel. The time evolutions of the x-component of the mass center velocity u for the above two cases are shown in Figures 4 and 5. A positive value of u implies that the center of mass is moving closer to the floor. In both figures this is the case for t < 0.02. This is due to the spreading and compaction of particles along the floor wall. From Figure 4 we note that u becomes negative at t ~0.035 due the lift off of the particles on the surface of the powder. This lift off is caused by the back flow of air in the closed channel as the particles fall downwards. Also note that some particles reach the top wall which partially blocks the back flow of air. This partial blockage leads to a larger pressure drop in the region occupied by the particles, and thus the downward acceleration of the mass center decreases. The increased pressure drop causes the particles in the top layer to move upwards, i.e., in a direction opposite to the bulk flow, and a clear channel close to the upper wall appears. This removal of particles causes the pressure drop to decrease and the center of mass again accelerates downwards. For $t > 0.1$ the x-component of the mass center velocity becomes negative as the particles begin to reach the bottom due to the piling of particles.

Finally, we note that u in Figure 5 becomes negative at an earlier time than in Figure 4. Since the density of particles in the front is smaller in Figure 5, they are lifted away at a smaller value of $|v|$. Therefore, in this case the channel is blocked at an earlier time. Also note that the maximum value of $|v|$ in Figure 5 is smaller than in Figure 4, as the weight of the powder is 25% smaller. This dependence of the settling velocity of the same sized particles on their weights causes the segregation shown in Figure 3b.Here, the less dense particles marked with a "+" eventually find themselves on the top of the heavier particles.

4.   Acknowledgement

This research was supported by grants from the SBR program and the PTC at NJIT, and the National Science Foundation KDI Grand Challenge grant (NSF/CTS-98-73236).

5.   References

1.   Akiyama, T., Kimura, N. and Iguchi, T.: Effects of reduced air pressure on the particle circulation rate in two-dimensional vibrating particle beds, *Particle Technol.*, **89** (1996),133-138.
2.   Bristeau, M. O., Glowinski, R. and Periaux, J.: Numerical methods for the Navier-Stokes equations. Applications to the simulation of compressible and incompressible flows. *Computer Physics Reports* **6** (1987), 73-187.
3.   Glowinski, R., Pan, T.W., Hesla, T.I and Joseph, D.D.: A distributed Lagrange multiplier/fictitious domain method for particulate flows, *Int. J. of Multiphase Flows* 25|5| (1988), 755-794.
4.   Glowinski, R., Pan, T.W., and Periaux: A Lagrange multiplier/fictitious domain method for the numerical simulation of incompressible viscous flow around moving rigid bodies, *C.R. Acad. Sci. Paris* **324** (1997), 361-369.
5.   Hesla, T.I.: The dynamical simulation of two-dimensional fluid/particle systems, Unpublished notes (1991).
6.   Lan, Y. and Rosato, A. D.: Macroscopic behavior of vibrating beds of smooth inelastic spheres. *Phys. Fluids* **9** (1995), 3615-3624.
7.   Lan, Y. and Rosato, A. D.: Convection related phenomena in granular dynamics simulations of vibrated beds, *Phys. Fluids*, **9** (1997), 3615-3624.
8.   Tsuji, Y. Tanka, T., Isdhida, T..: Langrangian numerical simulation of plug flow of cohesionless particles in a horizontal pipe, *Powder Technol.* **71** (1992), 239-250.
9.   Tsuji, Y. Kawaguchi, T., Tanaka, T.: Discrete particle simulation of two dimensional fluidized bed. *Powder Technol.* **77** (1993), 79-87.
10. Hoooman, B.P.B., Kuiper, J.A.M., Breils, W.J., Van Swaiij, W.P.M.: Discrete particle simulation of bubble and slug formation in two-dimensional gas fluidized bed: hard sphere approach, *Chem. Eng. Sci.* **51** (1996), 99-118.
11. Kuipers, J.A.M.,Van Duin, K.J., Van Beckum, F.P.H., Van Swaiij, W.P.M.: Numerical model of gas fluidized beds, *Chem. Eng. Sci.* **47** (1992), 1913-1924.
12. Sochlichin. A., Eigenberg. G., Lapin, A., Lubert, A.: Dynamic numerical solution of gas liquid particle two phase flow Euler/Euler vs Euler/Lagrange, *Chem. Eng. Sci.* **52** (1997), 611-626.
13. D. Gera, M.Gaytman, D. W. Lyons, Y. Tsuji.: Numerical Simulation of large particles fluidized bed: TFM vs. DEM approach, *Fluidization IX*, Durango, CO, May 17-22, 1998.

14. Pak, H.K., Van Doorn, E. and Behringer, R.P.: Effects of ambient gases on granular materials under vibration. *Phys. Rev. Lett.* **74** (1995), 4643-4646.
15. Singh, P. and Joseph, D. D.: Dynamics of fluidized suspensions of spheres of finite size, *Int. J. of Multiphase Flows* **21**[1] (1995), 1-26.
16. Singh, P., Joseph, D. D., Hesla, T.I., Glowinski, R. Pan, T.W.: *J. of Non-Newtonian Fluid Mech.* (to appear).
17. Van Doorn, E. and Behringer, R.P.: Dilation of a vibrated granular layer, *Europhys. Lett.* **40** (1997), 387-392.

PARTICIPANTS

Mr. Birgir Arnarson
Theoretical & Applied Mechanics
Cornell University
212 Kimball Hall
Ithaca, NY 14853, USA
boa1@cornell.edu

Dr. John Baxter
Dept. of Chemical & Process Engineering
University of Surry
Guildford
Surrey GU2 5XH, U.K.
U.Tuzun@surrey.ac.uk

Prof. Denis Blackmore
Dept. of Mathematics
New Jersey Institute of Technology
University Heights
Newark, NJ 07102, USA
deblac@chaos.njit.edu

Prof. John Bridgwater
Dept. of Chemical Engineering
Cambridge University
Pembroke Street
Cambridge CB2 3RA, United Kingdom
john_bridgwater@cheng.cam.ac.uk

Mrs. Luciana Bruno
Departamento de Fisica - Facultad de
Ingenieria
Unidersidad de Buenos Aires
Paseo Colon 850
1063 Buenos Aires, Argentina
lbruno@tron.fi.uba.ar

Dr. Arvind Caprihan
New Mexico Resonance, Inc.
2425 Ridgecrest Drive SE
Albuquerque, NM 87108, USA
arvind@nmr.org

Prof. Sunil de Silva
POSTEC
Telemark College
Kjoelnes Ring 56

3914 Porsgrunn, Norway
Sunil.deSilva@hit.no

Prof. Antonio Di Carlo
Dipartimento di Scienze dell'Ingegneria
Civile
Universita Degli Studi "Roma Tre"
Via Corrado Segre, 60
I-00146 Rome, Italy
dicarlo@fenice.dsic.uniroma3.it

Prof. Jon Eggert
Department of Physics
Colorado School of Mines
Golden, CO 80401, USA
Jeggert@mines.edu

Dr. Pierre Evesque
Laboratoire de Mecanique, Sols,
Structures, Materiaux
Ecole Centrale Paris
Garande Voie des Vignes
92295 Chatenay-Malabry Cedex, France
evesque@mssmat.ecp.fr

Dr. Eiichi Fukushima
New Mexico Resonance, Inc.
New Mexico Resonance, Inc.
Albuquerque, NM 87108, USA
eiichi@nmr.org

Dr. J. M. N. T. Gray
Department of Mathematics
University of Manchester
Oxford Road
Manchester M13 9PL, England
ngray@ma.man.ac.uk

Dr. Kimberly M. Hill
Department of Physics
University of Wisconsin
Upham Hall / 800 West Main Street
Whitewater, WI 53190

340

Prof. Daniel Hong
Department of Physics
Lehigh University
Lewis Laboratory
Bethlehem, PA 18015, USA
dh09@lehigh.edu

Dr. Shu-San Hsiau
Mechanical Engineering
National Central University
Chung-Li, Taiwan 32054, R.O.C.
sshsiau@cc.ncu.edu.tw

Dr. Cunkiu Huang
Division of Engineering
Colorado School of Mines
Golden, CO 80401, USA

Mr. Sunil Jain
Pharmaceutical R&D
Merck Research Labs
WP 78-110
Sumneytown Pike, PA 19486
sunil_jain@merck.com

Prof. James Jenkins
Theoretical & Applied Mechanics
Cornell University
212 Kimball Hall
Ithaca, NY 14853, USA
jtj2@cornell.edu

Ms. Anna Karion
Division of Engineering / Mech. Engr.
California Institute of Technology
Pasadena, CA 91125, USA
karion@caltech.edu

Prof. Arshad Kudrolli
Department of Physics
Clark University
Worcester
MA 01610, USA
A.Kudrolli@clarku.edu

Mr. Mark C. Leaper
The Wolfson Centre for Bulk Solids
Handling
University of Greenwich
Woolwich, London SE 18 6PF, UK
m.c.leaper@gre.ac.uk

Prof. Michel Louge
Mechanical Engineering Dept.
Cornell University
284 Grumman Hall
Ithaca, NY 14853, USA
myl3@cornell.edu

Dr. Stefan Luding
Institute for Computer Applications I
University of Stuttgart
Pfaffenwaldring 27
70569 Stuttgart, Germany
lui@ica1.uni-stuttgart.de

Prof. Joseph J. McCarthy
Dept. of Chemical & Petroleum
Engineering
University of Pittsburgh
Pittsburgh, PA 15261, USA
mccarthy@engrng.pitt.edu

Dr. James McElwaine
Institute of Low Temperature Science
Hokkaido University
Kita-19, Nishi-8
Sapporo 060-0819, Japan
jim@orange.lowtem.hokudai.ac.jp

Prof. Sean McNamara
The Levich Institute
City College of New York
New York, NY 10031, USA
mcnamara@levdev.engr.ccny.cuny.edu

Prof. Masami Nakagawa
Division of Engineering
Colorado School of Mines
Golden, CO 80401, USA
mnakagaw@mines.edu

Dr. Thomas O'Brien
U.S. Dept. of Energy (FETC)
Mail Stop NO4
3610 Collins Ferry Road/PO Box 880
Morgantown, , WV 26507-0880, USA
tobrie@fetc.doe.gov

Dr. William R. Paterson
Dept. of Chemical Engineering
Cambridge University
Pembroke Street
Cambridge CB3 0HE, U.K.
wrp1@cheng.cam.ac.uk

Prof. Bruce Pitman
Dept. of Mathematics
State Universitiy of New York
106 Diefendorf Hall
Buffalo, NY 14214-3093, USA
pitman@galileo.math.buffalo.edu

Dr. Patrice Porion
CRMD
1B Rue de la Ferollerie
45071 Orleans cedex 2, France
porion@admin.cnrs-orleans.fr

Dr. James Prescott
Jenike and Johanson, Inc.
Jenike and Johanson, Inc.
Westford, MA 01886, USA
jkprescott@jenike.com

Prof. Gerald Ristow
Institut für Theoretische Physik
Universität des Saarlandes
Postfach 15 11 50
D-66041 Saarbrücken, Germany
ristow@lusi.uni-sb.de

Prof. Anthony Rosato
Mechanical Engineering Dept.
New Jersey Institute of Technology
University Heights
Newark, NJ 07102, USA
rosato@njit.edu

Prof. Clara Saluena
Departament de Fisica Fonamental
University of Barcelona
Diagonal 647
Barcelona 08028, Spain
clara@physik.hu-berlin.de

Ms. Azadeh Samadani
Department of Physics
Clark University
Worcester
MA 01610, USA

Dr. Roman Samulyak
Dept. of Applied Science (DAS)
Brookhaven National Laboratory
DAS – Center for Data Intensive
Computing / Building 515
PO Box 5000
Upton, NY 11973-5000

Mr. Lenny Scarolla
Union Carbide Corporation
P. O. Box 670
Boundbrook, NJ 08805, USA
scarolls@ucarb.com

Dr. Troy Shinbrot
Chemical & Biochemical Engr. Dept.
Rutgers University
Room C228, Engineering Building
Piscataway, NJ 08855, USA
shinbrot@sol.rutgers.edu

Prof. Pushpendra Singh
Mechanical Engineering Dept.
Rutgers University
University Heights
Newark, NJ 07102, USA
singhp@adm.njit.edu

Ms. Jaimie Turner
Division of Engineering
Colorado School of Mines
Golden, CO 80401, USA

Dr. James Vallance
Dept. of Civil Engineering & Applied
Mechanics / Macdonald Engr. Bldg.
McGill University
817 Sherbrooke St. W.
Montreal, Quebec H3A 2K6, Canada
james@fuego.civil.mcgill.ca

Dr. D. F. Wang
Polymers R & D
Union Carbide Corporation
P.O. Box 670
Boundbrook, NJ 08805, USA

Prof. Carl Wassgren
School of Mechanical Engineering
Purdue University
1288 Mechanical Engr. Bldg.
West Lafayette, IN 47907, USA
cwassgren@ecn.purdue.edu

Prof. Leslie Woodcock
Dept. of Chemical Engineering
University of Bradford
Bradford BD7 1DP
W. Yorkshire, U.K.
L.V.Woodcock@Bradford.ac.uk

Prof. David Wu
Dept. of Chemical Engineering
Colorado School of Mines
Golden, CO 80401, USA
dwu@mines.edu

Prof. Peter Wypych
Centre for Bulk Solids & Particulate
Technologies
University of Wollongong
Mechanical Engineering Dept.
Wollongong 2522, New South Wales,
Australia
peter_wypych@uow.edu.au

Dr. Kenji Yamane
Taiho Pharmaceutical Co., Ltd.
Tokushima 771-0194
Japan
yamane@mupf.mech.eng.osaka-u.ac.jp

Mechanics

SOLID MECHANICS AND ITS APPLICATIONS
Series Editor: G.M.L. Gladwell

Aims and Scope of the Series

The fundamental questions arising in mechanics are: *Why?*, *How?*, and *How much?* The aim of this series is to provide lucid accounts written by authoritative researchers giving vision and insight in answering these questions on the subject of mechanics as it relates to solids. The scope of the series covers the entire spectrum of solid mechanics. Thus it includes the foundation of mechanics; variational formulations; computational mechanics; statics, kinematics and dynamics of rigid and elastic bodies; vibrations of solids and structures; dynamical systems and chaos; the theories of elasticity, plasticity and viscoelasticity; composite materials; rods, beams, shells and membranes; structural control and stability; soils, rocks and geomechanics; fracture; tribology; experimental mechanics; biomechanics and machine design.

1. R.T. Haftka, Z. Gürdal and M.P. Kamat: *Elements of Structural Optimization.* 2nd rev.ed., 1990
ISBN 0-7923-0608-2
2. J.J. Kalker: *Three-Dimensional Elastic Bodies in Rolling Contact.* 1990 ISBN 0-7923-0712-7
3. P. Karasudhi: *Foundations of Solid Mechanics.* 1991 ISBN 0-7923-0772-0
4. *Not published*
5. *Not published.*
6. J.F. Doyle: *Static and Dynamic Analysis of Structures.* With an Emphasis on Mechanics and Computer Matrix Methods. 1991 ISBN 0-7923-1124-8; Pb 0-7923-1208-2
7. O.O. Ochoa and J.N. Reddy: *Finite Element Analysis of Composite Laminates.*
ISBN 0-7923-1125-6
8. M.H. Aliabadi and D.P. Rooke: *Numerical Fracture Mechanics.* ISBN 0-7923-1175-2
9. J. Angeles and C.S. López-Cajún: *Optimization of Cam Mechanisms.* 1991
ISBN 0-7923-1355-0
10. D.E. Grierson, A. Franchi and P. Riva (eds.): *Progress in Structural Engineering.* 1991
ISBN 0-7923-1396-8
11. R.T. Haftka and Z. Gürdal: *Elements of Structural Optimization.* 3rd rev. and exp. ed. 1992
ISBN 0-7923-1504-9; Pb 0-7923-1505-7
12. J.R. Barber: *Elasticity.* 1992 ISBN 0-7923-1609-6; Pb 0-7923-1610-X
13. H.S. Tzou and G.L. Anderson (eds.): *Intelligent Structural Systems.* 1992
ISBN 0-7923-1920-6
14. E.E. Gdoutos: *Fracture Mechanics.* An Introduction. 1993 ISBN 0-7923-1932-X
15. J.P. Ward: *Solid Mechanics.* An Introduction. 1992 ISBN 0-7923-1949-4
16. M. Farshad: *Design and Analysis of Shell Structures.* 1992 ISBN 0-7923-1950-8
17. H.S. Tzou and T. Fukuda (eds.): *Precision Sensors, Actuators and Systems.* 1992
ISBN 0-7923-2015-8
18. J.R. Vinson: *The Behavior of Shells Composed of Isotropic and Composite Materials.* 1993
ISBN 0-7923-2113-8
19. H.S. Tzou: *Piezoelectric Shells.* Distributed Sensing and Control of Continua. 1993
ISBN 0-7923-2186-3
20. W. Schiehlen (ed.): *Advanced Multibody System Dynamics.* Simulation and Software Tools. 1993
ISBN 0-7923-2192-8
21. C.-W. Lee: *Vibration Analysis of Rotors.* 1993 ISBN 0-7923-2300-9
22. D.R. Smith: *An Introduction to Continuum Mechanics.* 1993 ISBN 0-7923-2454-4
23. G.M.L. Gladwell: *Inverse Problems in Scattering.* An Introduction. 1993 ISBN 0-7923-2478-1

Mechanics

Mechanics

SOLID MECHANICS AND ITS APPLICATIONS
Series Editor: G.M.L. Gladwell

49. J.R. Willis (ed.): *IUTAM Symposium on Nonlinear Analysis of Fracture*. Proceedings of the IUTAM Symposium held in Cambridge, U.K. 1997   ISBN 0-7923-4378-6
50. A. Preumont: *Vibration Control of Active Structures*. An Introduction. 1997
                                                                              ISBN 0-7923-4392-1
51. G.P. Cherepanov: *Methods of Fracture Mechanics: Solid Matter Physics*. 1997
                                                                              ISBN 0-7923-4408-1
52. D.H. van Campen (ed.): *IUTAM Symposium on Interaction between Dynamics and Control in Advanced Mechanical Systems*. Proceedings of the IUTAM Symposium held in Eindhoven, The Netherlands. 1997   ISBN 0-7923-4429-4
53. N.A. Fleck and A.C.F. Cocks (eds.): *IUTAM Symposium on Mechanics of Granular and Porous Materials*. Proceedings of the IUTAM Symposium held in Cambridge, U.K. 1997
                                                                              ISBN 0-7923-4553-3
54. J. Roorda and N.K. Srivastava (eds.): *Trends in Structural Mechanics*. Theory, Practice, Education. 1997   ISBN 0-7923-4603-3
55. Yu.A. Mitropolskii and N. Van Dao: *Applied Asymptotic Methods in Nonlinear Oscillations*. 1997   ISBN 0-7923-4605-X
56. C. Guedes Soares (ed.): *Probabilistic Methods for Structural Design*. 1997
                                                                              ISBN 0-7923-4670-X
57. D. François, A. Pineau and A. Zaoui: *Mechanical Behaviour of Materials*. Volume I: Elasticity and Plasticity. 1998   ISBN 0-7923-4894-X
58. D. François, A. Pineau and A. Zaoui: *Mechanical Behaviour of Materials*. Volume II: Viscoplasticity, Damage, Fracture and Contact Mechanics. 1998   ISBN 0-7923-4895-8
59. L.T. Tenek and J. Argyris: *Finite Element Analysis for Composite Structures*. 1998
                                                                              ISBN 0-7923-4899-0
60. Y.A. Bahei-El-Din and G.J. Dvorak (eds.): *IUTAM Symposium on Transformation Problems in Composite and Active Materials*. Proceedings of the IUTAM Symposium held in Cairo, Egypt. 1998   ISBN 0-7923-5122-3
61. I.G. Goryacheva: *Contact Mechanics in Tribology*. 1998   ISBN 0-7923-5257-2
62. O.T. Bruhns and E. Stein (eds.): *IUTAM Symposium on Micro- and Macrostructural Aspects of Thermoplasticity*. Proceedings of the IUTAM Symposium held in Bochum, Germany. 1999
                                                                              ISBN 0-7923-5265-3
63. F.C. Moon: *IUTAM Symposium on New Applications of Nonlinear and Chaotic Dynamics in Mechanics*. Proceedings of the IUTAM Symposium held in Ithaca, NY, USA. 1998
                                                                              ISBN 0-7923-5276-9
64. R. Wang: *IUTAM Symposium on Rheology of Bodies with Defects*. Proceedings of the IUTAM Symposium held in Beijing, China. 1999   ISBN 0-7923-5297-1
65. Yu.I. Dimitrienko: *Thermomechanics of Composites under High Temperatures*. 1999
                                                                              ISBN 0-7923-4899-0
66. P. Argoul, M. Frémond and Q.S. Nguyen (eds.): *IUTAM Symposium on Variations of Domains and Free-Boundary Problems in Solid Mechanics*. Proceedings of the IUTAM Symposium held in Paris, France. 1999   ISBN 0-7923-5450-8
67. F.J. Fahy and W.G. Price (eds.): *IUTAM Symposium on Statistical Energy Analysis*. Proceedings of the IUTAM Symposium held in Southampton, U.K. 1999   ISBN 0-7923-5457-5
68. H.A. Mang and F.G. Rammerstorfer (eds.): *IUTAM Symposium on Discretization Methods in Structural Mechanics*. Proceedings of the IUTAM Symposium held in Vienna, Austria. 1999
                                                                              ISBN 0-7923-5591-1

Mechanics

Kluwer Academic Publishers – Dordrecht / Boston / London

ICASE/LaRC Interdisciplinary Series in Science and Engineering

1. J. Buckmaster, T.L. Jackson and A. Kumar (eds.): *Combustion in High-Speed Flows.* 1994
   ISBN 0-7923-2086-X
2. M.Y. Hussaini, T.B. Gatski and T.L. Jackson (eds.): *Transition, Turbulence and Combustion.* Volume I: Transition. 1994
   ISBN 0-7923-3084-6; set 0-7923-3086-2
3. M.Y. Hussaini, T.B. Gatski and T.L. Jackson (eds.): *Transition, Turbulence and Combustion.* Volume II: Turbulence and Combustion. 1994
   ISBN 0-7923-3085-4; set 0-7923-3086-2
4. D.E. Keyes, A. Sameh and V. Venkatakrishnan (eds): *Parallel Numerical Algorithms.* 1997
   ISBN 0-7923-4282-8
5. T.G. Campbell, R.A. Nicolaides and M.D. Salas (eds.): *Computational Electromagnetics and Its Applications.* 1997
   ISBN 0-7923-4733-1
6. V. Venkatakrishnan, M.D. Salas and S.R. Chakravarthy (eds.): *Barriers and Challenges in Computational Fluid Dynamics.* 1998
   ISBN 0-7923-4855-9
7. M.D. Salas, J.N. Hefner and L. Sakell (eds.): *Modeling Complex Turbulent Flows.* 1999
   ISBN 0-7923-5590-3